APPLIED GROUP THEORY

For Physicists & Chemists

GEORGE H. DUFFEY
South Dakota State University

Dover Publications, Inc.
Mineola, New York

Copyright

Copyright © 1992 by George H. Duffey
All rights reserved.

Bibliographical Note

This Dover edition, first published in 2015, is an unabridged republication of the work originally published in 1992 by Prentice-Hall, Inc., Englewood Cliffs, New Jersey.

Library of Congress Cataloging-in-Publication Data

Duffey, George H.
  Applied group theory : for physicists and chemists / George H. Duffey. — Dover edition.
    p. cm.
  Originally published: Englewood Cliffs, NJ : Prentice Hall, 1992.
  Includes bibliographical references and index.
  ISBN-13: 978-0-486-78314-7
  ISBN-10: 0-486-78314-6
  1. Symmetry (Physics) 2. Group theory. 3. Chemistry—Mathematics. 4. Chemistry, Physical and theoretical. I. Title.

QC174.17S9D84 2015
530.15'22—dc23

2014028695

Manufactured in the United States by Courier Corporation
78314601   2015
www.doverpublications.com

Contents

Preface		ix
Chapter 1/Symmetry Operations		
1.1	Causality and Symmetry	*1*
1.2	Common Symmetry Operations	*2*
1.3	Reorientation Matrices	*3*
1.4	Operations Involving Translations	*8*
1.5	Permutation Matrices	*12*
1.6	How Symmetry Operations for a System Make up a Group	*17*
1.7	Identifying Spatial Point Groups	*18*
1.8	Group Generators and Cayley Diagrams	*19*
1.9	Structures of Common Point Groups	*24*
1.10	Representing Group Elements with Complex Numbers	*26*
1.11	Occurences of Irreducible Matrix Representations	*33*
1.12	Representing Translations in Crystals	*36*
1.13	Key Concepts	*38*
	Discussion Questions	*39*
	Problems	*41*
	References	*43*
Chapter 2/Classes and Characters		
2.1	The Immediate Agenda	*45*
2.2	Classes within a Group	*46*
2.3	Adding the Elements in a Class	*51*
2.4	Products of Class Sums	*53*
2.5	Eigenvalues and Eigenoperators of the Class Sums	*56*
2.6	Key Orthogonality Conditions	*63*
2.7	Restrictions on the Classes	*66*
2.8	An Alternate Procedure	*68*
2.9	Main Features	*69*
	Discussion Questions	*70*
	Problems	*72*
	References	*73*

Chapter 3/Expedient Arrangements of Displacements

3.1	Vibrating Arrays	*74*
3.2	Suitable Bases and Coordinates	*75*
3.3	Generating Symmetry-Adapted Arrays	*77*
3.4	Procedural Details	*78*
3.5	Separating the Modes	*87*
3.6	Symmetry-Adapted Coordinates	*91*
3.7	Interactions between Modes	*94*
3.8	Reducing the Symmetry	*97*
3.9	Highlights	*98*
	Discussion Questions	*100*
	Problems	*101*
	References	*104*

Chapter 4/Symmetry-Adapted Stresses and Strains

4.1	Condensed Phases	*106*
4.2	Stress Arrays	*107*
4.3	Elements of Strain as Generalized Coordinates	*111*
4.4	Dependence of Stress on Strain	*114*
4.5	Constitutive Relations for the Different Crystal Types	*115*
4.6	Constitutive Relations for Other Materials	*122*
4.7	Basic Ideas	*123*
	Discussion Questions	*124*
	Problems	*125*
	References	*126*

Chapter 5/Matrix Representations

5.1	Explicit Forms for the Elements of a Group	*128*
5.2	Generating Representations from Bases	*129*
5.3	Traces of the Matrices in a Representation	*139*
5.4	Character Vectors for Representations	*142*
5.5	Symmetry Species in a Representation	*147*
5.6	Freedom in Representing a Group	*148*
5.7	Schur's Lemma	*149*
5.8	An Overview	*154*
	Discussion Questions	*155*
	Problems	*156*
	References	*159*

Contents v

Chapter 6/Symmetry-Adapted Kets and Bras

6.1	Quantum Mechanical States	*161*
6.2	Symmetry Properties of the Eigenstates	*162*
6.3	Matrix Elements for Scalar Physical Operators	*172*
6.4	Generating Representations of Groups with Kets	*175*
6.5	Reduction of Symmetry Induced by Degeneracy	*181*
6.6	Summary	*184*
	Discussion Questions	*185*
	Problems	*186*
	References	*189*

Chapter 7/Combinations of Products of Bases

7.1	Significant Combinations	*190*
7.2	Product Representations	*191*
7.3	Product Groups	*195*
7.4	Quantum Angular Momenta	*198*
7.5	Clebsch-Gordon Coefficients	*202*
7.6	Relating Tensorial-Operator Matrix Elements	*205*
7.7	Synopsis	*208*
	Discussion Questions	*209*
	Problems	*210*
	References	*212*

Chapter 8/Permutation Groups

8.1	The Role Played by Permutations	*213*
8.2	Elements of a Full Permutation Group	*213*
8.3	Symmetry Species for a Symmetric Group	*218*
8.4	Trace for the Reorientation of a Polynomial Ket with Simple Permutational Symmetry	*224*
8.5	Combining Spin and Orbital Kets	*227*
8.6	States Assumed by Systems of Equivalent Particles	*229*
8.7	Electronic states	*233*
8.8	Recapitulation	*234*
	Discussion Questions	*237*
	Problems	*238*
	References	*239*

Chapter 9/Continuous Groups

9.1	Group-Element Continua	240
9.2	Infinitesimal Operators for Single-Parameter Groups	241
9.3	The Canonical Parameter	244
9.4	Form-Preserving Transformations of Functions	246
9.5	Canonical Coordinates	249
9.6	Elements of Single-Piece Multiparameter Groups	251
9.7	Important Lie Groups	253
9.8	Commutators of the Infinitesimal Operators	258
9.9	Lie Algebras	260
9.10	The $U(1, c)$ Group	262
9.11	The $U(2, c)$ Group	263
	Discussion Questions	265
	Problems	267
	References	269

Chapter 10/Rotation Groups

10.1	Introduction	271
10.2	Two-Dimensional Rotation Groups	272
10.3	The $SO(3)$ Spherical Group	276
10.4	Key Eigenvalue Equations for the Spherical Group	280
10.5	Behavior of Irreducible Tensor Components	284
10.6	The Relationship of $SU(2)$ to $SO(3)$	284
10.7	Finite Operators of the $SU(2)$ Group	285
10.8	Commutation Relations for a Step Operator	287
10.9	Conditions Defining Related Groups	289
	Discussion Questions	290
	Problems	292
	References	293

Chapter 11/Physical Lie Algebras

11.1	General Considerations	294
11.2	Characterizing Creation and Annihilation Operators	295
11.3	Commutators of Bilinear Products of the Operators	297
11.4	Proton-Neutron Systems	298
11.5	The Quark Lattice	301
11.6	Classifying Fundamental Particles	311
11.7	Commutators of the $SU(3)$ Operators	313
11.8	$SU(3)$ Young Diagrams	314

11.9	The Octet Degeneracy	*317*
11.10	A Representation of the Color Field	*318*
	Discussion Questions	*320*
	Problems	*322*
	References	*323*

Appendix/Characters and Bases for the Primitive Symmetry Species of Important Groups

A.1	Bases for Representations of Geometric Symmetry Groups	*325*
A.2	Properties of the Primitive Symmetry Species of Geometric Symmetry Groups	*326*
A.3	Characters for the Primitive Symmetry Species Introduced into Rotation Groups by Extending the Period of Rotation from 2π to 4π	*352*
A.4	Characters and Spin Bases for the Primitive Symmetry Species of Full Permutation Groups	*353*
	Permutation Groups	*353*
	Answers to Problems	*355*

Index *367*

Preface

This work introduces students to the aspects of group theory that are most important in applications. As such, it can serve as a text for a senior or first-year graduate course taken by physicists, chemists, and applied mathematicians. In the development, concrete examples accompany the abstract development. Thus, understanding and usefulness are enhanced.

Since the primary applications of group theory are to symmetric structures, the nature of symmetry operations is considered first. How these make up a group is noted. Then the structures of key groups are described using generators and Cayley diagrams.

Classes of group elements are defined, class sums formed, and the characters for symmetry species determined. Their use in constructing symmetry-adapted structures for physical systems is then developed. Applications are made to vibrating systems, to continuum mechanics, to quantum structures.

The text then contains considerable material on product systems. Some special techniques needed for permutation groups are developed. These include the development of Young tables and diagrams. How spin and orbital states combine is thus elucidated.

Each piece of a continuous group is generated by infinitesimal operators. These combine to form a Lie algebra. In this algebra commutation relations play a key role. Thus, we obtain a basis for the use of these relations in angular-momentum theory, in fundamental-particle theory, in general quantum mechanics.

How bilinear products of creation and annihilation operators lead to Lie algebras is considered. Thus, a basis is laid for quark theory, for interpreting particle multiplets.

CHAPTER 1 / *Symmetry Operations*

1.1
Causality and Symmetry

The universe is not a single indivisible whole. Instead, it consists of parts that can act independently in spite of interactions binding the parts together. This feature allows observers and observing instruments to exist. It also permits both analysis and synthesis to proceed.

An observer first notes that events in his life fall in order; he experiences *time* locally. Second, whatever he observes can be located at points or small regions in a 3-dimensional *space* based on his own position at the time of observation. Similar relationships presumably prevail for an observing instrument.

Time is not observed as a global entity but as an independent property of the observing point or small region. It behaves as a directed coordinate orthogonal to the three spatial coordinates of the point. Furthermore, an interval of time can be measured by the distance traveled by a photon in the interval. Consequently, the arena in which phenomena occur is a 4-dimensional continuum in which a displacement may be oriented to be either timelike or spacelike.

In constructing science, one seeks out the *patterns* that exist among the observations. One presumes that the material world is not capricious or lawless—that it is not governed by spirits as primitive man believed. If certain events appear to follow as a consequence of particular conditions, these events are said to be *caused* by the conditions. Thus in Newtonian mechanics, one says that the acceleration of a body is caused by the net force acting on it.

In principle, uniqueness need not prevail. A given set of conditions, a given cause, may lead to various possible results rather than to a single result. Then *degeneracy* is said to obtain. For instance, the radioactive nuclei in a sample may be shown to be identical by statistical tests. Nevertheless, they will disintegrate at random times with a definite half life.

In general, we will call the part of the universe under study a *system*. The system may be subdivided in various ways. And the resulting parts may be further subdivided. Each of the subsystems may be considered a system in its own right in the approximation that it behaves as an entity.

Now, an operation performed on a system may yield an equivalent system with the same spectrum of properties. The entity under study is then said to possess *symmetry*. The operation is called a symmetry operation.

When these conditions are only approximately satisfied, one says that a near-symmetry exists. The system may then be considered as a pertubation of a corresponding symmetric entity. When the perturbation is small, it may be neglected.

Symmetry operations may act in position space, or in the space-time continuum. They may act in momentum space, or in phase space. Alternatively, they may act in a more general space or plot.

Symmetry operations may also involve other attributes besides position and momentum. Examples of these include particle spin, isotopic spin, hypercharge, color.

1.2
Common Symmetry Operations

For certain properties, the behavior of a system may be governed by a particular function or operator. In classical mechanics, the discriminating function may be a potential, a Lagrangian, or a Hamiltonian. In quantum mechanics, the discriminating operator may be that for some angular momentum or energy.

Now, any system for which distinct operations fail to alter the form of a discriminating function or operator is said to possess symmetry. The operations that leave the pertinent function or operator unchanged are called *symmetry operations*.

Some of the processes that transform a symmetric region into an equivalent region are geometric, while some are not. Others consist of a geometric change combined with a nongeometric change. Each geometric symmetry operation occurs with respect to a structure in the system, a *base*.

A nongeometric alteration is called a *conversion*. In magnetic systems, a conversion involves changing the magnetic state of a particle (as when spins are reversed). In colored systems, a conversion involves changing one color to another in a cycle. In particle systems, a conversion may involve changing one particle into another.

Symmetry operations often met in dealing with the mechanics of macroscopic and microscopic systems are described in Table 1.1. The geometric processes include reorientations about a base, translations, and translations combined with reorientations. In symmetric systems, these appear as permutation of like parts.

1.3
Reorientation Matrices

The reorientations in Table 1.1 may be carried out on the physical system or on the coordinate axes. The first kind is said to be *active*, the second kind *passive*. Both kinds are described by homogenous linear transformations of appropriate Cartesian coordinates. Such transformations can be represented by linear matrix equations.

Where there is a point about which the reorientation is executed, this point is chosen as the origin. Where there is more than one such invariant point, a representative point from these points is chosen as the origin. The axes are drawn in appropriate directions.

The coordinates of a typical point of the system before and after the transformation under discussion are designated (x, y, z) and (x', y', z'), respectively. These are related by the equation

$$\begin{pmatrix} x' \\ y' \\ z' \end{pmatrix} = \begin{pmatrix} A_{11} & A_{12} & A_{13} \\ A_{21} & A_{22} & A_{23} \\ A_{31} & A_{32} & A_{33} \end{pmatrix} \begin{pmatrix} x \\ y \\ z \end{pmatrix} \qquad (1.1)$$

which may be abbreviated as

$$\mathbf{r'} = \mathbf{Ar}. \qquad (1.2)$$

In the *identity* operation I there is no change and formula (1.1)

Table 1.1 Physically Important Symmetry Operations

Reference Structure or Base	Corresponding Geometric Process	Symbols*	The Geometric Process Followed by Some Other Change	Symbols*
All space	Identity	1 or $\underline{1}$	Conversion only	θ or $\underline{1}$
Mirror plane	Reflection in the plane	$\sigma \quad m$	Reflection, then conversion	$\theta\sigma \quad \underline{m}$
Straight line	Rotation by $1/n$ turn about the line	$C_n \quad n$ $(2,3,\ldots)$	Rotation followed by the conversion	$\theta C_n \quad \underline{n}$ $(2,\underline{3},\ldots)$
Point	Inversion	$i \quad \bar{1}$	Inversion, then conversion	$\theta i \quad \underline{\bar{1}}$
Point on straight line	Rotation by $1/n$ turn followed by inversion	$iC_n \quad \bar{n}$	The rotoinversion followed by the conversion	$\theta i C_n \quad \underline{\bar{n}}$
Point on plane	Rotation by $1/n$ turn, then reflection in the plane	$S_n \quad n/m$	The rotoreflection followed by the conversion	$\theta S_n \quad \underline{n/m}$
Vector	Translation	$T \quad t$	Translation, then conversion	$\theta T \quad \underline{t}$
Glide plane	Translation followed by reflection in the plane	$\sigma T \quad a,b,$ c,n,d	Translation-reflection, then conversion	$\theta\sigma T \quad \underline{a},\underline{b},$ $\underline{c},\underline{n},\underline{d}$
Screw axis	Translation, then rotation by $1/n$ turn about the axis	$C_n T \quad n_j$	The screw motion followed by conversion	$\theta C_n T \quad \underline{n}_j$

*The notation in the first column is modeled after that of Schönflies, in the second column after that of Hermann and Mauguin.

reduces to

$$\begin{pmatrix} x' \\ y' \\ z' \end{pmatrix} = \begin{pmatrix} 1 & 0 & 0 \\ 0 & 1 & 0 \\ 0 & 0 & 1 \end{pmatrix} \begin{pmatrix} x \\ y \\ z \end{pmatrix}. \qquad (1.3)$$

The symbolic form for equation (1.3) is

$$\mathbf{r'} = \mathbf{I}\mathbf{r}. \qquad (1.4)$$

The base for a 1-dimensional *reflection* may be made a coordinate plane. When the plane is the yz plane, the operation changes the sign of coordinate x and

$$\begin{pmatrix} x' \\ y' \\ z' \end{pmatrix} = \begin{pmatrix} -1 & 0 & 0 \\ 0 & 1 & 0 \\ 0 & 0 & 1 \end{pmatrix} \begin{pmatrix} x \\ y \\ z \end{pmatrix}. \qquad (1.5)$$

A symbolic form for equation (1.5) is

$$\mathbf{r'} = \sigma_v \mathbf{r}. \qquad (1.6)$$

Reflection by a vertical plane is often labeled σ_v. Reflection by the horizontal xy plane is labeled σ_h. Operation σ_d is reflection through a vertical plane that forms part of a dihedral angle of symmetry, or that bisects the angle between successive σ_v planes.

Reflection through a point is called *inversion i*. With the origin this point, the matrix equation for the operation is

$$\begin{pmatrix} x' \\ y' \\ z' \end{pmatrix} = \begin{pmatrix} -1 & 0 & 0 \\ 0 & -1 & 0 \\ 0 & 0 & -1 \end{pmatrix} \begin{pmatrix} x \\ y \\ z \end{pmatrix} \qquad (1.7)$$

or

$$\mathbf{r'} = \mathbf{i}\mathbf{r}. \qquad (1.8)$$

Symmetry Operations

Rotation of a physical body counterclockwise by angle ϕ about the z axis produces the same change in the coordinates of a point in the body as rotation of the coordinate axes clockwise by angle ϕ about the z axis. From definitions of the sine and cosine, the change is described by the equation

$$\begin{pmatrix} x' \\ y' \\ z' \end{pmatrix} = \begin{pmatrix} \cos\phi & -\sin\phi & 0 \\ \sin\phi & \cos\phi & 0 \\ 0 & 0 & 1 \end{pmatrix} \begin{pmatrix} x \\ y \\ z \end{pmatrix}, \quad (1.9)$$

which may be abbreviated as

$$\mathbf{r'} = \mathbf{C}_n \mathbf{r} \quad (1.10)$$

if

$$\phi = 2\pi/n. \quad (1.11)$$

and

$$\mathbf{r'} = \mathbf{C}_n^m \mathbf{r} \quad (1.12)$$

when

$$\phi = 2\pi(m/n). \quad (1.13)$$

When all symmetry rotations about a given axis have the form C_n^m where n is an integer and

$$m = 1, 2, \ldots, n-1, n, \quad (1.14)$$

then the axis is said to be an n-fold axis. Rotations by $\frac{1}{2}$ turn about axes perpendicular to the n-fold axis are labeled C_2', C_2'', ... Operation C_n followed by inversion i through the origin is called *rotoinversion*. Operation C_n followed by reflection σ_h with respect to a plane perpendicular to the axis of rotation is called a *rotoreflection* S_n.

Since a symmetry operation changes a physical system into an equivalent system, it does not introduce any distortion. However, it

Example 1.1

Construct a reorientation matrix for the C_3 operation.

Equation (1.9) is the form equation (1.1) assumes when the operation consists of rotation by angle ϕ about the z axis. When ϕ is ⅓ turn, the cosine is $-\tfrac{1}{2}$ and the sine is $\sqrt{3}/2$. Then the square matrix in equation (1.9) becomes

$$\mathbf{C}_3 = \begin{pmatrix} -\tfrac{1}{2} & -\tfrac{\sqrt{3}}{2} & 0 \\ \tfrac{\sqrt{3}}{2} & -\tfrac{1}{2} & 0 \\ 0 & 0 & 1 \end{pmatrix}.$$

Example 1.2

Construct a matrix that represents reorientation S_3.

From the definition, S_3 equals rotation C_3 followed by reflection σ_h. When the axis of rotation is the z axis, the reflection changes the sign of z. We have

$$\mathbf{r}' = \begin{pmatrix} x' \\ y' \\ z' \end{pmatrix} = \begin{pmatrix} 1 & 0 & 0 \\ 0 & 1 & 0 \\ 0 & 0 & -1 \end{pmatrix} \begin{pmatrix} x \\ y \\ z \end{pmatrix} = \sigma_h \mathbf{r}$$

and

$$\mathbf{S}_3 = \sigma_h \mathbf{C}_3 = \begin{pmatrix} 1 & 0 & 0 \\ 0 & 1 & 0 \\ 0 & 0 & -1 \end{pmatrix} \begin{pmatrix} -\tfrac{1}{2} & -\tfrac{\sqrt{3}}{2} & 0 \\ \tfrac{\sqrt{3}}{2} & -\tfrac{1}{2} & 0 \\ 0 & 0 & 1 \end{pmatrix}$$

$$= \begin{pmatrix} -\frac{1}{2} & -\frac{\sqrt{3}}{2} & 0 \\ \frac{\sqrt{3}}{2} & \frac{1}{2} & 0 \\ 0 & 0 & -1 \end{pmatrix}.$$

1.4 Operations Involving Translations

The reorientations just considered are described by homogenous linear transformations. When the homogeneity is dropped by adding a constant vector to the right side of equation (1.2), the operation includes a translation.

A pure *translation* entails displacing each point of the system by a constant vector $\mathbf{a}$:

$$\mathbf{r}' = \mathbf{r} + \mathbf{a}. \tag{1.15}$$

When the translation is followed by a reorientation effected by matrix $\mathbf{R}$, we have

$$\mathbf{r}'' = \mathbf{R}\mathbf{r}' = \mathbf{R}(\mathbf{r} + \mathbf{a}) = \mathbf{R}\mathbf{r} + \mathbf{R}\mathbf{a} = \mathbf{R}\mathbf{r} + \mathbf{b}. \tag{1.16}$$

Note that $\mathbf{b}$ is the translation $\mathbf{R}\mathbf{a}$.

When a system repeats itself at regular intervals in a certain direction, it is said to possess translational symmetry and to be crystalline in that direction. When it is periodic in three independent directions, the unit that is repeated again and again is called a *unit cell*. One of these cells can be picked as reference and its edges labeled.

$$\mathbf{a}_1 = a_1 \mathbf{e}_1, \quad \mathbf{a}_2 = a_2 \mathbf{e}_2, \quad \mathbf{a}_3 = a_3 \mathbf{e}_3, \tag{1.17}$$

where $\mathbf{e}_1$, $\mathbf{e}_2$, and $\mathbf{e}_3$ are the appropriate unit vectors. Then the symmetry translations for the system involve

$$\mathbf{a} = \Sigma n_j \mathbf{a}_j = \Sigma n_j a_j \mathbf{e}_j, \tag{1.18}$$

where n_1, n_2, and n_3 are integers. Equivalent to the point at the origin is the point

$$\mathbf{r} = \Sigma n_j \mathbf{a}_j. \tag{1.19}$$

Note how $\mathbf{a}_1$, $\mathbf{a}_2$, and $\mathbf{a}_3$ serve as base vectors.

Designate the angle between $\mathbf{a}_2$ and $\mathbf{a}_3$ as α, that between $\mathbf{a}_3$ and $\mathbf{a}_1$ as β, and that between $\mathbf{a}_1$ and $\mathbf{a}_2$ as γ. Then the unit cells needed to fit observed crystals satisfy the conditions listed in Table 1.2.

If a sinusoidal disturbance

$$\mathbf{F} = \mathbf{F}_0 \sin (\mathbf{k} \cdot \mathbf{r} - \phi) \sin \omega t \tag{1.20}$$

is to affect equivalent positions in a crystal in the same manner, wavevector $\mathbf{k}$ must be chosen so that $\mathbf{k} \cdot \mathbf{r}$ increases by an integral number of 2π radians for each symmetry translation $\mathbf{a}$. This condition is satisfied when the wavevector is an integral combination

$$\mathbf{k} = \Sigma h_j \mathbf{A}_j \tag{1.21}$$

of the *reciprocal vectors*

$$\mathbf{A}_j = 2\pi \frac{\mathbf{a}_{j+1} \times \mathbf{a}_{j+2}}{\mathbf{a}_1 \cdot \mathbf{a}_2 \times \mathbf{a}_3}. \tag{1.22}$$

Here h_1, h_2, and h_3 are integers and numbers 1, 2, 3 are considered to be in cyclic order (so $3 + 1 = 1, \dots$).

Table 1.2 Properties of 3-Dimensional Unit Cells

Lattice	Unit-cell Edges	Angles
Triclinic	$a_1 \neq a_2 \neq a_3$	$\alpha \neq \beta \neq \gamma \neq 90°$
Monoclinic	$a_1 \neq a_2 \neq a_3$	$\alpha = \gamma = 90°, \beta \neq 90°$
Orthorhombic	$a_1 \neq a_2 \neq a_3$	$\alpha = \beta = \gamma = 90°$
Tetragonal	$a_1 = a_2 \neq a_3$	$\alpha = \beta = \gamma = 90°$
Cubic	$a_1 = a_2 = a_3$	$\alpha = \beta = \gamma = 90°$
Hexagonal	$a_1 = a_2 \neq a_3$	$\alpha = \beta = 90°, \gamma = 120°$
Rhombohedral	$a_1 = a_2 = a_3$	$\alpha = \beta = \gamma \neq 90°$

For then

$$\mathbf{A}_j \cdot \mathbf{a}_k = 2\pi\delta_{jk} \qquad (1.23)$$

and at equivalent points from equation (1.19) we have

$$\mathbf{k} \cdot \mathbf{r} = (\Sigma h_j \mathbf{A}_j) \cdot (\Sigma n_k \mathbf{a}_k) = \Sigma h_j n_k 2\pi\delta_{jk} = 2\pi\Sigma h_j n_j$$

$$= 2\pi \text{ (an integer).} \qquad (1.24)$$

The wavevectors

$$\mathbf{k} = \Sigma h_j \mathbf{A}_j, \qquad (1.25)$$

with each h_j an integer, define an array of points called the *reciprocal lattice*. This lattice appears in a plot of the wavevectors, that is, in **k**-space.

In either **k**-space or **r**-space (physical space) the smallest possible unit cell is called a *primitive cell*. This is not necessarily the same as the conventional unit cell. Thus, the face-centered and the body-centered cubic lattices have rhombohedral primitive cells. Edges $\mathbf{a}_1$, $\mathbf{a}_2$, $\mathbf{a}_3$ of these are illustrated in Figures 1.1 and 1.2.

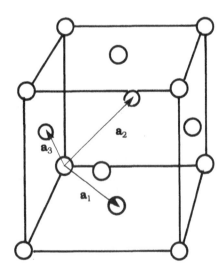

Figure 1.1 Primitive base vectors for a face-centered cubic lattice.

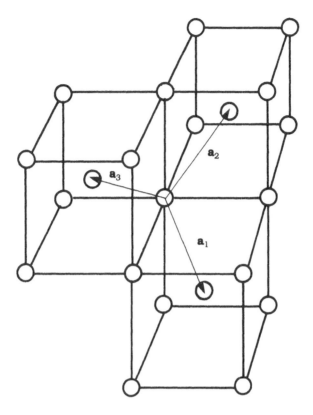

Figure 1.2 Primitive base vectors for a body-centered cubic lattice.

Example 1.3

A primitive cell of a face-centered cubic lattice is bounded by the vectors

$$\mathbf{a}_1 = \frac{d}{2}(\mathbf{i} + \mathbf{j}), \qquad \mathbf{a}_2 = \frac{d}{2}(\mathbf{j} + \mathbf{k}), \qquad \mathbf{a}_2 = \frac{d}{2}(\mathbf{i} = \mathbf{k}),$$

where d is the length of an edge of the unit cube. What lattice is reciprocal to this lattice?

From the determinant representation of the triple scalar product, we obtain

$$\mathbf{a}_1 \cdot \mathbf{a}_2 \times \mathbf{a}_3 = \frac{d^3}{8} \begin{vmatrix} 1 & 1 & 0 \\ 0 & 1 & 1 \\ 1 & 0 & 1 \end{vmatrix} = \frac{d^3}{4}.$$

Then using the determinant representation of the vector product in equation (1.22) leads to

$$\mathbf{A}_1 = \frac{2\pi}{d^3/4} \frac{d^2}{4} \begin{vmatrix} \mathbf{i} & \mathbf{j} & \mathbf{k} \\ 0 & 1 & 1 \\ 1 & 0 & 1 \end{vmatrix} = \frac{2\pi}{d}(\mathbf{i} + \mathbf{j} - \mathbf{k}),$$

$$\mathbf{A}_2 = \frac{2\pi}{d^3/4} \frac{d^2}{4} \begin{vmatrix} \mathbf{i} & \mathbf{j} & \mathbf{k} \\ 1 & 0 & 1 \\ 1 & 1 & 0 \end{vmatrix} = \frac{2\pi}{d}(-\mathbf{i} + \mathbf{j} + \mathbf{k}),$$

$$\mathbf{A}_3 = \frac{2\pi}{d^3/4} \frac{d^2}{4} \begin{vmatrix} \mathbf{i} & \mathbf{j} & \mathbf{k} \\ 1 & 1 & 0 \\ 0 & 1 & 1 \end{vmatrix} = \frac{2\pi}{d}(\mathbf{i} - \mathbf{j} + \mathbf{k}),$$

Vectors $\mathbf{A}_1$, $\mathbf{A}_2$, $\mathbf{A}_3$ bound a primitive cell of a body-centered cubic lattice of edge length $4\pi/d$. So the reciprocal lattice for an **r**-space face-centered cubic lattice is a **k**-space body-centered cubic lattice.

1.5
Permutation Matrices

Many physical systems are composed of equivalent parts. A symmetry operation then acts by permuting these parts, yielding an equivalent system.

Permutation Matrices

To represent such a process, a person may locate equivalent positions in the equivalent parts and establish a reference point that is not moved during the transformation. The vector drawn from the reference point to the chosen position for the jth part is labeled r_j before, and r_j' after, the transformation. One then forms the column matrices

$$\mathbf{r} = \begin{pmatrix} r_1 \\ r_2 \\ \cdot \\ \cdot \\ \cdot \\ r_n \end{pmatrix}, \quad \mathbf{r}' = \begin{pmatrix} r_1' \\ r_2' \\ \cdot \\ \cdot \\ \cdot \\ r_n' \end{pmatrix}, \quad (1.26)$$

in which n is the total number of parts permuted.

From the changes that a given operation causes, one relates the vectors:

$$r_j' = r_k. \quad (1.27)$$

These relations are inserted into matrix $\mathbf{r}'$ and matrix $\mathbf{r}$ is factored out to give

$$\begin{pmatrix} r_1' \\ r_2' \\ \cdot \\ \cdot \\ \cdot \\ r_n' \end{pmatrix} = \begin{pmatrix} \cdot & \cdot & \cdot & 0 & 1 & 0 & \cdot & \cdot & \cdot \\ \cdot & & & & & & & & \cdot \\ \cdot & & & & & & & & \cdot \\ 1 & 0 & \cdot & \cdot & \cdot & \cdot & \cdot & \cdot & 0 \\ \cdot & & & & & & & & \cdot \\ \cdot & & & & & & & & \cdot \end{pmatrix} \begin{pmatrix} r_1 \\ r_2 \\ \cdot \\ \cdot \\ \cdot \\ r_n \end{pmatrix}. \quad (1.28)$$

This may be abbreviated as

$$\mathbf{r}' = \mathbf{A}\mathbf{r} \quad (1.29)$$

with $\mathbf{A}$ being called a *permutation matrix* for the operation.

In a *cyclic* permutation, the transformation does not affect the order of the radius vectors. The matrix $\mathbf{A}$ representing such an operation may consist of a sequence of 1's along one diagonal and 0's everywhere else.

A general permutation can be broken down into cyclic permutations. The corresponding matrix partitions into null matrices and matrices representing the cyclic actions.

Example 1.4

Construct a permutation matrix representing the C_4 operation.

A system for which C_4 is a symmetry operation contains four parts that are permuted cyclicly by the operation. (A simple example appears in Figure 1.3). Initially, the parts are arranged as shown, with vectors r_1, r_2, r_3, r_4, r_5 drawn from a point on the axis (an invariant point) to the first, second, third, fourth, and fifth parts. After a transformation, the parts have moved, together with their vectors, which are now designated $r_1', r_2', r_3', r_4', r_5'$.

Under the C_4 operation, the first body moves to the second position, the second body to the third position, the third body to the fourth position, the fourth body to the first position, and the sixth body is merely rotated. Consequently,

$$r_1' = C_4 r_1 = r_2$$
$$r_2' = C_4 r_2 = r_3,$$
$$r_3' = C_4 r_3 = r_4,$$
$$r_4' = C_4 r_4 = r_1,$$
$$r_5' = C_4 r_5 = r_5,$$

and we have

$$\begin{pmatrix} r_1' \\ r_2' \\ r_3' \\ r_4' \end{pmatrix} = \begin{pmatrix} C_4 r_1 \\ C_4 r_2 \\ C_4 r_3 \\ C_4 r_4 \end{pmatrix} = \begin{pmatrix} r_2 \\ r_3 \\ r_4 \\ r_1 \end{pmatrix} = \begin{pmatrix} 0 & 1 & 0 & 0 \\ 0 & 0 & 1 & 0 \\ 0 & 0 & 0 & 1 \\ 1 & 0 & 0 & 0 \end{pmatrix} \begin{pmatrix} r_1 \\ r_2 \\ r_3 \\ r_4 \end{pmatrix}.$$

In the last step, the matrix

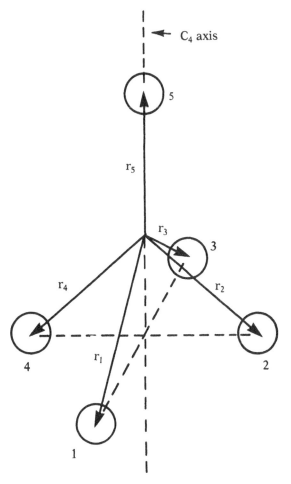

Figure 1.3 Five equivalent bodies or kets arranged about a 4-fold axis of symmetry.

$$\begin{pmatrix} r_1 \\ r_2 \\ r_3 \\ r_4 \end{pmatrix} \text{ is factored out of } \begin{pmatrix} r_2 \\ r_3 \\ r_4 \\ r_1 \end{pmatrix}.$$

Comparing the overall equation with

Symmetry Operations

$$\mathbf{r}' = \mathbf{C}_4 \mathbf{r},$$

we obtain the permutation matrix

$$\mathbf{C}_4 = \begin{pmatrix} 0 & 1 & 0 & 0 \\ 0 & 0 & 1 & 0 \\ 0 & 0 & 0 & 1 \\ 1 & 0 & 0 & 0 \end{pmatrix}.$$

Example 1.5

Construct a permutation matrix representing a σ_d operation on the system in Figure 1.3.

Under the σ_d operation, r_1 and r_2 are interchanged; also r_3 and r_4 are interchanged. Thus

$$r_1' = \sigma_d r_1 = r_2$$

$$r_2' = \sigma_d r_2 = r_1,$$

$$r_3' = \sigma_d r_3 = r_4,$$

$$r_4' = \sigma_d r_4 = r_3,$$

and

$$\begin{pmatrix} r_1' \\ r_2' \\ r_3' \\ r_4' \end{pmatrix} = \begin{pmatrix} \sigma_d r_1 \\ \sigma_d r_2 \\ \sigma_d r_3 \\ \sigma_d r_4 \end{pmatrix} = \begin{pmatrix} r_2 \\ r_1 \\ r_4 \\ r_3 \end{pmatrix} = \begin{pmatrix} 0 & 1 & 0 & 0 \\ 1 & 0 & 0 & 0 \\ 0 & 0 & 0 & 1 \\ 0 & 0 & 1 & 0 \end{pmatrix} \begin{pmatrix} r_1 \\ r_2 \\ r_3 \\ r_4 \end{pmatrix}$$

With

$$\mathbf{r}' = \sigma_d \mathbf{r},$$

we have the representation

$$\sigma_d = \begin{pmatrix} 0 & 1 & 0 & 0 \\ 1 & 0 & 0 & 0 \\ 0 & 0 & 0 & 1 \\ 0 & 0 & 1 & 0 \end{pmatrix}.$$

1.6
How Symmetry Operations for a System Constitute a Group

The symmetry operations for a particular aspect of a system constitute a set with notable properties, which we will now consider.

Among the symmetry operations, there is one that reproduces the original configuration, regardless of what it is. This is the identity operation I. Furthermore, no transformation in the set destroys information about the system. Hence, each is reversible. Corresponding to each operation A is an *inverse* operation A^{-1}.

Since no operation in the set has an appreciable effect on the governing function or operator, potential or Hamiltonian, each of the transformed systems is like a possible form of the untransformed system. And each of the operations changes a transformed system to a configuration that is still like a possible form of the untransformed system.

Since the set of operations is complete, by assumption, it contains an operation that takes the system directly from the original state to the twice-transformed state. Therefore, any two of the operations, A, and B, combine to form an operation in the set C,

$$BA = C \tag{1.30}$$

and the set is said to be *closed*. Each operation in equation (1.30) may act on a following expression. So in product BA, we consider that operation A would act first, then operation B.

In a series of three transformations, the first two operations can be replaced by the equivalent single operation, or the last two by the equivalent one, without affecting the result:

$$CBA = C(BA) = (CB)A. \qquad (1.31)$$

As long as the sequence of operations is maintained, how the factors are associated has no influence; the *associative law* is obeyed.

Let us abstract these properties of the symmetry operations for a given system. We thus come to the following formulation.
Any set of elements

$$A_1, A_2, \ldots, A_g \qquad (1.32)$$

with a prescription for combining pairs that meets the four conditions,

   (a) presence of the identity,
   (b) presence of a complete set of inverses,
   (c) closure,
   (d) associatativity,

is called a *group*. The *order* of the group is the number of elements g in it. If part of the set also meets these conditions, it forms a subgroup of the group.

For convenience, the combining operation is often called *multiplication*.

1.7
Identifying Spatial Point Groups

The operations that transform a physical system into equivalent forms, systems with the same potential function or Hamiltonian operator as the original system, and complete subsets thereof, form groups. For reorientations about a point in physical space, the pertinent groups are subgroups of the *full rotation group*. This group consists of the complete set of operations that transform a sphere erected on the point as center into itself.

A convenient way to identify such a group is to ask whether certain bases for symmetry operations are present, in the sequence given in Figure 1.4. The symbol at the end of a line labels the corresponding group.

Errors in identification arise if a key base is missed. In particular, the C_2' axes perpendicular to the C_n axis are difficult to see in a $\mathbf{D}_n$ or a $\mathbf{D}_{nd}$ molecule. But when such axes are present, S_{2n-a} and S_{2n+a} are symmetry operations and these are easy to spot. If they are found together with C_n, then $\mathbf{D}_n$ is a covering group. If in addition, a is zero, the molecule possess vertical planes of symmetry (σ_d's) and $\mathbf{D}_{nd}$ is a covering group.

As an alternative procedure, a person can check possible generators for the given group against a standard list. Such generators will be considered next.

1.8
Group Generators and Cayley Diagrams

To describe a group, a person need only specify how each pair of elements in it combine. The results may be collected in a multiplication table. However, a description in terms of generators is more succinct. A diagram may well be constructed in which points representing the elements are linked by lines representing application of the pertinent generators.

Because of the closure property, combining any given element of a group with itself any number of times must yield an element of the group. Such an element is called a *generator* for the elements produced. If the first m powers of generator A,

$$A, A^2, \ldots, A^{m-1}, A^m, \qquad (1.33)$$

are all different and the mth one is the identity,

$$A^m = I, \qquad (1.34)$$

higher powers merely reproduce the elements in the first cycle. For, if we let the power be $nm + p$ where n is any integer and p is a positive integer less than m, we have

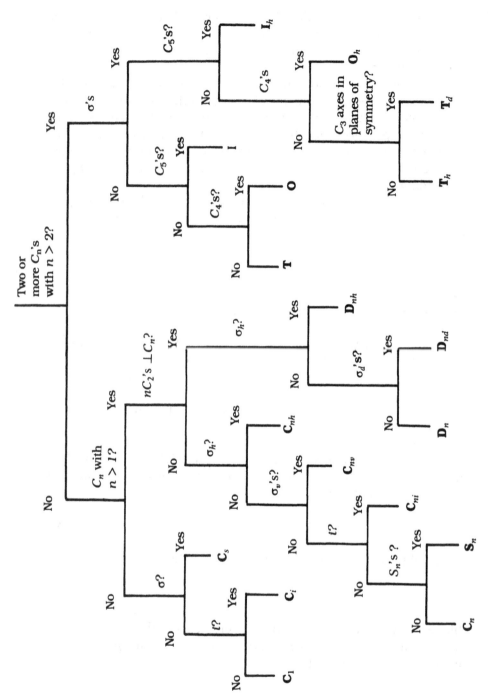

Figure 1.4 Flow chart for identifying subgroups of the full rotation group.

$$A^{mn+p} = (A^m)^n A^p = A^p. \tag{1.35}$$

Number m is called the *order* of A while condition (1.34) is called a *defining relationship*. When a group contains two or more independent generators, A, B, ..., relations defining how the different generators combine to yield the identity also exist. If a group is *finite* (that is, if it contains only a finite number of elements) the order of each generator is, of necessity, finite.

Diagrams, or graphs, illustrating how the elements combine were originated by Arthur *Cayley*. In a typical plot, each element of the group under discussion is assigned a different point. A line directed from one point to another implies application of a generator to the first element to give the second one. When the order of the generator is 2, the inverse of the generator equals the generator and no direction is indicated.

A solid directed line stands for application of generator A, a dashed directed line application of B, a dotted directed line application of C, ... (See Figure 1.5) The arrow on the line is omitted when the direction of travel is immaterial. (See Figure 1.6.) Since each element combines with each generator and its inverse to give elements of the group, the same number and kinds of lines must emanate from each point.

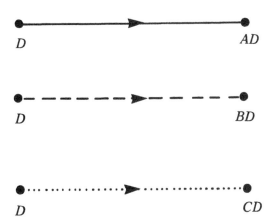

Figure 1.5 Directed lines representing application of operations A, B, and C, respectively, after operation D.

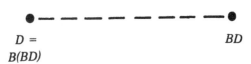

$$D = \qquad\qquad BD$$
$$B(BD)$$

Figure 1.6 Line for which passage in either direction represents the same operation B.

The simplest groups are generated by a single element A of order m, where m is any integer. The element may be combined with itself an infinite number of times. According to equation (1.35), however, combinations of more than m A's merely repeat the elements in the first set of m elements, Thus, each of these groups is *cyclic*.

The next most simple groups are those constructed from one generator of order m and another generator of order 2, with four laps in the smallest circuits involving both generators twice. Diagrams of six-element groups with one generator and with two generators obeying these conditions appear in Figures 1.7 through 1.9.

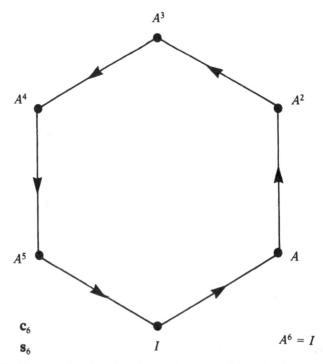

Figure 1.7 Diagram for the six-element group with a single generator A.

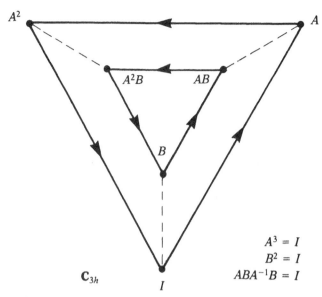

Figure 1.8 Combination of the three-element cyclic group with a second-order generator that does not alter the direction in which the first generator acts.

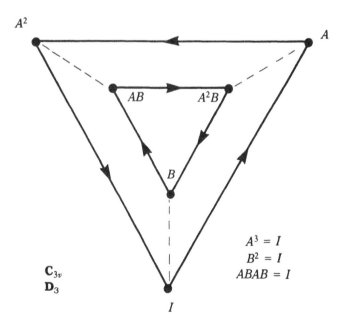

Figure 1.9 Combination of the three-element cyclic group with a second-order generator that reverses the direction in which the first generator acts.

1.9
Structures of Common Point Groups

Generators and defining relations for the most common point groups are listed in Table 1.3. Cayley diagrams for the cyclic groups C_n, S_1, and S_{2n} are constructed like Figure 1.7. Twin cyclic groups C_{nh} yield diagrams like Figure 1.8. Twin cyclic groups C_{nv} and dihedral groups D_n, D_{nd} have structures like the one in Figure 1.9.

Dihedral groups D_{nh} are formed from one generator of order n and two oppositely acting generators of order 2, with four laps in a circuit involving each pair of generators. An example appears in Figure 1.10.

Tetrahedral group **T** contains four threefold cycles linked by twofold cycles. The smallest circuit involving both generators is six laps long. (See Figure 1.11) Tetrahedral group T_h contains eight threefold cycles linked by twofold cycles. The smallest circuit involving the second-order generator is eight laps long. (See Figure 1.12)

Table 1.3 Generating Operations for Common Groups

Symbol for Group	Generators A	B	C	Defining Relations
C_1	I			$A = I$
C_n	C_n			$A^n = I$
$C_s = S_1$	σ_h			$A^2 = I$
$C_i = S_2$	i			$A^2 = I$
S_{2n}	S_{2n}			$A^{2n} = I$
C_{nv}	C_n	σ_v		$A^n = B^2 = ABAB = I$
C_{nh}	C_n	σ_h		$A^n = B^2 = ABA^{-1}B = I$
D_n	C_n	C_2'		$A^n = B^2 = ABAB = I$
D_{nd}	S_{2n}	σ_d		$A^{2n} = B^2 = ABAB = I$
D_{nh}	C_n	σ_v	σ_h	$A^n = B^2 = C^2 = I$
				$(AB)^2 = (BC)^2 = ACA^{-1}C = I$
T	C_3	C_2		$A^3 = B^2 = (AB)^3 = I$
T_h	C_3	C_2		$A^3 = B^2 = (AB)^6 = I$
T_d	S_4	C_3		$A^4 = B^3 = (AB)^4 = I$
O	C_4	C_3		$A^4 = B^3 = (AB)^4 = I$
O_h	C_4	σ_d		$A^4 = B^2 = (AB)^6 = I$
I	C_5	C_2		$A^5 = B^2 = (AB)^3 = I$

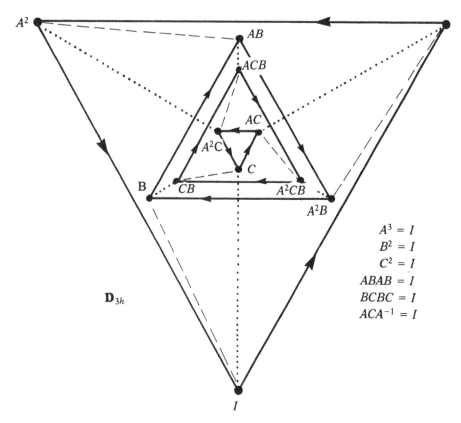

Figure 1.10 Combination of the three-element cyclic group with a second-order generator that does not alter the direction in which the first generator acts and with a second-order generator that reverses this direction.

Tetrahedral group T_d and octahedral group O are constructed from a fourth-order generator and a third-order generator, with the smallest circuit including both, four laps long. (See Figure 1.13) Octahedral group O_h is constructed from a fourth- order and a second-order generator, with the smallest circuit including both generators alternately twelve laps long. (See Figure 1.14.) Icosahedral group I contains twelve fivefold cycles linked by twofold cycles. The smallest circuit involving both generators is six laps long. (See Figure 1.15.)

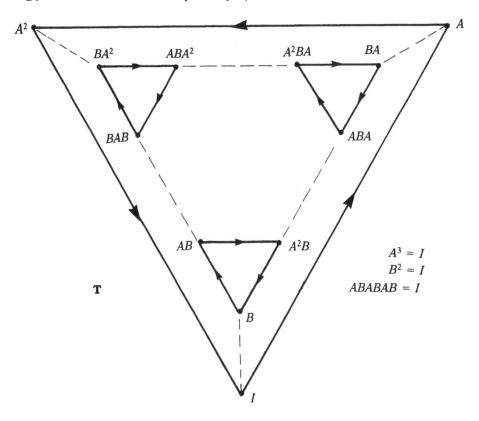

Figure 1.11 Combination of the three-element cyclic group with a second-order generator that shifts the cycle to another triangular face of a truncated tetrahedron.

1.10
Representing Group Elements with Complex Numbers

A cyclic group may be characterized by sets of numbers whose members combine as the group elements do. Each such set is said to provide a *representation* of the group.

Consider a group with one generator A of order n, that is, with the defining relation

$$A^n = I. \tag{1.36}$$

Representing Group Elements with Complex Numbers

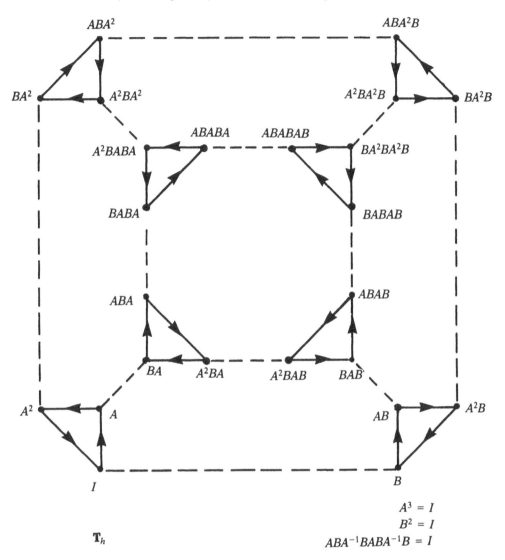

$A^3 = I$
$B^2 = I$
$ABA^{-1}BABA^{-1}B = I$

Figure 1.12 Combination of the three-element cyclic group with a second-order generator that shifts the cycle to another triangular face of a truncated cube.

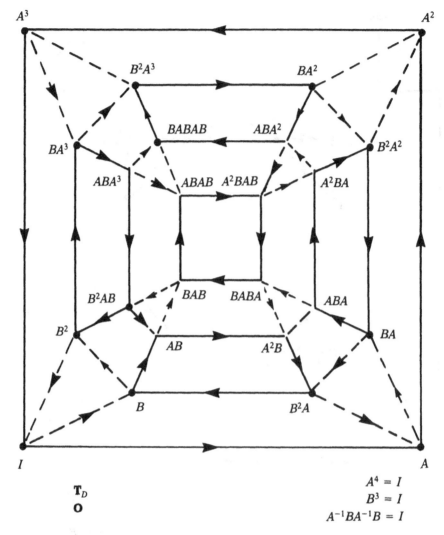

Figure 1.13 Combination of a fourth-order generator with a third-order generator in a rhombicuboctahedral arrangement.

Representing Group Elements with Complex Numbers

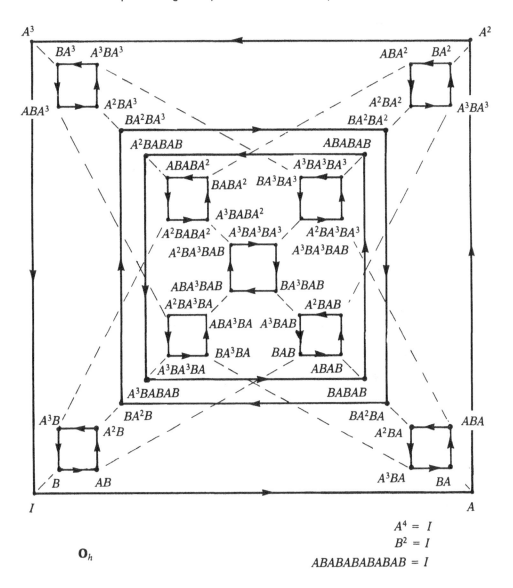

$A^4 = I$
$B^2 = I$
$ABABABABABAB = I$

Figure 1.14 Twelve equivalent four-element cycles linked by a second-order generator.

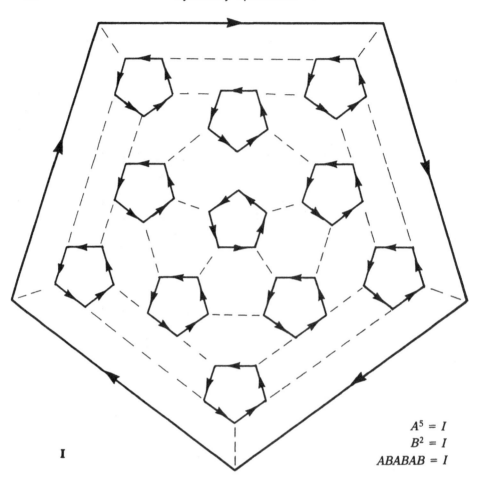

Figure 1.15 Twelve equivalent five-element cycles linked by a second-order generator.

Then any number λ representing A would satisfy the equation

$$\lambda^n = 1. \tag{1.37}$$

But the n roots of (1.37) are

$$\lambda = 1, \omega, \omega^2, \ldots, \omega^{n-1}$$

where

$$\omega = \exp(2\pi i/n). \quad (1.38)$$

Each of these roots may be used to represent A. Element A^r is then the rth power of the one chosen. We thus obtain the n distinct 1-dimensional representations listed in Table 1.4. When $n = 2$, Table 1.4 reduces to Table 1.5.

Table 1.4 Numerical Representations for C_n and Isomorphic Groups

	I	A	...	A^r	...	A^{n-1}
Γ_0	1	1	...	1	...	1
Γ_1	1	ω	...	ω^r	...	ω^{n-1}
Γ_2	1	ω^2	...	ω^{2r}	...	$\omega^{2(n-1)}$
...	...	...	...	...	...	...
Γ_{n-1}	1	ω^{n-1}	...	$\omega^{(n-1)r}$	...	$\omega^{(n-1)(n-1)}$

Table 1.5 Numerical Representations for C_2

	I	B
Γ_0	1	1
Γ_1	1	-1

Now, consider a group with two generators, A of order n and B of order 2. Also consider that B commutes with A. The defining relations are then

$$A^n = I, \ B^2 = I, \ AB = BA. \quad (1.39)$$

The subgroup involving only powers of A is represented by Table 1.4; the subgroup involving only powers of B, by Table 1.5. Since A commutes with B, these elements combine as numbers. The resulting representations are found by combining the two tables as in Table 1.6.

Table 1.6 Numerical Representations for C_{nh} and Isomorphic Groups

	I	A	$\cdots$	A^r	$\cdots$	B	$BA = AB$	$\cdots$	$BA^r = A^rB$	$\cdots$
Γ_0'	1	1	$\cdots$	1	$\cdots$	1	1	$\cdots$	1	$\cdots$
Γ_1'	1	ω	$\cdots$	ω^r	$\cdots$	1	ω	$\cdots$	ω^r	$\cdots$
Γ_2'	1	ω^2	$\cdots$	ω^{2r}	$\cdots$	1	ω^2	$\cdots$	ω^{2r}	$\cdots$
$\vdots$	$\vdots$	$\vdots$	$\ddots$	$\vdots$		$\vdots$	$\vdots$	$\ddots$	$\vdots$	
Γ_{n-1}'	1	ω^{n-1}	$\cdots$	$\omega^{(n-1)r}$	$\cdots$	1	ω^{n-1}	$\cdots$	$\omega^{(n-1)r}$	$\cdots$
Γ_0''	1	1	$\cdots$	1	$\cdots$	-1	-1	$\cdots$	-1	$\cdots$
Γ_1''	1	ω	$\cdots$	ω^r	$\cdots$	-1	$-\omega$	$\cdots$	$-\omega^r$	$\cdots$
Γ_2''	1	ω^2	$\cdots$	ω^{2r}	$\cdots$	-1	$-\omega^2$	$\cdots$	$-\omega^{2r}$	$\cdots$
$\vdots$	$\vdots$	$\vdots$	$\ddots$	$\vdots$		$\vdots$	$\vdots$	$\ddots$	$\vdots$	
Γ_{n-1}''	1	ω^{n-1}	$\cdots$	$\omega^{(n-1)r}$	$\cdots$	-1	$-\omega^{n-1}$	$\cdots$	$-\omega^{(n-1)r}$	$\cdots$

1.11
Occurrences of Irreducible Matrix Representations

Groups with two noncommuting generators have some matrix representations that cannot be reduced to diagonal form throughout. Thus, these cannot be broken down into numerical representations.

Consider a group with only two generators, A of order n and B of order p, with $p < n$. The defining relations include

$$A^n = I, \; B^p = I. \tag{1.40}$$

Also, suppose that A and B do not commute; so that

$$C = BAB^{-1} \neq A. \tag{1.41}$$

But taking the n-tuple of each side of the equality yields

$$C^n = BAB^{-1} \ldots BAB^{-1} = BA^nB^{-1} = BB^{-1} = I. \tag{1.42}$$

Element C is of order n.

Let us here limit ourselves to groups that do not contain any element of order n except A and its powers; so

$$C = A^t. \tag{1.43}$$

In any case, the elements may be represented by distinct operators acting on vectors in an abstract space. Indeed when vector **v** is the eigenvector of A with eigenvalue λ, we have

$$A\mathbf{v} = \lambda\mathbf{v}. \tag{1.44}$$

And when operator B acts on equation (1.44), we get

$$BA\mathbf{v} = \lambda B\mathbf{v}. \tag{1.45}$$

But from equation (1.41), we have

$$BA = CB; \tag{1.46}$$

so equation (1.45) becomes

$$CB\mathbf{v} = \lambda B\mathbf{v}. \qquad (1.47)$$

Thus, vector $B\mathbf{v}$ is an eigenvector of C with the eigenvalue λ. Substituting equation (1.43) into equation (1.47) leads to

$$A^t B\mathbf{v} = \lambda B\mathbf{v}. \qquad (1.48)$$

We see that $\lambda^{1/t}$ is also an eigenvalue of A. This process can be repeated again and again. We obtain an infinite sequence of eigenvalues

$$\lambda, \lambda^{1/t}, \lambda^{1/t^2}, \ldots, \qquad (1.49)$$

which is not allowed unless

$$t = \pm 1. \qquad (1.50)$$

The upper sign is ruled out by the inequality in equation (1.41); so

$$C = A^{-1} \qquad (1.51)$$

and equation (1.46) yields

$$BA = A^{-1}B \qquad (1.52)$$

whence

$$AB = BA^{-1}. \qquad (1.53)$$

Letting B act on equation (1.45) now gives us

$$B^2 A\mathbf{v} = BA^{-1}B\mathbf{v} = AB^2\mathbf{v} = \lambda B^2\mathbf{v}, \qquad (1.54)$$

a relationship satisfied with

$$B^2 = I. \qquad (1.55)$$

Thus, generator B is of order 2.

Equations (1.44) and (1.48) with $t = -1$ make up the matrix equation

$$A \begin{pmatrix} \mathbf{v} \\ B\mathbf{v} \end{pmatrix} = \begin{pmatrix} \lambda & 0 \\ 0 & \lambda^{-1} \end{pmatrix} \begin{pmatrix} \mathbf{v} \\ B\mathbf{v} \end{pmatrix}. \qquad (1.56)$$

With equation (1.55), we also have

$$B \begin{pmatrix} \mathbf{v} \\ B\mathbf{v} \end{pmatrix} = \begin{pmatrix} 0 & 1 \\ 1 & 0 \end{pmatrix} \begin{pmatrix} \mathbf{v} \\ B\mathbf{v} \end{pmatrix}. \qquad (1.57)$$

Thus, elements A and B are represented by the matrices

$$A = \begin{pmatrix} \lambda & 0 \\ 0 & \lambda^{-1} \end{pmatrix}, \qquad B = \begin{pmatrix} 0 & 1 \\ 1 & 0 \end{pmatrix}. \qquad (1.58)$$

Because the nth power of A is the identity,

$$\lambda^n = 1, \qquad (1.59)$$

and λ may be any of the nth roots of 1 as in formula (1.38).

When λ is not ± 1, the matrix representations are not reducible to numbers. For odd n, λ may be $+1$ but not -1. For even n, λ may be either $+1$ or -1. When λ is $+1$, it equals its reciprocal,

$$\lambda = \lambda^{-1}, \qquad (1.60)$$

the two eigenvectors are equivalent,

$$B\mathbf{v} = \pm \mathbf{v}, \qquad (1.61)$$

and B is represented by factors 1 and -1, while A is represented by the number 1. When λ is -1, equations (1.60) and (1.61) still hold and B is represented by factors 1 and -1, while A is represented by the number -1.

These results are summarized in Tables 1.7 and 1.8 where order n has been replaced by $2n + 1$ when the order is odd and by $2n$ when it is even. Note that the independent products of powers of A and B are not listed, but these have the same dimensionality as the pertinent A and B. For more complicated groups, where equation (1.43) does not hold, higher dimensional irreducible representations occur. The theory in Chapter 5 will cover such cases.

Table 1.7 Simplest Matrix Representations for $C_{2n+1\,v}$ and Isomorphic Groups

	I	A	A^{-1}	$\ldots$	B
Γ_0'	1	1	1	$\ldots$	1
Γ_0''	1	1	1	$\ldots$	-1
$\Gamma_{\pm 1}$	$\begin{pmatrix} 1 & 0 \\ 0 & 1 \end{pmatrix}$	$\begin{pmatrix} \omega & 0 \\ 0 & \omega^{-1} \end{pmatrix}$	$\begin{pmatrix} \omega^{-1} & 0 \\ 0 & \omega \end{pmatrix}$	$\ldots$	$\begin{pmatrix} 0 & 1 \\ 1 & 0 \end{pmatrix}$
$\ldots$					
$\Gamma_{\pm n}$	$\begin{pmatrix} 1 & 0 \\ 0 & 1 \end{pmatrix}$	$\begin{pmatrix} \omega^n & 0 \\ 0 & \omega^{-n} \end{pmatrix}$	$\begin{pmatrix} \omega^{-n} & 0 \\ 0 & \omega^n \end{pmatrix}$	$\ldots$	$\begin{pmatrix} 0 & 1 \\ 1 & 0 \end{pmatrix}$

Table 1.8 Simplest Matrix Representations for $C_{2n\,v}$ and Isomorphic Groups

	I	A	A^{-1}	$\ldots$	B
Γ_0'	1	1	1	$\ldots$	1
Γ_0''	1	1	1	$\ldots$	-1
$\Gamma_{\pm 1}$	$\begin{pmatrix} 1 & 0 \\ 0 & 1 \end{pmatrix}$	$\begin{pmatrix} \omega & 0 \\ 0 & \omega^{-1} \end{pmatrix}$	$\begin{pmatrix} \omega^{-1} & 0 \\ 0 & \omega \end{pmatrix}$	$\ldots$	$\begin{pmatrix} 0 & 1 \\ 1 & 0 \end{pmatrix}$
$\ldots$					
$\Gamma_{\pm(n-1)}$	$\begin{pmatrix} 1 & 0 \\ 0 & 1 \end{pmatrix}$	$\begin{pmatrix} \omega^{n-1} & 0 \\ 0 & \omega^{-n+1} \end{pmatrix}$	$\begin{pmatrix} \omega^{-n+1} & 0 \\ 0 & \omega^{n-1} \end{pmatrix}$	$\ldots$	$\begin{pmatrix} 0 & 1 \\ 1 & 0 \end{pmatrix}$
Γ_0'''	1	-1	-1	$\ldots$	1
Γ_0''''	1	-1	-1	$\ldots$	-1

1.12
Representing Translations in Crystals

Complex numbers also serve to represent translatory operations. In crystals these form groups.

Consider a system that possesses translational symmetry in a particular direction. Let a be the fundamental period in the direction while operation T translates entities by this period. If the radius vector locating a physical point is $\mathbf{r}$ before and $\mathbf{r}'$ after the operation, then

$$\mathbf{r}' = T\mathbf{r}. \tag{1.62}$$

The group of translations is generated by T. However, no power of T yields the identity operations. So we have no definite defining relation and the argument in Section 1.10 based on equation (1.36) does not apply. But since all powers of T (positive and negative) commute, all the simplest representations are numerical.

Let us consider a complex vector $\mathbf{v}$ associated with a point on the axis. Under translation T, its direction is unchanged; but its phase and magnitude may be altered.

$$T\mathbf{v} = \lambda\mathbf{v}. \tag{1.63}$$

This eigenvalue equation has a solution for any complex value of λ. But when $|\lambda| \neq 1$, repetition of T in

$$T^n\mathbf{v} = \lambda^n\mathbf{v} \tag{1.64}$$

or in

$$T^{-n}\mathbf{v} = \lambda^{-n}\mathbf{v} \tag{1.65}$$

leads to a divergence of $\mathbf{v}$. Such divergence would not be allowed in an infinite crystal.

The suitable eigenvalues require

$$|\lambda| = 1. \tag{1.66}$$

These are generally written as

$$\lambda = e^{ika} \tag{1.67}$$

and

38 Symmetry Operations

$$\lambda = e^{-ika}. \tag{1.68}$$

Parameter k is the wavevector described in Section 1.4 following equation (1.2). By equation (1.63), operation T is represented as multiplication by λ. So equations (1.67) and (1.68) lead to the 1-dimensional representations listed in Table 1.7.

Table 1.9 Numerical Representations for Key Elements of the Translation Group

	I	T	...	T^r	...
Γ_0	1	1	...	1	...
Γ_k	1	e^{ika}	...	e^{ikra}	...
Γ_{-k}	1	e^{-ika}	...	e^{-ikra}	...
...	..	..	..	..	..

1.13
Key Concepts

Every operation that transforms a given system into an equivalent system is called a symmetry operation. By equivalent system, we mean one that is indistinguishable from the original one with respect to the properties under study. All the symmetry operations for a system considered in a certain way form a group. Also, certain subsets of the operations may form groups.

To be a group with respect to a law for combining pairs, a set of elements

$$A, B, \ldots P, Q \tag{1.69}$$

must contain

(a) the identity,
(b) an inverse for each element.

Furthermore, the result of each combination of elements must lie

(c) within the set

and must obey

(d) the associative law.

All finite and denumerable groups can be characterized by their generators and defining relations. Important examples appear in Table 1.3. In physically significant groups, the elements are symmetry operations. Such groups are characterized by the bases for the operations. (Recall how Figure 1.2 is used.)

Elements of cyclic groups correlate with complex numbers in a sum of ways equal to the order of the group. The numbers in each set combine by multiplication as the group elements.

Adding a commuting second-order generator doubles the size of each set and the number of sets. Adding a noncommuting generator leads to representation sets of matrices.

Discussion Questions

1.1 Characterize the arena in which physical phenomena occur.
1.2 What is causality, degeneracy?
1.3 What is a system?
1.4 What is symmetry? Where may it exist?
1.5 Cite possible symmetry operations.
1.6 Distinguish between active and passive operations.
1.7 What is a homogenous linear transformation? How can such a transformation be represented by a matrix equation?
1.8 Explain the matrix equations representing the actions of (a) I, (b) σ_v, (c) σ_h, (d) i, (e) C_n, (f) S_n.
1.9 A given symmetry operation may be effected as reorientation R followed by translation $\mathbf{b}$ or as translation $\mathbf{a}$ followed by reorientation R.. How are $\mathbf{a}$ and $\mathbf{b}$ related?
1.10 What restricts the kind of unit cells one can use in crystallography?
1.11 How is the reciprocal of a vector defined?
1.12 Explain how the reciprocal lattice is defined.
1.13 Is information lost on going from the spatial lattice to the reciprocal lattice?
1.14 When can a reorientation be described as a permutation?

1.15 How can such a permutation be represented by a matrix equation?

1.16 How can a general permutation be broken down into cyclic permutations?

1.17 What combinatory characteristics do the symmetry operations for a given physical system exhibit?

1.18 Why is the identity operation included among the elements in a group?

1.19 Why does a physically occurring set of symmetry operations always include an inverse for each operation in the set?

1.20 Why do two symmetry operations applied in succession produce the same effect as a single symmetry operation?

1.21 What common binary operation does not obey the associative law?

1.22 What structure does the closure property impose on a group?

1.23 How does a Cayley diagram display the structure of a group?

1.24 What kind of groups are formed from a single generator?

1.25 How can a second-order generator be combined with an nth-order generator?

1.26 In a Cayley diagram, why must the same number and kinds of lines emanate from each vertex?

1.27 How does one represent (a) a second-order generator, (b) an nth-order generator?

1.28 What is the full rotation group?

1.29 How can one identify a group by noting what bases for symmetry operations are present?

1.30 How does a person identify a group from its generators?

1.31 Explain the structure of a cyclic group.

1.32 What kinds of twin cyclic groups, groups with two nth-order cycles, arise? Explain.

1.33 Explain the structure of a dihedral D_{nh} group.

1.34 Describe how the different tetrahedral and octahedral groups are constructed.

1.35 How can a fifth-order and a second- order generator combine to give a third-order generator? Where do such combinations appear?

1.36 How can sets of numbers represent a group?
1.37 When must some representations of a group involve matrices?
1.38 When can a matrix representation be broken down into numerical representations? Explain.

Problems

1.1 Construct a 3×3 matrix representing reorientation C_6 about the z axis.

1.2 Construct in standard form the matrix that rotates a body by angle ϕ about the y axis.

1.3 By multiplying the corresponding matrices, show that rotation by angle A succeeded by rotation by angle B about the same axis is equivalent to rotation by angle $A + B$ about the axis.

1.4 A unit cell of a lattice is bounded by the vectors

$$\mathbf{a}_1 = \frac{d}{2}(\mathbf{i} + \sqrt{3}\,\mathbf{j}), \quad \mathbf{a}_2 = \frac{d}{2}(-\mathbf{i} + \sqrt{3}\,\mathbf{j}), \quad \mathbf{a}_3 = e\mathbf{k}.$$

Construct vectors bounding a unit cell of the reciprocal lattice.

1.5 Formulate permutation matrices for C_3 and C_3^2. Show that the matrix for the latter is the square of the matrix for the former.

1.6 To what groups do the symmetry operations for (a) CH_4, (b) CO_2, (c) $H_2C=C=CH_2$, (d) PCl_5 belong?

1.7 Explain why Figure 1.16 does not represent a group. Alter the figure to make it a suitable diagram for a group.

1.8 Construct a Cayley diagram for C_{2v}. What other common groups does the figure represent?

1.9 What do the defining relations

$$B^2 = I, \quad (AB)^4 = I, \quad ABA^{-1}B = I$$

imply about the order of A?

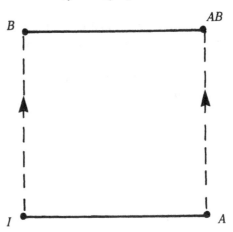

Figure 1.16 A Cayley diagram that does not represent a group.

1.10 Construct a 3 × 3 matrix representing the rotoinversion $\bar{6}$.

1.11 Using the corresponding matrices, relate C_4 to σ_h. Explain the result with a figure.

1.12 By multiplying the corresponding reorientation matrices, show that

$$C_3^3 = I.$$

1.13 Construct the lattice that is reciprocal to the lattice for which

$$\mathbf{a}_1 = d\mathbf{i}, \qquad \mathbf{a}_2 = e\mathbf{i} + f\mathbf{j}, \qquad \mathbf{a}_3 = g\mathbf{k}.$$

1.14 For what symmetry operation is

$$\begin{pmatrix} 0 & 0 & 0 & 0 & 1 \\ 1 & 0 & 0 & 0 & 0 \\ 0 & 1 & 0 & 0 & 0 \\ 0 & 0 & 1 & 0 & 0 \\ 0 & 0 & 0 & 1 & 0 \end{pmatrix}$$

a permutation matrix? Explain with a figure. What is the order of this operation as a generator?

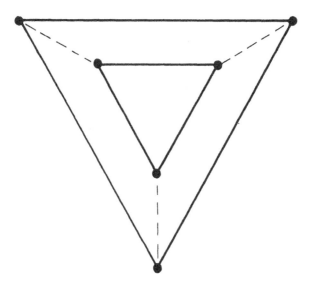

Figure 1.17 Erroneous Cayley diagram.

1.15 Explain why Figure 1.17 does not represent a group. Alter the figure so that it becomes suitable.

1.16 Draw a Cayley diagram for the D_{4h} group.

1.17 To what groups do the symmetry operations for (a) Y_{p_z} (b) $Y_{d_{3z^2-r^2}}$, (c) $Y_{d_{x^2-y^2}}$, (d) $Y_{f_{xyz}}$ belong?

1.18 What is the order of A, if

$$B^2 = I, \quad (AB)^3 = I, \quad ABA^{-1}BA^{-1}BA = I?$$

References

Books

Cotton, F. A.: 1963, *Chemical Applications of Group Theory*, Wiley-Interscience, New York, pp. 6-49. In Chapter 2, basic definitions and theorems are presented in a very simple manner. In Chapter 3, Cotton interprets the symmetries of representative molecules. A possible source of confusion is his use of the term symmetry element for the reference structure or base.

Duffey, G. H.: 1980, *Theoretical Physics: Classical and Modern Views*, Krieger, Melbourne, Fla., pp. 105-119, 203-210. More information on the use of matrices appears in Chapter 4. Symmetry in physical systems is discussed in Chapter 7.

Hahn, T. (editor): 1985, *International Tables for Crystallography*, Brief Teaching Edition of Volume A *Space-Group Symmetry*, Reidel, Dordrecht, pp. 1-119. These tables describe the most frequently occurring crystallographic space groups. Here again, a symmetry element is a reference structure or base for symmetry operations.

Nussbaum, A.: 1971, *Applied Group Theory for Chemists, Physicists and Engineers*, Prentice-Hall, Englewood Cliffs, N.J., pp. 1-78. As an experimental solid-state physicist, Nussbaum presents key concepts in a detailed descriptive manner. Chapter 1 contains many useful diagrams, tables, and descriptions of symmetric systems.

Articles

Breneman, G. L.: 1987, "Crystallographic Point Group Notation Flow Chart," *J. Chem. Educ.* **64,** 216-217.

Burrows, E. L., and Clark, M. J.: 1974, "Pictures of Point Groups," *J. Chem. Educ.* **51,** 87-91.

Craig, N. C.: 1959, "Molecular Symmetry Models," *J. Chem. Educ.* **46,** 23-26.

Donohue, J.: 1969, "A Key to Point Group Classification," *J. Chem. Educ.* **46,** 27.

Fackler, J. P., Jr.: 1978, "Symmetry in Molecular Structure—Facts, Fiction and Fun," *J. Chem. Educ.* **55,** 79-83.

Quane, D.: 1976, "Systematic Procedures for the Classification of Molecules into Point Groups: The Problem of the $\mathbf{D}_{nd}$ Group," *J. Chem. Educ.* **53,** 190.

CHAPTER 2 / Classes and Characters

2.1
The Immediate Agenda

In a particular group, the elements are distinguished by how they combine with other elements. If the result of each combination does not depend on the order of the factors, the group is said to be *Abelian*.

In Example 2.2 we will find that an Abelian group does not contain any element that can serve to correlate two different elements. As a consequence, each element will be considered in a class by itself.

In a group that is not Abelian, the products of some combinations do depend on the order of the factors. Then, some elements are correlated by the presence of another element and, as we will see in Section 2.2, such elements act similarly when they are operators.

Each set of similar elements makes up a *class*. We will find that the elements of a group can be placed uniquely into distinct classes. But, the elements in a group may be represented by distinct operators acting in an abstract space. Consequently, they can be summed. Of particular interest are the sums of the elements within classes, the *class sums*.

Since by assumption a law for combining elements exists, class sums can be similarly combined. This process does not take one outside the system of sums. Consequently, each product of class sums can be reduced to a linear combination of class sums or of inverse class sums. And, eigenvalue equations in the space of the sums exist.

Each common solution to the eigenvalue equations can be expressed as a linear combination of inverse class sums. The coefficients obtained characterize the corresponding symmetry species. After normalization and choice of phase, they are said to constitute a *character* vector. The components of this vector are used in formulating physically significant functions with elementary symmetry properties—those belonging to the chosen primitive symmetry species.

2.2
Classes within a Group

The presence of certain elements in a group can make a given element behave like one or more of the other elements. Thus, each σ_v operation in $\mathbf{C}_{nv}$ removes the distinction between counterclockwise and clockwise rotation, making C_n similar to C_n^{-1}. Also in $\mathbf{C}_{nv}$, the powers of C_n remove the distinctions between the different reflection planes, making the σ_v's similar.

The formal rule for determining what elements are similar can be induced from how they act in geometric groups. Consider that elements $A, B, \ldots, P$ are represented by operations that change vectors $\mathbf{x}$ to $\mathbf{y}$, $\mathbf{x}'$ to $\mathbf{y}'$, ..., $\mathbf{x}$ to $\mathbf{x}'$, $\mathbf{y}$ to $\mathbf{y}'$,

$$\mathbf{A}\mathbf{x} = \mathbf{y}, \tag{2.1}$$

$$\mathbf{B}\mathbf{x}' = \mathbf{y}', \tag{2.2}$$

$$\mathbf{P}\mathbf{x} = \mathbf{x}', \tag{2.3}$$

$$\mathbf{P}\mathbf{y} = \mathbf{y}'. \tag{2.4}$$

(See Figure 2.1.) Then

$$\mathbf{x} = \mathbf{P}^{-1}\mathbf{x}', \tag{2.5}$$

$$\mathbf{y} = \mathbf{P}^{-1}\mathbf{y}', \tag{2.6}$$

and equation (2.1) can be rewritten in the form

$$\mathbf{A}\mathbf{P}^{-1}\mathbf{x}' = \mathbf{P}^{-1}\mathbf{y}' \tag{2.7}$$

whence

$$\mathbf{P}\mathbf{A}\mathbf{P}^{-1}\mathbf{x}' = \mathbf{y}'. \tag{2.8}$$

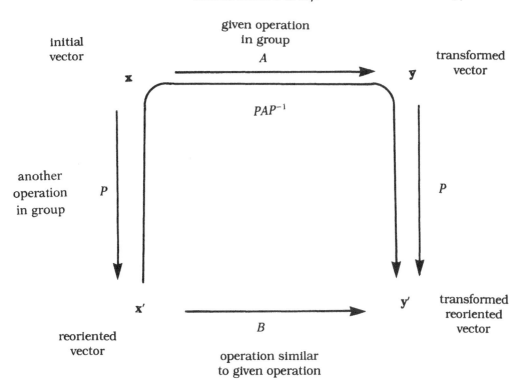

Figure 2.1 How similar symmetry operations are related.

For equations (2.8) and (2.2) to be true, regardless of how the original vector **x** is chosen, we must have

$$\mathbf{B} = \mathbf{PAP}^{-1}. \qquad (2.9)$$

Consequently, the corresponding elements in the group satisfy

$$B = PAP^{-1}. \qquad (2.10)$$

Since the P operation imposes the same reorientation on both **x** and **y**, operation B is similar to operation A. Formula (2.10) tells us how these are related. To get the other operations that are similar to A, a person merely replaces P in formula (2.10) with each of the other elements in turn and lists the different results.

Formula (2.10) can be applied whether the group has a geometric realization or not. Whenever a group contains the element P that transforms one element A into another element B by the formula

PAP^{-1}, A and B are said to belong to the same *class*. The change of A to B by formula (2.10) is called a *similarity transformation*.

Since the inverse of an element combined with the element equals the identity, by definition, we have

$$(AB)^{-1}AB = I. \tag{2.11}$$

Combining this with B^{-1} from the right,

$$(AB)^{-1}ABB^{-1} = (AB)^{-1}A = B^{-1}, \tag{2.12}$$

and then with A^{-1} from the right yields

$$(AB)^{-1}AA^{-1} = (AB)^{-1} = B^{-1}A^{-1}. \tag{2.13}$$

Thus, the inverse of a combination of elements equals the combination of the inverse elements in reverse order.

Suppose that both

$$B = PAP^{-1} \tag{2.14}$$

and

$$C = QAQ^{-1} \tag{2.15}$$

with $A, B, C, \ldots, P, Q$ all in the group. Then B and C are in the class or classes of A. But does this fact imply that B and C are in the same class? Are the classes mutually exclusive?

From equation (2.15), we obtain the relationship

$$Q^{-1}CQ = A \tag{2.16}$$

which converts equation (2.14) to

$$B = PQ^{-1}CQP^{-1} = PQ^{-1}C(PQ^{-1})^{-1}. \tag{2.17}$$

Since PQ^{-1} is an element of the group, B and C are related through a similarity transformation by an element of the group. Consequently, B does belong to the class of C. Different elements that belong to the class of a given element are in the same class and the classes within a group are mutually exclusive.

Example 2.1

Identify the classes in a cyclic group.

The cyclic group of order n is generated by C_n. The similarity transformation of the mth power of this generator by the lth power is

$$C_n^l C_n^m C_n^{-l} = C_n^{l+m-l} = C_n^m.$$

Since this result applies regardless of l, all similarity transforms of C_n^m are merely C_n^m. Each integral power of the generator up to and including the nth power is in a class by itself.

Example 2.2

Show that each element in an Abelian group is in a class by itself.

Let A be the given element and P any other element in an Abelian group. Then from the definition of Abelian

$$PA = AP.$$

Combine this with P^{-1} from the right to get

$$PAP^{-1} = APP^{-1} = A.$$

The similarity transform of A by any P in the group yields A. Thus A is in a class by itself.

Example 2.3

Show that C_3 and C_3^{-1} belong to the same class when σ_v is also present.

Place three equivalent bodies or kets symmetrically around an axis so that it becomes a C_3 axis with three σ_v planes intersecting there. Next inverse-reflect (that is, reflect) the system in one of these planes, rotate the result by ⅓ turn around the C_3 axis and reflect in the plane initially chosen, as Figure 2.2 illustrates. From the figure, the same final result is obtained on rotating the initial system $-⅓$ turn around the C_3 axis. Therefore,

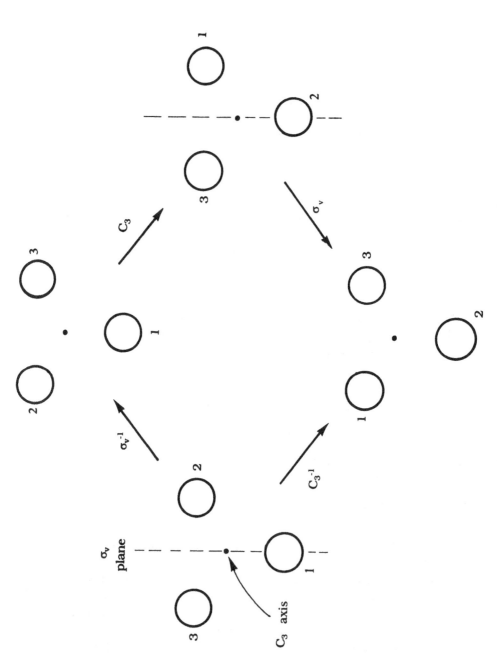

Figure 2. How reflection operation σ_v makes C_3^{-1} similar to C_3.

$$\sigma_v C_3 \sigma_v^{-1} = C_3^{-1},$$

C_3 and C_3^{-1} are in the same class.

2.3
Adding the Elements in a Class

Many physically significant groups consist of elements with representations that act on various operands, transforming them. The processes can be described by equations. Adding multiples of these equations produces forms that suggest how a second combinatory operation should be defined. Applying this operation to the elements of each class of a group and manipulating the resulting expressions leads to a deeper knowledge of group structure. Certain quantities and relationships obtained will enable us to construct symmetry-adapted expressions quite simply.

So far, we have combined elements to obtain single elements of a group. Thus, we have written

$$BA = C \tag{2.18}$$

where A, B, and C are all elements in the given group. Because this combining operation acts as the multiplication process in algebra, we say that in equation (2.18) we have *multiplied* A by B from the left to get C. Combination BA is called the product of B and A.

In geometric groups, the elements are represented by operations that transform position vectors. When **A** and **B** are such operations and **y**, **z** the corresponding transformations of vector **x**, we have

$$\mathbf{A}\mathbf{x} = \mathbf{y}, \tag{2.19}$$

$$\mathbf{B}\mathbf{x} = \mathbf{z}. \tag{2.20}$$

A person can multiply the first equation by scalar α, the second equation by scalar β, and add the results to obtain

$$\alpha \mathbf{A}\mathbf{x} + \beta \mathbf{B}\mathbf{x} = \alpha \mathbf{y} + \beta \mathbf{z} \tag{2.21}$$

or

$$(\alpha \mathbf{A} + \beta \mathbf{B})\mathbf{x} = \alpha \mathbf{y} + \beta \mathbf{z}. \qquad (2.22)$$

The operator acting on **x** in equation (2.22) is the sum of $\alpha \mathbf{A}$ and $\beta \mathbf{B}$. Analgously, one can form the sum,

$$\alpha \mathbf{A} + \beta \mathbf{B} + \ldots, \qquad (2.23)$$

of multiples of elements in a group.

Of particular interest is the simple sum of elements in each class. For the class consisting of the identity element, we write

$$\mathscr{C}_1 = I. \qquad (2.24)$$

For the jth class, consisting of $C, D, \ldots, F$, we write

$$\mathscr{C}_j = C + D + \ldots + F. \qquad (2.25)$$

Expression (2.25) is called the jth *class sum*. The sum for the class containing the inverses of these elements is written

$$\mathscr{C}_{\bar{j}} = C^{-1} + D^{-1} + \ldots + F^{-1}. \qquad (2.26)$$

When $C^{-1}, D^{-1}, \ldots, F^{-1}$ are included in set $C, D, \ldots, F$, then $\mathscr{C}_{\bar{j}}$ is the same as $\mathscr{C}_j$.

Since a similarity transformation by any other element in the group replaces, in order, each element in a class by another element in the class, the transformation has the effect of rearranging the elements in the sum. Consequently,

$$G\mathscr{C}_j G^{-1} = \mathscr{C}_j \qquad (2.27)$$

whence

$$G\mathscr{C}_j = \mathscr{C}_j G \qquad (2.28)$$

as long as G is in the group. Each element in another class commutes with a given class sum. Consequently, the kth class sum commutes with the jth class sum:

$$\mathscr{C}_k \mathscr{C}_j = \mathscr{C}_j \mathscr{C}_k. \qquad (2.29)$$

2.4
Products of Class Sums

Multiplying class sums does not take one outside the space of the sums because the various products equal linear combinations of class sums in the given group. Since $\mathscr{C}_j$ and $\mathscr{C}_k$ are the sums of group elements, the product $\mathscr{C}_j\mathscr{C}_k$ equals the sum of products of group elements. But each product of group elements is a group element and so $\mathscr{C}_j\mathscr{C}_k$ must be a linear combination of group elements. These can be collected in their particular classes.

No one of these classes can be incomplete. We know from equation (2.28), that

$$G\mathscr{C}_j\mathscr{C}_k G^{-1} = \mathscr{C}_j G \mathscr{C}_k G^{-1} = \mathscr{C}_j \mathscr{C}_k G G^{-1} = \mathscr{C}_j\mathscr{C}_k, \qquad (2.30)$$

and thus that a similarity transformation with any element in the group leaves the sum $\mathscr{C}_j\mathscr{C}_k$ unchanged. If one of the classes were incomplete, at least one of these similarity transformations would yield a sum containing a missing element and formula (2.30) would be violated.

Consequently, product $\mathscr{C}_j\mathscr{C}_k$ equals a linear combination of class sums; and alternatively, it equals a linear combination of inverse class sums. Thus we can write

$$\mathscr{C}_j\mathscr{C}_k = \sum_l c_{jkl}\mathscr{C}_{\bar{l}} = c_{jk1}\mathscr{C}_1 + c_{jk2}\mathscr{C}_{\bar{2}} + \ldots. \qquad (2.31)$$

Coefficient c_{jkl} gives the distribution of $\mathscr{C}_{\bar{l}}$ through the multiplication table for $\mathscr{C}_j$ and $\mathscr{C}_k$.

The identity element $I = \mathscr{C}_1$ appears only if $\mathscr{C}_k$ contains the inverse of an element in $\mathscr{C}_j$. But when $\mathscr{C}_k$ contains one inverse element, all of its elements must be inverses of elements in $\mathscr{C}_j$ and

$$k = \bar{j}. \qquad (2.32)$$

If h_j is the number of elements in $\mathscr{C}_j$, then h_j identities are obtained and

$$c_{jkl} = h_j \delta_{j\bar{k}}. \qquad (2.33)$$

Formulas (2.31) and (2.33) reduce higher products of class sums to linear combinations. For the triple product, we have

$$\begin{aligned}\mathscr{C}_j\mathscr{C}_k\mathscr{C}_l &= \sum c_{jkm}\mathscr{C}_{\bar{m}}\mathscr{C}_l = \sum\sum c_{jkm}c_{\bar{m}ln}\mathscr{C}_{\bar{n}} \\ &= \sum c_{jkm}(h_{\bar{m}}\delta_{\bar{m}l}\mathscr{C}_1 + \ldots) \\ &= h_l c_{jkl}\mathscr{C}_1 + \ldots .\end{aligned} \quad (2.34)$$

Since all class sums commute, equation (2.34) implies that

$$h_l c_{jkl} = h_l c_{kjl} = h_j c_{lkj} = h_j c_{klj} = h_k c_{jlk} = h_k c_{ljk} \quad (2.35)$$

or

$$\frac{c_{jkl}}{h_j h_k} = \frac{c_{kjl}}{h_k h_j} = \frac{c_{lkj}}{h_l h_k} = \frac{c_{klj}}{h_k h_l} = \frac{c_{jlk}}{h_j h_l} = \frac{c_{ljk}}{h_l h_j}. \quad (2.36)$$

Example 2.4

Determine how the class sums of $\mathbf{C}_{3v}$ multiply.

Group $\mathbf{C}_{3v}$ is formed on combining the third-order generator $C_3 = A$ with the second-order generator $\sigma_v = B$ as Figure 1.9 shows. From the defining relations, and also from a construction like that in Figure 2.2, we find that

$$C_3\sigma_v = AB = B^{-1}A^{-1} = BA^2 = \sigma_v C_3^2.$$

The construction also shows that

$$\sigma_v C_3^2 = \sigma_v'',$$

where σ_v'' causes reflection in the plane obtained on rotating the σ_v plane ⅔ turn. Similarly,

$$\sigma_v C_3 = BA = A^2B = C_3^2 \sigma_v = \sigma_v'$$

where σ_v' causes reflection in the plane obtained on rotating the σ_v plane ⅓ turn.

In $\mathbf{C}_{3v}$ the reflections cause the rotations to belong to the same class, while the rotations cause the reflections to belong to a single class. The class sums are therefore

$$\mathscr{C}_1 = I,$$
$$\mathscr{C}_2 = C_3 + C_3^2 = A + A^2,$$
$$\mathscr{C}_3 = \sigma_v + \sigma_v' + \sigma_v'' = B + A^2B + AB.$$

The results of combining these in the various possible pairs appear in Table 2.1.

Table 2.1 Class-Sum Multiplication Table for $\mathbf{C}_{3v}$

$\mathbf{C}_{3v}$	$\mathscr{C}_1$	$\mathscr{C}_2$	$\mathscr{C}_3$
$\mathscr{C}_1$	$\mathscr{C}_1$	$\mathscr{C}_2$	$\mathscr{C}_3$
$\mathscr{C}_2$	$\mathscr{C}_2$	$2\mathscr{C}_1 + \mathscr{C}_2$	$2\mathscr{C}_3$
$\mathscr{C}_3$	$\mathscr{C}_3$	$2\mathscr{C}_3$	$3\mathscr{C}_1 + 3\mathscr{C}_2$

Example 2.5

Obtain the c_{jkl} coefficients for $\mathbf{C}_{3v}$ and arrange them in matrices.

By definition, c_{jkl} times $\mathscr{C}_{\bar{l}}$ is the contribution of the inverse class sum $\mathscr{C}_{\bar{l}}$ to product $\mathscr{C}_j\mathscr{C}_k$. But in the $\mathbf{C}_{3v}$ group, the inverse of each class sum is the same class sum; so the bar over an index does not introduce a change. The contribution of $\mathscr{C}_{\bar{l}}$ then equals the contribution of $\mathscr{C}_l$.

The composition of each $\mathscr{C}_j\mathscr{C}_k$ for $\mathbf{C}_{3v}$ appears in Table 2.1. From this table, we find that

$$(c_{jk1}) = \begin{pmatrix} 1 & 0 & 0 \\ 0 & 2 & 0 \\ 0 & 0 & 3 \end{pmatrix},$$

$$(c_{jk2}) = \begin{pmatrix} 0 & 1 & 0 \\ 1 & 1 & 0 \\ 0 & 0 & 3 \end{pmatrix},$$

$$(c_{jk3}) = \begin{pmatrix} 0 & 0 & 1 \\ 0 & 0 & 2 \\ 1 & 2 & 0 \end{pmatrix}.$$

Alternative matrices containing the same information are

$$(c_{1jk}) = \begin{pmatrix} 1 & 0 & 0 \\ 0 & 1 & 0 \\ 0 & 0 & 1 \end{pmatrix},$$

$$(c_{2jk}) = \begin{pmatrix} 0 & 1 & 0 \\ 2 & 1 & 0 \\ 0 & 0 & 2 \end{pmatrix},$$

$$(c_{3jk}) = \begin{pmatrix} 0 & 0 & 1 \\ 0 & 0 & 2 \\ 3 & 3 & 0 \end{pmatrix}.$$

2.5
Eigenvalues and Eigenoperators of the Class Sums

Since one class sum acting on any fraction of another class sum of a given group yields a linear combination of class sums in the group, a person can construct an eigenvalue equation for each class sum. A number of eigenvalues, and normalized independent eigenoperators, equal to the number n of class sums exist for each $\mathscr{C}_j$. Since the class sums commute, n eigenoperators common to all the class sums can be constructed. The coefficients in the normalized, properly phased, expansions are characteristic of the group.

Let us formally write

$$\mathscr{C}_j \mathscr{E} = \lambda_j \mathscr{E} \tag{2.37}$$

with

$$\mathscr{E} = \Sigma y_k \mathscr{C}_{\bar{k}}. \tag{2.38}$$

Here $\mathscr{C}_j$ is the jth class sum while $\mathscr{E}$ is a trial solution constructed as a linear combination of the inverse class sums with y_k the kth coefficient.

Expression (2.38) converts eigenvalue equation (2.37) to the form

$$\mathscr{C}_j \sum y_k \mathscr{C}_{\bar{k}} = \lambda_j \sum y_k \mathscr{C}_{\bar{k}} \tag{2.39}$$

which equation (2.31) reduces to

$$\sum_k \sum_l c_{j\bar{k}l} y_k \mathscr{C}_{\bar{l}} = \lambda_j \sum_k y_k \mathscr{C}_{\bar{k}} = \lambda_j \sum_k \sum_l y_k \delta_{kl} \mathscr{C}_{\bar{l}}. \tag{2.40}$$

For equation (2.40) to put no restriction on the class sums, the overall equation must be an identity in these sums. Then

$$\sum_k c_{j\bar{k}l} y_k = \lambda_j \sum_k y_k \delta_{kl} \tag{2.41}$$

or

$$\sum_k (c_{j\bar{k}l} - \lambda_j \delta_{kl}) y_k = 0. \tag{2.42}$$

On applying Cramer's rule, we find that all y_k's are zero unless the determinant of the coefficients equals zero:

$$\begin{vmatrix} c_{j\bar{1}1} - \lambda_j & c_{j\bar{2}1} & \cdot & \cdot & \cdot & c_{j\bar{n}1} \\ c_{j\bar{1}2} & c_{j\bar{2}2} - \lambda_j & \cdot & \cdot & \cdot & c_{j\bar{n}2} \\ \cdot & \cdot & & & & \cdot \\ \cdot & \cdot & & & & \cdot \\ \cdot & \cdot & & & & \cdot \\ c_{j\bar{1}n} & c_{j\bar{2}n} & \cdot & \cdot & \cdot & c_{j\bar{n}n} - \lambda_j \end{vmatrix} = 0. \tag{2.43}$$

Relationship (2.43) is referred to as a *secular equation*. From equation (2.31), coefficient $c_{j\bar{k}l}$ equals the contribution of $\mathscr{C}_{\bar{l}}$ to the product $\mathscr{C}_j \mathscr{C}_{\bar{k}}$.

In analyzing a group, a person may first determine these coefficients and calculate the λ_j's. Each of the eigenvalues is then substituted back into equation (2.42) and conditions on the y_k's are deduced. A similar calculation is carried out for the other classes. The common sets of y_k/y_1 ratios are identified and labled by a superscript (r).

The phase of each set of $y_k^{(r)}$'s is fixed by requiring the coefficient of the identity, $y_1^{(r)}$, to be real and positive. All magnitudes are then fixed by requiring that

$$\sum_{m=1}^{n} h_m y_{\overline{m}}^{(r)} y_m^{(r)} = g. \tag{2.44}$$

Free index r identifies the eigenoperator, number g is the order of the group, and h_m is the order of the mth class.

We call the resulting $y_m^{(r)}$ the mth component of the rth *character vector* $\chi^{(r)}$ for the given group. Table 2.2 illustrates how such components may be arranged. Such a listing is called a *character table*.

Table 2.2 Components of the Character Vectors of a Group

	$\mathscr{C}_1$	$\mathscr{C}_2$	...	$\mathscr{C}_n$
$\chi^{(1)}$	$y_1^{(1)}$	$y_2^{(1)}$	...	$y_n^{(1)}$
$\chi^{(2)}$	$y_1^{(2)}$	$y_2^{(2)}$	...	$y_n^{(2)}$
.	.	.	.	.
.	.	.	.	.
.	.	.	.	.
$\chi^{(n)}$	$y_1^{(n)}$	$y_2^{(n)}$	...	$y_n^{(n)}$

From equation (2.41), one obtains

$$\sum c_{\bar{j}k1} y_k^{(r)} = \lambda_j^{(r)} y_1^{(r)}. \tag{2.45}$$

Formula (2.33) reduces this to

$$\sum_k h_j \delta_{jk} y_k^{(r)} = \lambda_j^{(r)} y_1^{(r)} \tag{2.46}$$

whence

$$\frac{\lambda_j^{(r)}}{h_j} = \frac{y_j^{(r)}}{y_1^{(r)}}. \tag{2.47}$$

Remember, $\lambda_j^{(r)}$ is the rth eigenvalue of the jth class sum, h_j is the number of group elements in the jth class sum, and $y_j^{(r)}$ is the jth component in the expansion of $\mathscr{E}^{(r)}$ in terms of the inverse class sums. By inverse class sum $\mathscr{C}_{\bar{j}}$, we mean the class sum composed of the inverses of the elements in the class sum $\mathscr{C}_j$.

Example 2.6

Determine the eigenvalues for the class sums of the $\mathbf{C}_{3v}$ group.

In the $\mathbf{C}_{3v}$ group, the bar over an index does not introduce any change since each class sum does not differ from the sum of its inverse elements. So we can substitute c_{jkl} for $c_{\bar{j}kl}$ in equation (2.43).

Using the c_{1jk} coefficients from Example 2.5 leads to

$$\begin{vmatrix} 1 - \lambda_1 & 0 & 0 \\ 0 & 1 - \lambda_1 & 0 \\ 0 & 0 & 1 - \lambda_1 \end{vmatrix} = 0,$$

whence

$$(1 - \lambda_1)^3 = 0$$

and

$$\lambda_1 = 1, \quad \lambda_1 = 1, \quad \lambda_1 = 1.$$

Next, substitute the c_{2jk} coefficients from Example 2.5 into the secular equation (2.43),

$$\begin{vmatrix} -\lambda_2 & 2 & 0 \\ 1 & 1 - \lambda_2 & 0 \\ 0 & 0 & 2 - \lambda_2 \end{vmatrix} = 0,$$

multiply out the determinant,

$$-\lambda_2(1 - \lambda_2)(2 - \lambda_2) - 2(2 - \lambda_2) = 0,$$

and solve for the roots:

$$\lambda_2 = 2, \qquad \lambda_2 = 2, \qquad \lambda_2 = -1.$$

The c_{3jk} coefficients from Example 2.5 similarly give us the secular equation

$$\begin{vmatrix} -\lambda_3 & 0 & 3 \\ 0 & -\lambda_3 & 3 \\ 1 & 2 & -\lambda_3 \end{vmatrix} = 0$$

whence

$$-\lambda_3^3 + 3\lambda_3 + 6\lambda_3 = 0$$

and

$$\lambda_3 = 3, \qquad \lambda_3 = -3, \qquad \lambda_3 = 0.$$

Example 2.7

Construct the characters for the C_{3v} group.

The simultaneous equations behind the first secular equation in Example 2.6 are

$$(1 - \lambda_1)y_1 + (0)y_2 + (0)y_3 = 0,$$
$$(0)y_1 + (1 - \lambda_1)y_2 + (0)y_3 = 0,$$
$$(0)y_1 + (0)y_2 + (1 - \lambda_1)y_3 = 0.$$

The threefold root $\lambda_1 = 1$ makes each $(1 - \lambda_1)$ factor zero so that no conditions on y_1, y_2, or y_3 result.

The simultaneous equations behind the second secular equation in Example 2.6 are

$$-\lambda_2 y_1 + 2y_2 + (0)y_3 = 0,$$
$$y_1 + (1 - \lambda_2)y_2 + (0)y_3 = 0,$$
$$(0)y_1 + (0)y_2 + (2 - \lambda_2)y_3 = 0.$$

Substituting the twofold root $\lambda_2 = 2$ into these equations yields

$$-2y_1 + 2y_2 = 0,$$
$$y_1 - y_2 = 0,$$
$$(0)y_3 = 0,$$

whence

$$y_2 = y_1, \qquad y_3 = ?$$

Since this root does not restrict y_3, it does not lead to unique solutions.

Introducing the root $\lambda_2 = -1$, on the other hand, gives us the relationships

$$y_1 + 2y_2 = 0,$$
$$3y_3 = 0,$$

from which

$$y_2 = -\frac{1}{2} y_1, \qquad y_3 = 0.$$

We now require y_1 to be real and positive, and introduce equation (2.44) to obtain

$$y_1 = 2, \qquad y_2 = -1, \qquad y_3 = 0.$$

The simultaneous equations behind the third secular equation in Example 2.6 are

$$-\lambda_3 y_1 + (0)y_2 + 3y_3 = 0,$$
$$(0)y_1 - \lambda_3 y_2 + 3y_3 = 0,$$
$$y_1 + 2y_2 - \lambda_3 y_3 = 0.$$

Substituting the root $\lambda_3 = 3$ into these equations yields

$$-3y_1 + 3y_3 = 0,$$

$$-3y_2 + 3y_3 = 0,$$
$$y_1 + 2y_2 - 3y_3 = 0,$$

whence

$$y_3 = y_1, \qquad y_2 = y_3.$$

Imposition of condition (2.44), together with the requirement that y_1 be real and positive, leads to the components

$$y_1 = y_2 = y_3 = 1.$$

The root $\lambda_3 = -3$ similarly yields

$$3y_1 + 3y_3 = 0,$$
$$3y_2 + 3y_3 = 0,$$
$$y_1 + 2y_2 + 3y_3 = 0,$$

whence

$$y_3 = -y_1, \qquad y_2 = -y_3.$$

Condition (2.44), together with the phase requirement, then leads to

$$y_1 = 1, \qquad y_2 = 1, \qquad y_3 = -1.$$

Finally, the root $\lambda_3 = 0$ reduces the third set of simultaneous equations to

$$3y_3 = 0,$$
$$y_1 + 2y_2 = 0,$$

whence

$$y_2 = -\frac{1}{2}y_1, \qquad y_3 = 0.$$

As before, we require y_1 to be real and positive, and introduce equation (2.44), obtaining

$$y_1 = 2, \quad y_2 = -1, \quad y_3 = 0.$$

Since the $\mathbf{C}_{3v}$ group contains three class sums, it possesses three normalized, properly phased, independent eigenoperators. Consequently, there are three character vectors (one for each eigenoperator). All three vectors have been obtained from the third secular equation. One of these and a condition satisfied by the other two have been obtained from the second secular equation. The first secular equation did not impose any conditions on the character vectors.

Components of the character vectors are commonly presented as in Table 2.3. Each class sum for the group is identified by a typical term preceded by the number of terms when this number exceeds 1. Each character vector is identified by a letter. The one to which all class sums contribute equally is labeled A. The others to which $\mathscr{C}_1$ contributes singly are labeled B, C, D. The others to which $\mathscr{C}_1$ contributes doubly are labeled E; those to which $\mathscr{C}_1$ contributes triply, quadruply, quintuply, F, G, H, respectively. Subscripts are added where necessary to distinguish characters of the same order.

Table 2.3 Characters for the $\mathbf{C}_{3v}$ Group

	I	$2C_3$	$3\sigma_v$
A	1	1	1
B	1	1	−1
E	2	−1	0

2.6
Key Orthogonality Conditions

It can now be determined how (a) the eigenoperators and (b) the character vectors multiply.

Because each eigenoperator for a given group is a linear combination of the inverse class sums, from equation (2.38), and because the class sums commute, from equation (2.29), the *j*th class sum commutes with the *r*th eigenoperator:

$$\mathscr{C}_j \mathscr{E}^{(r)} = \mathscr{E}^{(r)} \mathscr{C}_j. \tag{2.48}$$

Let us multiply equation (2.48) by the sth eigenoperator,

$$\mathscr{C}_j \mathscr{E}^{(r)} \mathscr{E}^{(s)} = \mathscr{E}^{(r)} \mathscr{C}_j \mathscr{E}^{(s)}. \tag{2.49}$$

Introduce the eigenvalue equation

$$\mathscr{C}_j \mathscr{E}^{(r)} = \lambda_j^{(r)} \mathscr{E}^{(r)}. \tag{2.50}$$

to convert equation (2.49) to

$$\lambda_j^{(r)} \mathscr{E}^{(r)} \mathscr{E}^{(s)} = \mathscr{E}^{(r)} \lambda_j^{(s)} \mathscr{E}^{(s)} = \lambda_j^{(s)} \mathscr{E}^{(r)} \mathscr{E}^{(s)}. \tag{2.51}$$

For equation (2.51) to hold, we must have either

$$\lambda_j^{(r)} = \lambda_j^{(s)} \tag{2.52}$$

or

$$\mathscr{E}^{(r)} \mathscr{E}^{(s)} = 0. \tag{2.53}$$

Either the eigenoperators belong to the same eigenvalue or they are orthogonal.

When they have the same eigenvalue, relations (2.38), (2.37), (2.47), and (2.44) tell us that

$$\mathscr{E}^{(r)} \mathscr{E}^{(r)} = \sum_k y_k^{(r)} \mathscr{C}_k \mathscr{E}^{(r)} = \sum_k y_k^{(r)} \lambda_k^{(r)} \mathscr{E}^{(r)}$$

$$= \sum_k y_k^{(r)} y_{\bar{k}}^{(r)} \frac{h_k}{y_1^{(r)}} \mathscr{E}^{(r)} = \frac{g}{y_1^{(r)}} \mathscr{E}^{(r)}. \tag{2.54}$$

Combining equations (2.53) and (2.54) yields

$$\mathscr{E}^{(r)} \mathscr{E}^{(s)} = \frac{g}{y_1^{(r)}} \delta_{rs} \mathscr{E}^{(r)}, \tag{2.55}$$

a formula for calculating binary products of the common eigenoperators of the class sums.

Multiplying equation (2.41) in the form

Key Orthogonality Conditions

$$\sum_k c_{j\bar{k}l} y_k^{(r)} = \lambda_j^{(r)} y_l^{(r)} \tag{2.56}$$

by

$$h_l y_{\bar{l}}^{(s)} \tag{2.57}$$

and summing over l yields

$$\sum_k \sum_l y_k^{(r)} h_l c_{j\bar{k}l} y_{\bar{l}}^{(s)} = \lambda_j^{(r)} \sum_l h_l y_l^{(r)} y_{\bar{l}}^{(s)}. \tag{2.58}$$

Multiplying equation (2.41) in the form

$$\sum_l c_{jl\bar{k}} y_{\bar{l}}^{(s)} = \lambda_j^{(s)} y_{\bar{k}}^{(s)} \tag{2.59}$$

by

$$h_k y_k^{(r)} \tag{2.60}$$

and summing over k yields

$$\sum_k \sum_l y_k^{(r)} h_k c_{jl\bar{k}} y_{\bar{l}}^{(s)} = \lambda_j^{(s)} \sum_k h_k y_k^{(r)} y_{\bar{k}}^{(s)}. \tag{2.61}$$

Because of equation (2.35), the left sides of equation (2.58) and (2.61) are equal. Therefore, the right sides also have to be equal. Consequently, either

$$\lambda_j^{(r)} = \lambda_j^{(s)} \tag{2.62}$$

or

$$\sum h_k y_k^{(r)} y_{\bar{k}}^{(s)} = 0. \tag{2.63}$$

Combining equation (2.63) with equation (2.44) yields the *orthogonality relationship*

$$\sum h_k y_k^{(r)} y_{\bar{k}}^{(s)} = g \delta_{rs} \tag{2.64}$$

for the character vectors.

Example 2.8

Show how the characters for the C_{3v} group obey the orthogonality relationship.

In the C_{3v} group, the inverse of I is I, the inverse of C_3 is C_3^2, the inverse of σ_v is σ_v, of σ_v', it is σ_v', and of σ_v'', it is σ_v''. So each of the class sums is its own inverse and

$$y_{\bar{k}}^{(s)} = y_k^{(s)}.$$

From Table 2.3, the number h_1 of elements in the I class is 1, the number h_2 in the C_3 class sum is 2, the number h_3 in the σ_v class sum is 3. Applying equation (2.64) separately to the first row, to the second row, and to the third row yields

$$1(1)(1) + 2(1)(1) + 3(1)(1) = 6,$$
$$1(1)(1) + 2(1)(1) + 3(-1)(-1) = 6,$$
$$1(2)(2) + 2(-1)(-1) + 3(0)(0) = 6,$$

the number of elements in the group. But when r is 1, s is 2, equation (2.64) leads to

$$1(1)(1) + 2(1)(1) + 3(1)(-1) = 0;$$

when r is 1, s is 3,

$$1(1)(2) + 2(1)(-1) + 3(1)(0) = 0;$$

when r is 2, s is 3,

$$1(1)(2) + 2(1)(-1) + 3(-1)(0) = 0.$$

2.7
Restrictions on the Classes

The orthogonality and normalization conditions on the rows in a character table lead to similar conditions on the columns.

Let the number of group elements in the kth class h_k multiplied by character component $y_k^{(r)}$ be the rkth element of a matrix. Further-

Restrictions on the Classes

more, let the character component $y_{\bar{k}}^{(s)}$ be the $k$sth element of a second matrix. Then all rs forms of equation (2.64) appear in the product

$$\begin{pmatrix} h_1 y_1^{(1)} & h_2 y_2^{(1)} & \cdots & h_n y_n^{(1)} \\ h_1 y_1^{(2)} & h_2 y_2^{(2)} & \cdots & h_n y_n^{(2)} \\ \cdot & \cdot & \cdot & \cdot \\ \cdot & \cdot & \cdot & \cdot \\ \cdot & \cdot & \cdot & \cdot \\ h_1 y_1^{(n)} & h_2 y_2^{(n)} & \cdots & h_n y_n^{(n)} \end{pmatrix} \begin{pmatrix} y_{\bar{1}}^{(1)} & y_{\bar{1}}^{(2)} & \cdots & y_{\bar{1}}^{(n)} \\ y_{\bar{2}}^{(1)} & y_{\bar{2}}^{(2)} & \cdots & y_{\bar{2}}^{(n)} \\ \cdot & \cdot & \cdot & \cdot \\ \cdot & \cdot & \cdot & \cdot \\ \cdot & \cdot & \cdot & \cdot \\ y_{\bar{n}}^{(1)} & y_{\bar{n}}^{(2)} & \cdots & y_{\bar{n}}^{(n)} \end{pmatrix}$$

$$= \begin{pmatrix} g & 0 & \cdots & 0 \\ 0 & g & \cdots & 0 \\ \cdot & \cdot & \cdot & \cdot \\ \cdot & \cdot & \cdot & \cdot \\ \cdot & \cdot & \cdot & \cdot \\ 0 & 0 & \cdots & g \end{pmatrix}. \qquad (2.65)$$

Now, multiply both sides of equation (2.65) from the left by the inverse of the first matrix and from the right by the first matrix itself. Since the matrix on the right commutes with either multiplicand, the inverse matrix cancels the original matrix on the right and one obtains

$$\begin{pmatrix} y_{\bar{1}}^{(1)} & y_{\bar{1}}^{(2)} & \cdots & y_{\bar{1}}^{(n)} \\ y_{\bar{2}}^{(1)} & y_{\bar{2}}^{(2)} & \cdots & y_{\bar{2}}^{(n)} \\ \cdot & \cdot & \cdot & \cdot \\ \cdot & \cdot & \cdot & \cdot \\ \cdot & \cdot & \cdot & \cdot \\ y_{\bar{n}}^{(1)} & y_{\bar{n}}^{(2)} & \cdots & y_{\bar{n}}^{(n)} \end{pmatrix} \begin{pmatrix} h_1 y_1^{(1)} & h_2 y_2^{(1)} & \cdots & h_n y_n^{(1)} \\ h_1 y_1^{(2)} & h_2 y_2^{(2)} & \cdots & h_n y_n^{(2)} \\ \cdot & \cdot & \cdot & \cdot \\ \cdot & \cdot & \cdot & \cdot \\ \cdot & \cdot & \cdot & \cdot \\ h_1 y_1^{(n)} & h_2 y_2^{(n)} & \cdots & h_n y_n^{(n)} \end{pmatrix}$$

$$= \begin{pmatrix} g & 0 & \cdots & 0 \\ 0 & g & \cdots & 0 \\ \cdot & \cdot & \cdot & \cdot \\ \cdot & \cdot & & \cdot \\ \cdot & \cdot & & \cdot \\ 0 & 0 & \cdots & g \end{pmatrix}. \qquad (2.66)$$

Equation (2.66) implies that

$$\sum_r h_j y_j^{(r)} y_k^{(r)} = g\delta_{jk}. \qquad (2.67)$$

2.8
An Alternative Procedure

The characters for given symmetry species can also be obtained from the corresponding representations by numbers or matrices. Only the simpler aspects of the theory will be presented here; a detailed discussion appears in Chapter 5.

Consider a group whose elements have been divided into classes. Now, each element in a class can be represented by an operator that acts on vectors in an abstract space. Let the sum of the operators in the jth class be designated $\mathscr{C}_j$ while an eigenvector of this sum is designated $\mathbf{u}$. Then analogous to equation (2.37), we have the equation

$$\mathscr{C}_j \mathbf{u} = \lambda_j \mathbf{u} \qquad (2.68)$$

with the same eigenvalue λ_j. By equation (2.47) this is related to $y_j^{(r)}$, the jth component of the pertinent character.

When $y_1 = 1$ and $h_j = 1$, $\mathscr{C}_j$ reduces to a single operator A^r or B. Equation (2.68) is then satisfied by equation (1.38) and equation (2.47) reduces to

$$\lambda_j = y_j. \qquad (2.69)$$

Consequently, the numerical representations in Tables 1.4, 1.5, and 1.6 are character components.

Similarly, the numerical representations for I and B in Tables 1.7 and 1.8 are character components. For the other classes, h_j equals 2; however, $\mathscr{C}_j$ includes A^r and A^{-r}. When $y_1 = 1$, each of these acting on **u** yields the same eigenvalue, which is one-half λ_j. (These values are listed in Tables 1.7 and 1.8.) Thus, the numerical representations here also are character components.

When y_1 equals 2 or more, equation (2.68) is replaced by a matrix equation such as equation (5.16). In Chapter 5, we will find that each character component then equals the trace of the corresponding matrix. For those in Tables 1.7 and 1.8, one would employ reductions of the form

$$\begin{aligned} \omega + \omega^{-1} &= e^{2\pi i/n'} + e^{-2\pi i/n'} \\ &= 2 \cos(2\pi/n') \end{aligned} \quad (2.70)$$

with n' equal to $2n + 1$ and $2n$, respectively.

2.9
Main Features

A group of g elements is distinguished by how its elements combine. Those that behave similarly are considered together. Thus, any two elements A and B, linked through a similarity transformation

$$B = PAP^{-1} \quad (2.71)$$

by some other element P in the group, are said to belong to the same class.

An operator

$$\alpha A + \beta B \quad (2.72)$$

that produces the effect of number α times the effect of A plus number β times the effect of B is called the sum of αA and βB. The sum of elements in the jth class

$$\mathscr{C}_j = C + D + \ldots + F \quad (2.73)$$

is called the jth class sum.

Each product of class sums is a linear function of class sums, or more conveniently, of the inverse class sums:

$$\mathscr{C}_j \mathscr{C}_k = \sum c_{jkl} \mathscr{C}_{\bar{l}}. \tag{2.74}$$

As a consequence, certain linear combinations of inverse class sums

$$\mathscr{E}^{(r)} = \sum y_k^{(r)} \mathscr{C}_{\bar{k}}. \tag{2.75}$$

are eigenoperators of the class sums:

$$\mathscr{C}_j \mathscr{E}^{(r)} = \lambda_j^{(r)} \mathscr{E}^{(r)}. \tag{2.76}$$

Since the class sums commute, there is a common set of n independent eigenoperators, where n is the number of class sums. These eigenoperators can be chosen so that each $y_1^{(r)}$ is real and positive and so that the sum

$$\sum_{m=1}^{n} h_m y_{\bar{m}}^{(r)} y_m^{(r)} = g, \tag{2.77}$$

equals order g of the group. Here h_m is the number of elements in the mth class sum. For each r, the $y_m^{(r)}$'s constitute the components of character vector $\chi^{(r)}$ for the group.

The different characters satisfy the orthogonality condition

$$\sum h_k y_{\bar{k}}^{(r)} y_k^{(s)} = g \delta_{rs}. \tag{2.78}$$

Similarly, the eigenoperators for a group obey the orthogonality condition

$$\mathscr{E}^{(r)} \mathscr{E}^{(s)} = \frac{g}{y_1^{(r)}} \delta_{rs} \mathscr{E}^{(r)}. \tag{2.79}$$

Discussion Questions

2.1 What distinguishes the elements in a group?

2.2 How can certain elements in a group make a given element behave like one or more of the other elements?

2.3 How may the rule for determining whether elements are similar be induced?

2.4 Explain how the inverse of a binary combination of group elements is obtained.

2.5 What is a class? Why are the classes in a group mutually exclusive?

2.6 By diagrams, show that $\sigma_v C_3$ produces the same effect as reflection σ_v' in a plane obtained by rotating the σ_v plane by ⅓ turn.

2.7 Interpret (a) the addition and (b) the multiplication of group elements.

2.8 What is a class sum? Why does one class sum commute with another?

2.9 Why does the product of class sums lie in the space of the class sums of the group under consideration?

2.10 What is coefficient c_{jkl}?

2.11 Show that

$$c_{jk1} = h_j \delta_{j\bar{k}}.$$

2.12 Why can an eigenvalue equation be written for a class sum?

2.13 Why can a set of eigenoperators common to all class sums in a group be constructed?

2.14 How are the eigenoperators normalized and phased?

2.15 Identify the contribution of each inverse class sum to each of the resulting common eigenoperators.

2.16 How do the character vectors characterize a group?

2.17 Why does a class sum commute with an eigenoperator for the same group?

2.18 Why are the different normalized eigenoperators orthogonal?

2.19 What orthogonality condition relates the character vectors for a given group? How can this condition be established?

2.20 In what groups are some inverse class sums distinct from the corresponding class sums?

2.21 How can the orthogonality condition for characters be expressed in matrix form?

2.22 What orthogonality condition relates the classes for a given group? How can this condition be derived from the orthogonality condition relating character vectors?

Problems

2.1 How does the Cayley diagram for C_{3v} show that introducing σ_v makes C_3 and $C_3{}^2$ belong to the same class?

2.2 Determine the c_{jkl} coefficients for the C_3 group from how the class sums multiply.

2.3 Calculate the eigenvalues for the class sums of C_3.

2.4 Construct the characters for the C_3 group.

2.5 Show that the characters of the C_3 group obey the orthogonality relationship.

2.6 Assume that $\mathscr{C}_{\bar{2}} = \mathscr{C}_4$, $\mathscr{C}_{\bar{3}} = \mathscr{C}_3$ and complete the following character table:

	$\mathscr{C}_1$	$\mathscr{C}_2$	$\mathscr{C}_3$	$\mathscr{C}_4$
$\chi^{(1)}$	1	1	1	1
$\chi^{(2)}$	1	-1	1	-1
$\chi^{(3)}$	1			
$\chi^{(4)}$	1			

2.7 Show that when the symmetry operations for a system include the identity, rotation by $1/n$ turn about a principal axis, all powers of this rotation, and rotations by $\frac{1}{2}$ turn about n axes perpendicular to the principal axis, $C_n{}^m$ and $C_n{}^{-m}$ belong to the same class.

2.8 Construct the Cayley diagram for D_5. Then determine how the class sums for D_5 multiply.

2.9 Calculate the c_{jkl} coefficients for D_5 from the results of Problem 2.8.

2.10 Obtain the eigenvalues for the class sums of D_5.

2.11 Construct the components of the character vectors for the D_5 group.

2.12 Show that the character vectors for the D_5 group obey the orthogonality relationship.

2.13
Assume that $\mathscr{C}_{\bar{2}} = \mathscr{C}_3$ and complete the character table:

	$\mathscr{C}_1$ ($h_1 = 1$)	$\mathscr{C}_2$ ($h_2 = 4$)	$\mathscr{C}_3$ ($h_3 = 4$)	$\mathscr{C}_4$ ($h_4 = 3$)
$\chi^{(1)}$	1	1	1	1
$\chi^{(2)}$	1			
$\chi^{(3)}$	1			
$\chi^{(4)}$	3	0	0	-1

References

Books

Hall, G. G.: 1967, *Applied Group Theory*, American Elsevier, New York, pp. 20-52. Hall assembles the group elements in class sums, constructs their eigenvalue equations, and from the eigenvalues determines the character components. He maintains that his approach is simpler than the conventional one employing matrices.

Articles

Rudin, R. A.: 1973, "Sufficient Conditions for Construction of the Character System of a Finite Group," *Am. J. Phys.* **41,** 490-494.

CHAPTER 3 / *Expedient Arrangements of Displacements*

3.1 Vibrating Arrays

In classical theory, an oscillating mass element moves periodically about a point of equilibrium. A system of such mass elements moves multiperiodically about an array of equilibrium points. At a given instant of time, the displacements of the elements from these points form an arrangement of displacements. Furthermore, the equilibrium points may move with respect to a reference inertial frame.

Governing the displacements are the equations of motion. For small displacements, and for some large displacements, the equations are linear. Then the vibrations appear as a superposition of modes of definite frequencies. Nonlinearities, which become significant at larger amplitudes, introduce changes in the frequencies and, also, new frequencies.

Solving the equations, when more than a very few mass elements are present, is difficult. Any symmetry present, however, imposes very useful restrictions. These we will consider in this chapter. We will find that each pure mode of motion belongs to a single symmetry species and row.

A physical system possesses symmetry whenever there are distinct operations that do not alter a discriminating function or operator. These operations together form a group. Certain subsets of these operations may also form smaller groups called subgroups.

The essence of a group lies in how its elements combine. The results can be expressed either in a multiplication table or, more succinctly, in a Cayley diagram. The elements can be classified using formula (2.10) or (2.71). The sum of elements in each class is an operator with eigenvalues, eigenoperators, and (as we will find in this chapter) eigenarrays. Other characteristic expressions can be similarly constructed.

Each common eigenoperator yields expressions belonging to a *primitive symmetry species*, with a definite character vector. When the species is degenerate, the results can be separated into mutually exclusive rows equal in number to the degeneracy.

The pure modes for a system belong to single symmetry species and rows. The equations of motion for some modes are linear about the equilibrium configuration. The equations for other modes are only approximately linear. In the approximation that all the equations are linear, the different modes do not interact and the complete motion can be described as a superposition of these modes.

Whenever each possible expression for an aspect or property arises as a linear combination of a set of standard expressions, the standard expressions are said to form a *basis*. The displacements that are combined to form the symmetry-adapted species make up such a basis. Standard symmetry-adapted expressions, which belong to primitive symmetry species and definite rows, form an alternative basis.

3.2
Suitable Bases and Coordinates

A system of mass elements, such as we have in a molecule, exhibits translation, rotation, and interelement movement. These may all be considered together; or the internal motions by themselves may be considered. In the former procedure, one locates all the equilibrium positions in an inertial frame at a given time. In the latter, one employs a reference frame based on the center of mass, either nonrotating or rotating with the system.

Either way, a person may employ Cartesian axes erected on each standard equilibrium point. The corresponding unit vectors at the jth position are designated

$$\mathbf{u}_{3j-2}, \mathbf{u}_{3j-1}, \mathbf{u}_{3j}, \qquad (3.1)$$

while the instantaneous position of the jth mass element has the coordinates

$$\eta_{3j-2}, \eta_{3j-1}, \eta_{3j}. \qquad (3.2)$$

Vectors (3.1) form a basis for describing the motion.

An alternate basis is provided by the symmetry-adapted arrays designated

$$\mathbf{f}_1, \mathbf{f}_2, \ldots, \mathbf{f}_{3n} \qquad (3.3)$$

where n is the number of mass elements.

The generalized coordinates for the system make up the vector

$$\mathbf{q} = \sum \eta_k \mathbf{u}_k = \sum q_l \mathbf{f}_l. \qquad (3.4)$$

The generalized forces similarly constitute the vector

$$\mathbf{Q} = \sum \phi_k \mathbf{u}_k = \sum Q_l \mathbf{f}_l. \qquad (3.5)$$

Governing the motion are the Lagrange equations

$$\frac{d}{dt}\frac{\partial T}{\partial \dot{q}_l} - \frac{\partial T}{\partial q_l} = Q_l, \qquad (3.6)$$

in which T is the kinetic energy and $\dot{q}_l$ stands for the derivative of q_l with respect to the time t. When the amplitudes of vibration are small, equation (3.6) reduces to the linear form

$$\sum A_{lk} \ddot{q}_k = Q_l \qquad (3.7)$$

in which the A_{lk}'s are parameters. Also, some modes retain this linear form when their amplitudes are large.

Expressing a physical law, equation (3.7) has to be covariant under the symmetry operations. Thus whenever the right side belongs to a primitive symmetry species, the left side does also. The integrals $q_k(t)$ must then belong to the same row of the same symmetry species. With a symmetric system, the number of forms in such a row is much below the number of independent coordinates, so that use of the symmetry-adapted basis facilitates solution of the equations.

When only the internal motions are being considered, one may employ the interelement distances and angles as variables; however, not all of these are independent. But in the symmetry treatment, one may not wish to eliminate any one by itself. The redundant combinations would later be rejected. As the basis, one may initially employ the complete set of equilibrium distances and angles—a set of scalars.

3.3
Generating Symmetry-Adapted Arrays

Our discussion will employ a base vector system. If a different system of expressions were used as basis, these would replace the base vectors in the formulas. We will first see how the common eigenoperators of the class sums acting on the basis vectors yield common eigenvectors of the class sums. The principles needed to construct the symmetry-adapted arrays from these eigenvectors will then be formulated.

Let $\mathbf{u}_l$ be the lth basis vector or basis vector array. And suppose that $\mathscr{E}^{(r)}$ is the rth common eigenoperator of the class sums; so

$$\mathscr{C}_j \mathscr{E}^{(r)} = \lambda_j^{(r)} \mathscr{E}^{(r)}. \tag{3.8}$$

Since the operators may act on any basis vector or array, we may construct the vector array

$$\mathscr{C}_j \mathscr{E}^{(r)} \mathbf{u}_l = \lambda_j^{(r)} \mathscr{E}^{(r)} \mathbf{u}_l. \tag{3.9}$$

But the eigenoperator is the linear combination of class sums

$$\mathscr{E}^{(r)} = \sum_k y_k^{(r)} \mathscr{C}_k, \tag{3.10}$$

where $y_k^{(r)}$ is the kth component of character vector $\chi^{(r)}$. The various eigenvectors of the class sums are given by

$$\mathscr{E}^{(r)}\mathbf{u}_l = \sum_k y_k^{(r)} \mathscr{C}_k^r \mathbf{u}_l. \tag{3.11}$$

So we can construct

$$\mathbf{f}_j = N \sum_k y_k^{(r)} \mathscr{C}_k^r \mathbf{u}_l. \tag{3.12}$$

where N is a normalization factor.

If the N's can be chosen so that the same form arises from all the $\mathbf{u}_l$'s, then this is the symmetry-adapted array for the primitive symmetry species. For a degenerate species, a number of independent arrays equal to the degeneracy (the character component for I) are found. These must be made mutually orthogonal so that they represent physically separate arrays.

With a given basis, a symmetry species can occur two or more times. Then equation (3.12) yields superpositions of the pure combinations. These superpositions have to be combined linearly to form independent sets of combinations equal in number to the number of occurrences. A criterion that is employed is that the combinations in each of these sets mix only among themselves under operations of the group.

Contributions can be arbitrarily separated when they are known to be independent for some physical reasons. In a problem where a translational or rotational mode appears with vibrational modes, superpositions that yield the pure translation or rotation can be separated out as long as the translation or rotation is physically independent of the vibration.

The quadratic nonlinearities that can contribute to the left side of equation (3.7) are obtained by determining the base products that belong to the same row of the same symmetry species as Q_l does. Thus, one finds that Coriolis interaction exists between modes whose bases multiply to form bases for rotational motion.

3.4
Procedural Details

In applying symmetry considerations to a vibration problem, a person must first determine what group to employ. We have already noted that the mass elements in a vibrating system are associated with points and when the system is at equilibrium these points form a characteristic array. One examines this array to find out what bases for symmetry operations are present. Table 1.1 lists the various possibilities and Figure 1.2 identifies the group. Note that when one omits certain bases from consideration, one obtains subgroups of the full group for the system.

On changing from the full group to a subgroup, however, the distinctions between some primitive symmetry species disappear. Conversely, on introducing more symmetry into the calculations, new primitive symmetry species appear. Modes from the more symmetric system may be employed in a system with the lower symmetry.

Furthermore, on going to a subgroup, some degenerate species become split into smaller species. Modes for the split species may be used for the degenerate source. Indeed, this procedure can be employed in constructing modes for different rows.

In applying symmetry considerations to a vibration problem, a person must, secondly, introduce a suitable basis for the displacements. With vectors, one needs three on each equilibrium position. To be completely independent of each other, the three need to be mutually perpendicular. For convenience, the basis vectors may be oriented so that the symmetry operations merely permute them.

One would then determine the effect of each operation in the group on certain of the base vectors. The results may be arranged in a table. The components of the character vectors are then obtained. In practice, one would refer to a standard tabulation.

Formula (3.12) can next be applied to generate the independent f_j's. Those representing primitive symmetry species are identified and, where degeneracy exists, distinct rows of such species are identified. In constructing the latter, one introduces the necessary orthogonality.

For multiple occurrences of species, the forms produced by equation (3.12) are combined linearly so that combinations that the group operations mix only among themselves are obtained.

Formally, one can regard

$$N\sum y_k^{(r)} \mathscr{O}_{\bar{k}} \tag{3.13}$$

as a *projection operator*. It acts to project u_l onto the rth primitive symmetry species and to normalize the result.

Example 3.1

What group should a person employ in determining the normal modes for a square molecule of equivalent atoms?

Going down the flow chart in Figure 1.2, we find that a system consisting of four equivalent atoms at the corners of a square has the bases for the operations of group D_{4h}. Describing fewer aspects of symmetry of the molecule are subgroups C_{4h}, D_4, C_{4v}, C_4, and so on.

In determining the normal modes, a person needs to take into account the symmetries (a) among the corners and (b) at each corner in the plane. Groups C_4 and C_{4h} allow for (a) but not for (b). Groups C_{4v} and D_4 allow for both effects. If a person wants to distinguish the out-of-plane motions from related motions in the plane of the ring, one should go to the complete group D_{4h}.

Example 3.2

Construct an array of equivalent vectors that can serve as bases for the in-plane displacements of the four atoms in a square molecule. Combine these fixed vectors to form eigenvectors for the class sums of the C_{4v} group.

Choose a set of equilibrium positions for the four atoms. Join these by straight lines. From each intersection draw two unit vectors and label them as in Figure 3.1. Determine the effects of the operations of C_{4v} on u_1, u_2, u_5, u_6 and tabulate the results.

From equations (3.9) and (3.10), common eigenvectors for the class sums are given by any constant times

$$\mathscr{E}^{e(r)} u_j = \sum y_k^{(r)} \mathscr{E}_{\bar{k}} u_j.$$

Employing the results from Table 3.1 and the components from Table 3.2 in this formula leads to the symmetry-adapted vectors

A_1: $f_1 = u_1 + u_2 + u_3 + u_4 + u_5 + u_6 + u_7 + u_8$,

A_2: $f_2 = u_1 + u_2 + u_3 + u_4 - u_5 - u_6 - u_7 - u_8$,

B_1: $f_3 = u_1 - u_2 + u_3 - u_4 + u_5 - u_6 + u_7 - u_8$,

B_2: $f_4 = u_1 - u_2 + u_3 - u_4 - u_5 + u_6 - u_7 + u_8$.

Procedural Details

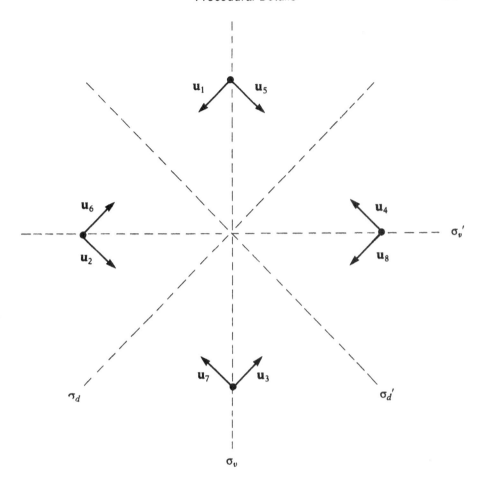

Figure 3.1 Base vectors for describing the oscillations in the plane of a square molecule.

Table 3.1 Effects of Pertinent Operations of the C_{4v} Group on Selected Base Vectors

I	C_4	C_2	C_4^{-1}	σ_v	σ_v'	σ_d	σ_d'
u_1	u_2	u_3	u_4	u_5	u_7	u_8	u_6
u_2		u_4				u_7	u_5
u_5	u_6	u_7	u_8	u_1	u_3	u_4	u_2
u_6		u_8					

For the E character, however, we obtain (after dividing by 2 and changing the sign of the last expression)

$$\mathbf{f}_5 = \mathbf{u}_1 - \mathbf{u}_3,$$
$$\mathbf{f}_6 = \mathbf{u}_2 - \mathbf{u}_4,$$
$$\mathbf{f}_7 = \mathbf{u}_5 - \mathbf{u}_7,$$
$$\mathbf{f}_8 = \mathbf{u}_8 - \mathbf{u}_6.$$

While the degeneracy of the E species is 2, four arrays that the operations mix among themselves have been generated. These have to be combined linearly to form two independent sets.

Table 3.2 Components of the Character Vectors for the $\mathbf{C}_{4v}$ Group

	I	$2C_4$	C_2	$2\sigma_v$	$2\sigma_d$
A_1	1	1	1	1	1
A_2	1	1	1	−1	−1
B_1	1	−1	1	1	−1
B_2	1	−1	1	−1	1
E	2	0	−2	0	0

Example 3.3

From the four arrays found for the E species, construct two independent sets.

Arrangements $\mathbf{f}_5$, $\mathbf{f}_6$, $\mathbf{f}_7$, and $\mathbf{f}_8$ are plotted in Figure 3.2. Applying the operations of the $\mathbf{C}_{4v}$ group to these yields the results in Table 3.3 Note that the first and fourth entries in each column are

$$\mathbf{f}_5, \mathbf{f}_8, \quad \text{or} \quad -\mathbf{f}_5, -\mathbf{f}_8, \quad \text{or} \quad \mathbf{f}_6, \mathbf{f}_7, \quad \text{or} \quad -\mathbf{f}_6, -\mathbf{f}_7.$$

Similarly, the second and third entries in each column are

$$\mathbf{f}_5, \mathbf{f}_8, \quad \text{or} \quad -\mathbf{f}_5, -\mathbf{f}_8, \quad \text{or} \quad \mathbf{f}_6, \mathbf{f}_7 \quad \text{or} \quad -\mathbf{f}_6, -\mathbf{f}_7.$$

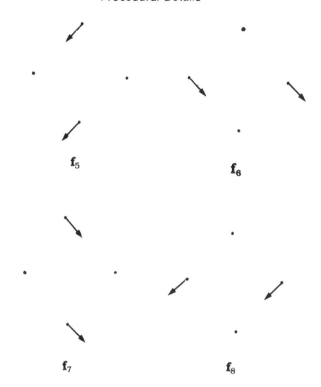

Figure 3.2 Displacements represented by $\mathbf{f}_5, \mathbf{f}_6, \mathbf{f}_7, \mathbf{f}_8$.

Table 3.3 Effects of Operations of the $\mathbf{C}_{4v}$ Group on Arrays $\mathbf{f}_5, \mathbf{f}_6, \mathbf{f}_7, \mathbf{f}_8$

I	C_4	C_2	C_4^{-1}	σ_v	σ_v'	σ_d	σ_d'
$\mathbf{f}_5$	$\mathbf{f}_6$	$-\mathbf{f}_5$	$-\mathbf{f}_6$	$\mathbf{f}_7$	$-\mathbf{f}_7$	$\mathbf{f}_8$	$-\mathbf{f}_8$
$\mathbf{f}_6$	$-\mathbf{f}_5$	$-\mathbf{f}_6$	$\mathbf{f}_5$	$\mathbf{f}_8$	$-\mathbf{f}_8$	$-\mathbf{f}_7$	$\mathbf{f}_7$
$\mathbf{f}_7$	$-\mathbf{f}_8$	$-\mathbf{f}_7$	$\mathbf{f}_8$	$\mathbf{f}_5$	$-\mathbf{f}_5$	$-\mathbf{f}_6$	$\mathbf{f}_6$
$\mathbf{f}_8$	$\mathbf{f}_7$	$-\mathbf{f}_8$	$-\mathbf{f}_7$	$\mathbf{f}_6$	$-\mathbf{f}_6$	$\mathbf{f}_5$	$-\mathbf{f}_5$

As a consequence, each operation causes

$$\mathbf{f}_5 + \mathbf{f}_8 \text{ to go into } \pm(\mathbf{f}_5 + \mathbf{f}_8) \text{ or } \pm(\mathbf{f}_6 + \mathbf{f}_7),$$
$$\mathbf{f}_6 + \mathbf{f}_7 \text{ to go into } \pm(\mathbf{f}_6 + \mathbf{f}_7) \text{ or } \pm(\mathbf{f}_5 + \mathbf{f}_8),$$

and causes

$$\mathbf{f}_5 - \mathbf{f}_8 \text{ to go into } \pm(\mathbf{f}_5 - \mathbf{f}_8) \text{ or } \pm(\mathbf{f}_6 - \mathbf{f}_7),$$
$$\mathbf{f}_6 - \mathbf{f}_7 \text{ to go into } \pm(\mathbf{f}_6 - \mathbf{f}_7) \text{ or } \pm(\mathbf{f}_5 - \mathbf{f}_8).$$

Under operations of the group, combinations $\mathbf{f}_5 + \mathbf{f}_8$ and $\mathbf{f}_6 + \mathbf{f}_7$ only mix among themselves; combinations $\mathbf{f}_5 - \mathbf{f}_8$ and $\mathbf{f}_6 - \mathbf{f}_7$ only mix among themselves. Therefore, the independent sets may be

$$E: \mathbf{f}_9 = \mathbf{f}_5 + \mathbf{f}_8 = \mathbf{u}_1 - \mathbf{u}_3 - \mathbf{u}_6 + \mathbf{u}_8,$$
$$\mathbf{f}_{10} = \mathbf{f}_6 + \mathbf{f}_7 = \mathbf{u}_2 - \mathbf{u}_4 + \mathbf{u}_5 - \mathbf{u}_7,$$

and

$$E: \mathbf{f}_{11} = \mathbf{f}_5 - \mathbf{f}_8 = \mathbf{u}_1 - \mathbf{u}_3 + \mathbf{u}_6 - \mathbf{u}_8,$$
$$\mathbf{f}_{12} = \mathbf{f}_6 - \mathbf{f}_7 = \mathbf{u}_2 - \mathbf{u}_4 - \mathbf{u}_5 + \mathbf{u}_7.$$

Example 3.4

Describe the normal modes for movements in their plane of four equivalent mass elements which have a square equilibrium configuration.

Each array of displacements belonging to a single occurrence of a primitive symmetry species describes a phase in a normal mode of motion. The arrays that we have found in Examples 3.2 and 3.3 are plotted in Figure 3.3.

We see that the A_1 mode of motion involves the mass elements moving together in and out along radii from the center. This is called a breathing movement. In the A_2 mode, the mass elements turn about the center in a rotation. In the B_1 mode of motion, the mass elements on one diagonal move in and out 180 degrees out of phase with those on the other diagonal. The square is alternately squashed on one diagonal, then on the other. In the B_2 mode of motion, the quadrilateral is alternately squashed between two opposite edges, then between the other two opposite edges. In the first set of type E, the mass elements are translating in the plane. In the second set of type E, opposite edges of the quadrilateral undergo scissors movements.

Procedural Details

The fact that translatory motion is independent of vibratory motion could have been used in constructing these sets. One would superpose (a) $\mathbf{f}_5$ and $\mathbf{f}_8$, (b) $\mathbf{f}_6$ and $\mathbf{f}_7$, so that only translations result. The orthogonal combinations would then yield the corresponding vibrations.

Example 3.5

Describe the normal modes for the out-of-plane movements of the four equivalent mass elements bound equivalently to the corners of a square.

To the unit vectors in Figure 3.1, add $\mathbf{u}_9, \mathbf{u}_{10}, \mathbf{u}_{11}, \mathbf{u}_{12}$ at the first, second, third, and fourth corners so that

$$\mathbf{u}_1 \times \mathbf{u}_5 = \mathbf{u}_9, \quad \mathbf{u}_2 \times \mathbf{u}_6 = \mathbf{u}_{10}, \quad \mathbf{u}_3 \times \mathbf{u}_7 = \mathbf{u}_{11}, \quad \mathbf{u}_4 \times \mathbf{u}_8 = \mathbf{u}_{12}.$$

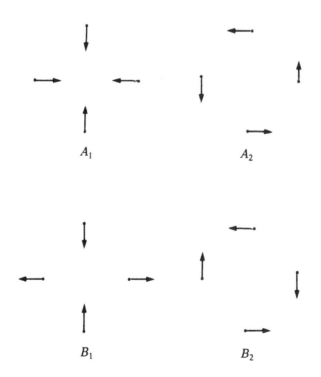

Figure 3.3 Coordinated displacements belonging to the primitive symmetry species of the C_{4v} group.

86 Expedient Arrangements of Displacements

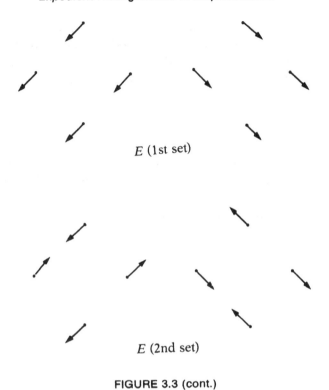

FIGURE 3.3 (cont.)

Then determine the effects of operations of C_{4v} on these new unit vectors and list the pertinent results in Table 3.4.

Table 3.4 Effects of Operations of the C_{4v} Group on Base Vectors u_9 and u_{10}

I	C_4	C_2	C_4^{-1}	σ_v	σ_v'	σ_d	σ_d'
u_9	u_{10}	u_{11}	u_{12}	u_9	u_{11}	u_{12}	u_{10}
u_{10}		u_{12}					

Employing these in equation (3.12), with $N = \tfrac{1}{2}$, yields

$$A_1: \mathbf{f}_{13} = \mathbf{u}_9 + \mathbf{u}_{10} + \mathbf{u}_{11} + \mathbf{u}_{12},$$
$$B_1: \mathbf{f}_{14} = \mathbf{u}_9 - \mathbf{u}_{10} + \mathbf{u}_{11} - \mathbf{u}_{12},$$

Separating the Modes

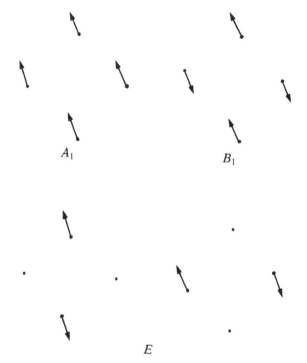

Figure 3.4 Coordinated displacements of the four equivalent masses perpendicular to the equilibrium plane, classified according to group C_{4v}.

$$E: \mathbf{f}_{15} = \mathbf{u}_9 - \mathbf{u}_{11},$$
$$\mathbf{f}_{16} = \mathbf{u}_{10} - \mathbf{u}_{12}.$$

This A_1 mode involves the elements translating perpendicular to the plane of the reference origins. The E mode describes rotation of the system about axes in the reference plane. In the B_1 mode, the two diagonals oscillate about each other; when two opposite mass elements are moving up, the other two are moving down, and vice versa. (See Figure 3.4.)

3.5
Separating the Modes

Two different pathways are now open. One may formulate the equations of motion in a coordinate system in which these equations are approximately linear. The linear combinations of coordinates belonging to the different primitive symmetry species are determined. These combinations are then applied to the equations of motion to gain separation of the modes. Alternatively, the kinetic energy and the potential energy may be formulated directly for each occurrence of a primitive symmetry species and the corresponding equations of motion constructed.

For the first procedure, the kinetic energy could be taken as

$$T = \sum \tfrac{1}{2} m_j (\dot{\eta}_{3j-2}^2 + \dot{\eta}_{3j-1}^2 + \dot{\eta}_{3j}^2). \tag{3.14}$$

This has the desired quadratic form

$$T = \sum \sum \dot{\eta}_j B_{jk} \dot{\eta}_k. \tag{3.15}$$

Here m_j is the mass of the jth element of the system while the η's are the Cartesian coordinates defined as in equations (3.1) and (3.2).

The interparticle distances equal the square root of a quadratic function of the η's. This relationship can be expanded as a Taylor series and higher terms than the linear dropped. The extensions in the distances would become linear in the η's. If the mass elements are held together by a force law approximately linear in the extensions, then the potential energy could be written in the form

$$V = \sum \sum \eta_j C_{jk} \eta_k, \tag{3.16}$$

which is also quadratic.

In general, the potential governing the forces depends on all the interparticle distances and angles. But as a first approximation, it appears as a quadratic function of the extensions of the interparticle distances and a quadratic function of the distortions of the angles.

In molecular systems, where the atoms are held together by covalent forces, both bond-stretching and bond-bending interactions need to be taken into account. But for masses held together by central forces or springs, the angular distortion terms vanish.

Example 3.6

Formulate the kinetic energy and the potential energy for the four equivalent mass elements held together by four equivalent springs

Separating the Modes

along the edges and two equivalent springs along the diagonals. Neglect the masses of the springs. Consider only the motion in the plane.

Employ the base vectors plotted in Figure 3.1. Let the mass of each element be m. Then in the η system the kinetic energy is

$$T = \tfrac{1}{2}m\left[(\dot{\eta}_1^2 + \dot{\eta}_5^2) + (\dot{\eta}_2^2 + \dot{\eta}_6^2) + (\dot{\eta}_3^2 + \dot{\eta}_7^2) + (\dot{\eta}_4^2 + \dot{\eta}_8^2) \right]$$

and the square matrix for equation (3.15) is

$$\mathbf{B} = \tfrac{1}{2}m \begin{pmatrix} 1 & 0 & 0 & 0 & 0 & 0 & 0 & 0 \\ 0 & 1 & 0 & 0 & 0 & 0 & 0 & 0 \\ 0 & 0 & 1 & 0 & 0 & 0 & 0 & 0 \\ 0 & 0 & 0 & 1 & 0 & 0 & 0 & 0 \\ 0 & 0 & 0 & 0 & 1 & 0 & 0 & 0 \\ 0 & 0 & 0 & 0 & 0 & 1 & 0 & 0 \\ 0 & 0 & 0 & 0 & 0 & 0 & 1 & 0 \\ 0 & 0 & 0 & 0 & 0 & 0 & 0 & 1 \end{pmatrix}.$$

Let the spring constant for each of the edge springs be k_1, that for each of the diagonal springs k_2. Also consider only the linear terms in the extensions. Then the potential energy reduces to

$$V = \tfrac{1}{2} k_1 \left[(\eta_1 + \eta_6)^2 + (\eta_2 + \eta_7)^2 + (\eta_3 + \eta_8)^2 + (\eta_4 + \eta_5)^2 \right]$$
$$\tfrac{1}{2} k_2 \left[\tfrac{1}{2}(\eta_1 + \eta_5 + \eta_3 + \eta_7)^2 + \tfrac{1}{2}(\eta_2 + \eta_6 + \eta_4 + \eta_8)^2 \right]$$

whence

$$\mathbf{C} = \tfrac{1}{2} \begin{pmatrix} k_1 + \tfrac{1}{2}k_2 & 0 & \tfrac{1}{2}k_2 & 0 & \tfrac{1}{2}k_2 & k_1 & \tfrac{1}{2}k_2 & 0 \\ 0 & k_1 + \tfrac{1}{2}k_2 & 0 & \tfrac{1}{2}k_2 & 0 & \tfrac{1}{2}k_2 & k_1 & \tfrac{1}{2}k_2 \\ \tfrac{1}{2}k_2 & 0 & k_1 + \tfrac{1}{2}k_2 & 0 & \tfrac{1}{2}k_2 & 0 & \tfrac{1}{2}k_2 & k_1 \\ 0 & \tfrac{1}{2}k_2 & 0 & k_1 + \tfrac{1}{2}k_2 & k_1 & \tfrac{1}{2}k_2 & 0 & \tfrac{1}{2}k_2 \\ \tfrac{1}{2}k_2 & 0 & \tfrac{1}{2}k_2 & k_1 & k_1 + \tfrac{1}{2}k_2 & 0 & \tfrac{1}{2}k_2 & 0 \\ k_1 & \tfrac{1}{2}k_2 & 0 & \tfrac{1}{2}k_2 & 0 & k_1 + \tfrac{1}{2}k_2 & 0 & \tfrac{1}{2}k_2 \\ \tfrac{1}{2}k_2 & k_1 & \tfrac{1}{2}k_2 & 0 & \tfrac{1}{2}k_2 & 0 & k_1 + \tfrac{1}{2}k_2 & 0 \\ 0 & \tfrac{1}{2}k_2 & k_1 & \tfrac{1}{2}k_2 & 0 & \tfrac{1}{2}k_2 & 0 & k_1 + \tfrac{1}{2}k_2 \end{pmatrix}.$$

Example 3.7

Construct equations of motion from the matrices obtained in Example 3.6.

From the **B** matrix, we find that

$$\frac{d}{dt}\frac{\partial T}{\partial \dot{\eta}_l} = \frac{d}{dt}(m\dot{\eta}_l) = m\ddot{\eta}_l, \quad \frac{\partial T}{\partial \eta_l} = 0,$$

while from **C**, we have

$$\frac{\partial V}{\partial \eta_l} = (k_1 + \tfrac{1}{2}k_2)\eta_l + \tfrac{1}{2}k_2\eta_{l+2}$$
$$+ \tfrac{1}{2}k_2\eta_{l+4} + \tfrac{1}{2}k_2\eta_{l+6} + k_1\eta_j.$$

The relationship between j and l is obtained from the written out matrix. Now,

$$\frac{\partial V}{\partial \eta_l} = -Q_l.$$

Substituting these results into the Lagrange equation yields

$$m\ddot{\eta}_1 + (k_1 + \tfrac{1}{2}k_2)\eta_1 + \tfrac{1}{2}k_2\eta_3 + \tfrac{1}{2}k_2\eta_5 + \tfrac{1}{2}k_2\eta_7 + k_1\eta_6 = 0,$$
$$m\ddot{\eta}_2 + (k_1 + \tfrac{1}{2}k_2)\eta_2 + \tfrac{1}{2}k_2\eta_4 + \tfrac{1}{2}k_2\eta_6 + \tfrac{1}{2}k_2\eta_8 + k_1\eta_7 = 0,$$
$$m\ddot{\eta}_3 + (k_1 + \tfrac{1}{2}k_2)\eta_3 + \tfrac{1}{2}k_2\eta_5 + \tfrac{1}{2}k_2\eta_7 + \tfrac{1}{2}k_2\eta_1 + k_1\eta_8 = 0,$$
$$m\ddot{\eta}_4 + (k_1 + \tfrac{1}{2}k_2)\eta_4 + \tfrac{1}{2}k_2\eta_6 + \tfrac{1}{2}k_2\eta_8 + \tfrac{1}{2}k_2\eta_2 + k_1\eta_5 = 0,$$
$$m\ddot{\eta}_5 + (k_1 + \tfrac{1}{2}k_2)\eta_5 + \tfrac{1}{2}k_2\eta_7 + \tfrac{1}{2}k_2\eta_1 + \tfrac{1}{2}k_2\eta_3 + k_1\eta_4 = 0,$$
$$m\ddot{\eta}_6 + (k_1 + \tfrac{1}{2}k_2)\eta_6 + \tfrac{1}{2}k_2\eta_8 + \tfrac{1}{2}k_2\eta_2 + \tfrac{1}{2}k_2\eta_4 + k_1\eta_1 = 0,$$
$$m\ddot{\eta}_7 + (k_1 + \tfrac{1}{2}k_2)\eta_7 + \tfrac{1}{2}k_2\eta_1 + \tfrac{1}{2}k_2\eta_3 + \tfrac{1}{2}k_2\eta_5 + k_1\eta_2 = 0,$$
$$m\ddot{\eta}_8 + (k_1 + \tfrac{1}{2}k_2)\eta_8 + \tfrac{1}{2}k_2\eta_2 + \tfrac{1}{2}k_2\eta_4 + \tfrac{1}{2}k_2\eta_6 + k_1\eta_3 = 0.$$

3.6
Symmetry-Adapted Coordinates

Since each of the symmetry-adapted arrays $\mathbf{f}_j$ represents a given phase of a pure mode and since these arrays are mutually orthogonal, the component of a general motion along $\mathbf{f}_j$ is constructed from the component displacements in the same manner.

Suppose that a given occurrence of a primitive symmetry species is represented by

$$\mathbf{f}_j = \sum a_{jk} \mathbf{u}_k. \tag{3.17}$$

On the other hand, a general displacement array is given by

$$\mathbf{q} = \sum \eta_l \mathbf{u}_l = \sum q_m \mathbf{f}_m. \tag{3.18}$$

Projecting $\mathbf{q}$ onto $\mathbf{f}_j$ leads to

$$q_j = \mathbf{q} \cdot \mathbf{f}_j = \sum \eta_l \mathbf{u}_l \cdot \sum a_{jk} \mathbf{u}_k = \sum \sum a_{jk} \eta_l \delta_{kl}$$
$$= \sum a_{jk} \eta_k. \tag{3.19}$$

The η_k's are combined as the $\mathbf{u}_k$'s are in equation (3.17).

Example 3.8

Construct and solve the equation of motion for the A_1 mode of the system in Example 3.6.

From Example 3.2, we have for this mode

$$\mathbf{f}_1 = \mathbf{u}_1 + \mathbf{u}_2 + \mathbf{u}_3 + \mathbf{u}_4 + \mathbf{u}_5 + \mathbf{u}_6 + \mathbf{u}_7 + \mathbf{u}_8 = \sum_1^8 \mathbf{u}_k.$$

So equation (3.19) tells us that

$$q_1 = \sum \eta_k, \quad \ddot{q}_1 = \sum \ddot{\eta}_k.$$

To separate out all the other modes, we therefore have to add the equations in Example 3.7 as they stand. We find that

$$m\ddot{q}_1 + 2(k_1 + k_2) q_1 = 0.$$

But for a harmonic oscillator with angular frequency ω, we have

$$\ddot{q} + \omega^2 q = 0.$$

So here

$$\omega_1 = \left[\frac{2(k_1 + k_2)}{m}\right]^{1/2}.$$

Example 3.9

Construct and solve the equation of motion for the B_1 mode of the system in Example 3.6.

From $\mathbf{f}_3$ in Example 3.2, proceeding as in Example 3.8, we obtain

$$q_3 = \eta_1 - \eta_2 + \eta_3 - \eta_4 + \eta_5 - \eta_6 + \eta_7 - \eta_8.$$

Combining the equations from Example 3.7 with alternating signs yields

$$m\ddot{q}_3 + 2k_2 q_3 = 0$$

whence

$$\omega_3 = \left[\frac{2k_2}{m}\right]^{1/2}.$$

Example 3.10

Construct and solve the equation of motion for the B_2 mode of the system in Example 3.6.

From $\mathbf{f}_4$ in Example 3.2, we obtain

$$q_4 = \eta_1 - \eta_2 + \eta_3 - \eta_4 - \eta_5 + \eta_6 - \eta_7 + \eta_8.$$

Combining the equations from Example 3.7 with alternating signs for the first four and reversed alternating signs for the second four yields

$$m\ddot{q}_4 + 2k_1 q_4 = 0$$

whence

$$\omega_4 = \left[\frac{2k_1}{m}\right]^{\frac{1}{2}}.$$

Example 3.11

Construct and solve the equations of motion for the E mode of the system in Example 3.6.

From $\mathbf{f}_{11}$ in Example 3.3, we obtain

$$q_{11} = \eta_1 - \eta_3 + \eta_6 - \eta_8.$$

Combining the first, third, sixth, and eighth equations from Example 3.7 with alternating signs yields

$$m\ddot{q}_{11} + 2k_1 q_{11} = 0$$

whence

$$\omega_{11} = \left[\frac{2k_1}{m}\right]^{\frac{1}{2}}.$$

From $\mathbf{f}_{12}$, we similarly obtain

$$q_{12} = \eta_2 - \eta_4 - \eta_5 + \eta_7.$$

The second equation minus the fourth minus the fifth plus the seventh yields

$$m\ddot{q}_{12} + 2k_1 q_{12} = 0$$

and

$$\omega_{12} = \left[\frac{2k_1}{m}\right]^{\frac{1}{2}}.$$

Both q_{11} and q_{12} satisfy the same linear differential equation; so any linear combination of them does also. All solutions for the E mode are of this form; so the degeneracy is 2. This degeneracy is caused by the symmetry in the group. Note that

$$\omega_{11} = \omega_{12} = \omega_4.$$

Thus the B_2 mode is degenerate with the oscillating E mode. This degeneracy would be removed, however, if there were a force associated with distortion of the angles. Insofar as the spatial symmetries for which we have allowed are concerned, this degeneracy is *accidental.*

3.7
Interactions between Modes

With a given vibrating system, a primitive symmetry species may occur more than once. The right side of equation (3.7) can then contain contributions from all of the occurrences, even when the left side is limited to each single one in turn. Thus, the different modes occupying the same row of the symmetry species interact with each other. The interactions tend to move the frequencies farther apart.

Consider a vibrating system in which two normal modes interact. Let the generalized coordinates for these be q_1 and q_2. In the linear approximation, the equations of motion may be written

$$\ddot{q}_1 + a_{11}q_1 + a_{12}q_2 = 0, \qquad (3.20)$$

$$a_{21}q_1 + \ddot{q}_2 + a_{22}q_2 = 0. \qquad (3.21)$$

A sinusoidal solution of angular frequency ω yields the relations

$$\ddot{q}_1 = -\omega^2 q_1, \qquad (3.22)$$

$$\ddot{q}_2 = -\omega^2 q_2. \qquad (3.23)$$

Substituting equations (3.22) and (3.23) into equations (3.20) and (3.21) produces

Interactions between Modes

$$(a_{11} - \omega^2)q_1 + a_{12}q_2 = 0, \qquad (3.24)$$

$$a_{21}q_1 + (a_{22} - \omega^2)q_2 = 0. \qquad (3.25)$$

These are homogenous linear equations. The Cramer-rule fractions for q_1 and q_2 have zero for their numerators. So the common denominator must equal zero. Indeed, for q_1 or q_2 to differ from zero, we must have

$$\begin{vmatrix} a_{11} - \omega^2 & a_{12} \\ a_{21} & a_{22} - \omega^2 \end{vmatrix} = 0, \qquad (3.26)$$

whence

$$\omega^4 - (a_{11} + a_{22})\omega^2 + a_{11}a_{22} - a_{12}a_{21} = 0 \qquad (3.27)$$

and

$$\omega^2 = \tfrac{1}{2}\left\{ a_{11} + a_{22} \pm \left[(a_{11} - a_{22})^2 + 4a_{12}a_{21} \right]^{\frac{1}{2}} \right\}. \qquad (3.28)$$

When there is no interaction, so that

$$a_{12} = 0, \qquad a_{21} = 0, \qquad (3.29)$$

we have the two roots

$$\omega_1^2 = a_{11}, \qquad \omega_2^2 = a_{22}, \qquad (3.30)$$

associated with the plus and minus signs in equation (3.28). Introducing interaction, with $a_{12}a_{21} > 0$, causes the upper root to move up and the lower root to move down.

Example 3.12

The symmetry in the system of Example 3.6 is reduced by altering the spring constant for the $\mathbf{u}_1 - \mathbf{u}_6$ and the $\mathbf{u}_3 - \mathbf{u}_8$ edges to k_0. How are the equations of motion changed? What happens to the $\mathbf{C}_{4v}$ group in the process?

The potential energy becomes

$$V = \tfrac{1}{2}k_0\left[(\eta_1 + \eta_6)^2 + (\eta_3 + \eta_8)^2\right]$$
$$+ \tfrac{1}{2}k_1\left[(\eta_2 + \eta_7)^2 + (\eta_4 + \eta_5)^2\right]$$
$$+ \tfrac{1}{2}k_2\left[\tfrac{1}{2}(\eta_1 + \eta_5 + \eta_3 + \eta_7)^2\right.$$
$$\left. + \tfrac{1}{2}(\eta_2 + \eta_6 + \eta_4 + \eta_8)^2\right].$$

Substituting this and the same kinetic energy into the Lagrange equations yields

$$m\ddot{\eta}_1 + (k_0 + \tfrac{1}{2}k_2)\eta_1 + \tfrac{1}{4}k_2\eta_3 + \tfrac{1}{4}k_2\eta_5 + \tfrac{1}{4}k_2\eta_7 + k_0\eta_6 = 0,$$
$$m\ddot{\eta}_2 + (k_1 + \tfrac{1}{2}k_2)\eta_2 + \tfrac{1}{4}k_2\eta_4 + \tfrac{1}{4}k_2\eta_6 + \tfrac{1}{4}k_2\eta_8 + k_1\eta_7 = 0,$$
$$m\ddot{\eta}_3 + (k_0 + \tfrac{1}{2}k_2)\eta_3 + \tfrac{1}{4}k_2\eta_5 + \tfrac{1}{4}k_2\eta_7 + \tfrac{1}{4}k_2\eta_1 + k_0\eta_8 = 0,$$
$$m\ddot{\eta}_4 + (k_1 + \tfrac{1}{2}k_2)\eta_4 + \tfrac{1}{4}k_2\eta_6 + \tfrac{1}{4}k_2\eta_8 + \tfrac{1}{4}k_2\eta_2 + k_1\eta_5 = 0,$$
$$m\ddot{\eta}_5 + (k_1 + \tfrac{1}{2}k_2)\eta_5 + \tfrac{1}{4}k_2\eta_7 + \tfrac{1}{4}k_2\eta_1 + \tfrac{1}{4}k_2\eta_3 + k_1\eta_4 = 0,$$
$$m\ddot{\eta}_6 + (k_0 + \tfrac{1}{2}k_2)\eta_6 + \tfrac{1}{4}k_2\eta_8 + \tfrac{1}{4}k_2\eta_2 + \tfrac{1}{4}k_2\eta_4 + k_0\eta_1 = 0,$$
$$m\ddot{\eta}_7 + (k_1 + \tfrac{1}{2}k_2)\eta_7 + \tfrac{1}{4}k_2\eta_1 + \tfrac{1}{4}k_2\eta_3 + \tfrac{1}{4}k_2\eta_5 + k_1\eta_2 = 0,$$
$$m\ddot{\eta}_8 + (k_0 + \tfrac{1}{2}k_2)\eta_8 + \tfrac{1}{4}k_2\eta_2 + \tfrac{1}{4}k_2\eta_4 + \tfrac{1}{4}k_2\eta_6 + k_0\eta_3 = 0.$$

Now C_4, C_4^3, σ_v, and σ_v' are no longer symmetry operations. Also, σ_d and σ_d' are no longer similar. For the new group, we employ $\mathbf{C}_{2v}$ and relabel σ_d and σ_d' as $\sigma(zx)$ and $\sigma(yz)$. Going from Table 3.2 to Table 3.5, symmetry species A_1 and B_2 coalesce into the new A_1, symmetry species A_2 and B_1 coalesce into the new A_2, symmetry species E splits into the new B_1 and B_2.

Table 3.5 Components of the Character Vectors for the $\mathbf{C}_{2v}$ Group

	I	C_2	$\sigma(zx)$	$\sigma(yz)$
A_1	1	1	1	1
A_2	1	1	-1	-1
B_1	1	-1	1	-1
B_2	1	-1	-1	1

3.8
Reducing the Symmetry

One may start with a highly symmetric system and then reduce the symmetry by altering some of the parameters. In such a process, certain class sums disappear and others may be split into separate sums. Correspondingly, the distinctions between some primitive symmetry species disappear and degenerate species may be split.

To determine what happens, a person compares the character tables for the initial and final groups. Wherever the change causes the characters of two or more species to become the same, a single species emerges.

Wherever the changes cause splitting, the character components for the final species add to give those that remain for the initial species. This relationship is evident on comparing Tables 3.2 and 3.5. It will be confirmed later by means of matrix representation theory.

Associated with the coalescing of primitive symmetry species is the introduction of interactions. These interactions tend to move the frequencies farther apart, as Section 3.7 indicates. Associated with the splitting of a species is a corresponding splitting of frequencies.

Example 3.13

Construct two independent symmetry-adapted vector arrays for the A_1 species of the system in Example 3.12.

Into equation (3.12) in the form

$$\mathbf{f}_j = \sum y_k^{(r)} \mathscr{C}_{\bar{k}} \mathbf{u}_b$$

let l be 1 and 2. Thus, introduce results from the first two rows of Table 3.1 and components from Table 3.5 to get

$$A_1: \quad \mathbf{f}_1 = \mathbf{u}_1 + \mathbf{u}_3 + \mathbf{u}_6 + \mathbf{u}_8,$$

$$\mathbf{f}_2 = \mathbf{u}_2 + \mathbf{u}_4 + \mathbf{u}_5 + \mathbf{u}_7.$$

Example 3.14

Construct and solve the equations of motion for the A_1 modes of the system in Example 3.12.

Following equation (3.19), the η_k's are combined as the $\mathbf{u}_k$'s are for each occurrence. So from Example 3.13, we have

$$q_1 = \eta_1 + \eta_3 + \eta_6 + \eta_8,$$
$$q_2 = \eta_2 + \eta_4 + \eta_5 + \eta_7.$$

Adding the first, third, sixth, and eighth equations in Example 3.12 yields

$$m\ddot{q}_1 + (2k_0 + k_2)q_1 + k_2 q_2 = 0,$$

while adding the second, fourth, fifth, and seventh equations in Example 3.12 yields

$$m\ddot{q}_2 + (2k_1 + k_2)q_2 + k_2 q_1 = 0.$$

Identifying the coefficients, substituting into equation (3.28), and reducing, leads to

$$\omega^2 = \frac{1}{m}\left[k_0 + k_1 + k_2 \pm (k_0^2 + k_1^2 + k_2^2 - 2k_0 k_1)^{\frac{1}{2}}\right].$$

3.9
Highlights

Each phase in the movement of a system of mass elements is described by an array of displacements from a set of equilibrium positions. When the equilibrium positions possess symmetry, it is advantageous to express the general array as a superposition of simple arrays belonging to primitive symmetry species of a covering group for the system.

When the amplitudes of critical modes of a given system are small, its equations of motion,

$$\frac{d}{dt}\frac{\partial T}{\partial \dot{q}_l} - \frac{\partial T}{\partial q_l} = Q_l, \qquad (3.31)$$

reduce to

$$\sum A_{lk}\ddot{q}_k = Q_l, \qquad (3.32)$$

with the A_{lk}'s parameters. Whenever Q_l belongs to a primitive symmetry species, the left sides of equations (3.31) and (3.32) do also. Thus a cause belonging to the species produces an effect belonging to the same species.

Each displacement is described by a vector bound to the pertinent equilibrium position. A given phase in a movement is described by an array of such vectors, each bound to its equilibrium position. This array can be represented by a single vector having the components of all the constiuent vectors.

When this array belongs to the rth primitive symmetry species, it is an rth common eigenvector of the class sums:

$$\mathscr{C}_j \mathbf{f}^{(r)} = \lambda_j^{(r)} \mathbf{f}^{(r)}. \tag{3.33}$$

When the species is not degenerate, each operation of the group acting on $\mathbf{f}^{(r)}$ yields a number times $\mathbf{f}^{(r)}$. When it is degenerate, there are a number of eigenvectors equal to the degeneracy and each operation of the group acting on $\mathbf{f}^{(r)}$ yields a linear combination of these eigenvectors. But we have

$$\mathscr{C}_j \sum y_k^{(r)} \mathscr{C}_{\bar{k}} = \lambda_j^{(r)} \sum y_k^{(r)} \mathscr{C}_{\bar{k}} \tag{3.34}$$

from Chapter 2. Letting each side act on a basis vector in the array yields

$$\mathscr{C}_j \sum y_k^{(r)} \mathscr{C}_{\bar{k}} \mathbf{u}_l = \lambda_j^{(r)} \sum y_k \mathscr{C}_{\bar{k}} \mathbf{u}_l. \tag{3.35}$$

Note that equations (3.33) and (3.35) have the same form. If N_l in

$$\mathbf{f}_i = N_l \sum y_k^{(r)} \mathscr{C}_{\bar{k}} \mathbf{u}_l \tag{3.36}$$

for all the base vectors $\mathbf{u}_1, \mathbf{u}_2, \ldots$, can be chosen so that the same expression is obtained, then $\mathbf{f}_i$ is identified with $\mathbf{f}^{(r)}$. Otherwise, one superposes the $\mathbf{f}_i$ that differ essentially to obtain expressions that behave under operations of the group as $\mathbf{f}^{(r)}$.

One obtains each occurrence of a primitive symmetry species as a linear combination of the base vectors:

$$\mathbf{f}_i = \sum a_{ij} \mathbf{u}_j. \tag{3.37}$$

If the displacement in the system along $\mathbf{u}_j$ is η_j at time t, the corresponding generalized coordinate at that time is

$$q_i = \sum a_{ij}\eta_j. \qquad (3.38)$$

Discussion Questions

3.1 When does a physical system possess symmetry? How is this symmetry characterized?

3.2 When do a set of expressions form a basis? Give examples.

3.3 How can a single variable number determine an array of displacements?

3.4 How can a single variable number determine an array of forces?

3.5 What is a primitive symmetry species?

3.6 Why should a force belonging to a primitive symmetry species cause an acceleration of the same species?

3.7 When is the complete solution to the equations of motion a linear combination of the solutions for a single species?

3.8 Why is an array belonging to the rth primitive symmetry species a common eigenvector of the class sums?

3.9 Why is

$$\sum_k y_k^{(r)}\mathscr{C}_{\bar{k}}\mathbf{u}_j$$

a common eigenvector of the class sums?

3.10 When does $\mathscr{C}_{\bar{k}}$ equal $\mathscr{C}_k$? When do they differ? Explain.

3.11 When does $y_{\bar{k}}$ differ from y_k?

3.12 If y_k equals $a + ib$, what should $y_{\bar{k}}$ equal so that the orthogonality relationship is satisfied?

3.13 When does the expression in Question 3.9 yield an array that represents a given phase in a normal mode of oscillation? When does the sum need to be combined with related sums to obtain such a phase?

3.14 Should one always employ the properties of the full group for a system in determining its normal modes? Why or why not?

3.15 When does the kinetic energy of a system have the quadratic form

$$T = \sum \sum \dot{\eta}_j B_{jk} \dot{\eta}_k?$$

3.16 With each C_{jk} a parameter, when does the potential energy of a system have the quadratic form

$$V = \sum \sum \eta_j C_{jk} \eta_k?$$

3.17 If a primitive symmetry species has the basis

$$\mathbf{f}_j = \sum a_{jk} \mathbf{u}_k$$

why does the corresponding generalized coordinate have the same form

$$q_j = \sum a_{jk} \eta_k?$$

3.18 Obtain the general solution for the harmonic oscillator equation

$$\ddot{q}_j + \omega^2 q_j = 0.$$

3.19 Show that parameter ω^2 is the square of the angular frequency of oscillation.

3.20 How do degeneracies arise?

3.21 What angular frequency does one obtain for (a) a rotational mode, (b), a translational mode?

3.22 When do different modes of motion interact?

3.23 How are frequencies altered by the interactions among modes?

Problems

3.1 What groups may a person employ in determining the normal modes for the molecule

$$O = C = O?$$

Why is it advantageous to use the $\mathbf{D}_2$ group?

3.2 Set up three parallel Cartesian systems on simultaneous equilibrium positions for the atoms in the CO_2 molecule. Introduce the corresponding base vectors. Then combine these to form eigenvectors for the class sums of the $\mathbf{D}_2$ group.

3.3 From the combinations obtained in Problem 3.2, separate out the translational and rotational motions and so obtain bases for the pure vibrational modes. Sketch these.

3.4 Construct the classical equations of motion for the CO_2 molecule, letting the kinetic energy be

$$T = \tfrac{1}{2}m_O(\dot{\eta}_1^2 + \dot{\eta}_2^2 + \dot{\eta}_3^2) + \tfrac{1}{2}m_C(\dot{\eta}_4^2 + \dot{\eta}_5^2 + \dot{\eta}_6^2)$$
$$+ \tfrac{1}{2}m_O(\dot{\eta}_7^2 + \dot{\eta}_8^2 + \dot{\eta}_9^2)$$

and the potential energy be

$$V = \tfrac{1}{2}k_1(\eta_4 - \eta_1)^2 + \tfrac{1}{2}k_1(\eta_7 - \eta_4)^2 + \tfrac{1}{2}k_2(\eta_5 - \eta_2)^2$$
$$+ \tfrac{1}{2}k_2(\eta_5 - \eta_8)^2 + \tfrac{1}{2}k_2(\eta_6 - \eta_3)^2 + \tfrac{1}{2}k_2(\eta_6 - \eta_9)^2.$$

Here η_1, η_4, η_7 are displacements of the first, second, and third atoms along the axis of the molecule in one direction while η_2, η_5, η_8 and η_3, η_6, η_9 are two sets of displacements perpendicular to the first set and perpendicular to each other.

3.5 Use the group theoretical results from Problem 3.3 to solve these classical equations for the A vibrational mode of CO_2.

3.6 Use the group theoretical results and solve for the B_1, B_2, and B_3 vibrational modes of CO_2.

3.7 At a given phase in a vibrational mode of CO_2, the oxygen atoms have moved distance S in one direction. How far has the carbon atom moved in the opposite direction if there is no net translation? Apply this result to the B_1 mode, constructing T and V in terms of $\dot{S}$ and S. Introduce an appropriate q and form the equation of motion. Obtain ω^2 as before.

3.8 In describing the vibrations of a square molecule of equivalent atoms, one may employ the internal coordinates of Figure 3.5. Combine these to form common eigenfunctions for the class sums of the $\mathbf{C}_{4v}$ group.

3.9 What internal coordinates would one need to describe the B_1 out-of-plane vibration? Introduce these and derive the corresponding eigenfunction for the class sums.

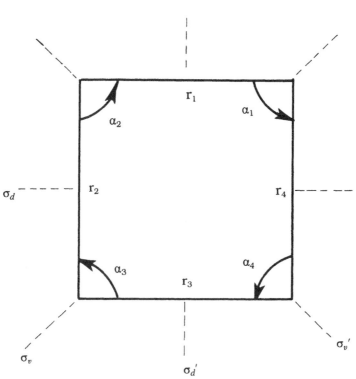

Figure 3.5 Internal coordinates for the square molecule. Mass m is at each corner; r_j is the length of the jth edge; α_j is the size of the jth angle.

3.10 Rotate each set of axes in Figure 3.1 by 45 degrees, so that $\mathbf{u}_5$, $\mathbf{u}_6$, $\mathbf{u}_7$, $\mathbf{u}_8$ point radially inward while $\mathbf{u}_1$, $\mathbf{u}_2$, $\mathbf{u}_3$, $\mathbf{u}_4$ point perpendicular to the radii in the counterclockwise direction. Then combine these to form common eigenvectors for the class sums of the $\mathbf{C}_{4v}$ group.

3.11 Identify the pure translational, pure rotational, and pure vibrational modes in the results from Problem 3.10. Indicate how the translational E mode and the vibrational E mode are obtained.

3.12 What groups may be used in analyzing the planar movements of an equilateral triangular molecule of three equivalent atoms? What is a good arrangement for the base vectors?

3.13 Combine the radial and tangential base vectors for the system in Problem 3.12 to form common eigenvectors for the group. For the doubly degenerate species, separate the translational modes from the vibrational modes. List the vibrational modes.

3.14 Construct the classical equations of motion for the equilateral triangular molecule of three equivalent atoms. Let the kinetic energy be

$$T = \tfrac{1}{2}m(\dot\rho_1^2 + \dot\tau_1^2 + \dot\rho_2^2 + \dot\tau_2^2 + \dot\rho_3^2 + \dot\tau_3^2)$$

and the potential energy be

$$V = \tfrac{1}{2}k\left\{\left[\frac{\sqrt{3}}{2}(\rho_1 + \rho_2) + \tfrac{1}{2}(\tau_1 - \tau_2)\right]^2 \right.$$
$$+ \left[\frac{\sqrt{3}}{2}(\rho_2 + \rho_3) + \tfrac{1}{2}(\tau_2 - \tau_3)\right]^2$$
$$\left. + \left[\frac{\sqrt{3}}{2}(\rho_3 + \rho_1) + \tfrac{1}{2}(\tau_3 - \tau_1)\right]^2\right\}$$

where ρ_j is the displacement inward along radial base vector $\mathbf{r}_j$ while τ_j is the displacement perpendicular to this direction along base vector $\mathbf{t}_j$.

3.15 Use the group theoretical results from Problem 3.13 to solve these classical equations for the vibrational modes. Also, employ them to show that the rotational and translational modes are zero-frequency modes.

References

Books

Duffey, G. H.: 1980, *Theoretical Physics: Classical and Modern Views*, Krieger, Melbourne, Fla., pp. 265-292. The treatment of modes of motion in classical systems is similar to that in this chapter.

Nakamoto, K.: 1986, *Infrared and Raman Spectra of Inorganic and Coordination Compounds*, 4th ed., Wiley, New York, pp. 3-97. Nakamoto develops the theory of normal vibrations in more detail. The behavior of representative molecules is considered.

Nussbaum, A.: 1971, *Applied Group Theory for Chemists, Physicists and Engineers*, Prentice-Hall, Englewood Cliffs, N.J., pp. 79-160. Nussbaum describes the relevant theory and applies it to constructing normal modes of representative symmetric systems.

Articles

Darensbourg, D. J., and Darensbourg, M. Y.: 1974, "Infrared Determination of Stereochemistry in Metal Complexes," *J. Chem. Educ.* **51,** 787-789.

Ermer, O.: 1990, "Independent Coordinates of Molecular Structures and Group Theory, " *J. Chem. Educ.* **67**, 209-210.

Hsu, C.-Y., and Orchin, M.: 1974, "Ligand Group Orbitals and Normal Molecular Vibrations," *J. Chem. Educ.* **51,** 725-729.

Nussbaum, A.: 1968, "Group Theory and Normal Modes," *Am. J. Phys.* **36,** 529-539.

Strommen, D. P., and Lippincott, E. R.: 1972, "Comments on Infinite Point Groups," *J. Chem. Educ.* **49,** 341-342.

Thomas, C. H.: 1974, "The Use of Group Theory to Determine Molecular Geometry from IR Spectra," *J. Chem. Educ.* **51,** 91-93.

CHAPTER 4 / *Symmetry-Adapted Stresses and Strains*

4.1
Condensed Phases

Solids and liquids consist of atoms, molecules, and/or ions vibrating about equilibrium positions. In a solid, these positions tend to remain fixed. Furthermore, the resulting material may be uniform in structure, or grainy, heterogeneous, composite. In a liquid, the equilibrium positions are mobile. Our concern here will be with uniform regions.

A crystal is characterized by equilibrium positions arranged on a regular lattice. Each unit of such a lattice is transformed into an equivalent unit by certain rotations, reflections, rotoreflections, and inversions. The complete set of these geometric operations forms a group. Also, certain subsets of the operations may form groups.

In the continuum approximation, independent attributes and effects are averaged over each small region of space and time so that microscopic uniformity is achieved. In this process, however, a person need not destroy the essential symmetry properties. Instead, we will suppose that each unit cell still exhibits the same transformation behavior about its center as it did before the averaging and that any appropriately shaped element of the continuum exhibits this behavior. The symmetry operations that occur form the groups listed in table 4.1.

A fiber uniform along its length is characterized by macroscopic isotropy along lines paralleling its axis. When the fiber is straight,

Table 4.1 Point Groups for the Most Symmetric Unit Cells of Possible Crystals

Lattice	Group
Triclinic	C_1, C_i
Monoclinic	C_s, C_2, C_{2h}
Orthorhombic	C_{2v}, D_2, D_{2h}
Tetragonal	$S_4, C_4, C_{4h}, C_{4v}, D_4, D_{2d}, D_{4h}$
Trigonal	$C_3, S_6, C_{3v}, D_3, D_{3d}$
Hexagonal	$C_6, C_{3h}, C_{6h}, C_{6v}, D_6, D_{3h}, D_{6h}$
Cubic	T, T_h, T_d, O, O_h

Table 4.2 Symmetry Groups for Partially and Completely Isotropic Materials

Isotropy	Group
Parallel lines*	C_s, C_2, C_{2h}
Parallel planes	$C_\infty, C_{\infty v}, D_{\infty h}$
All space	I, I_h, N_{3p}, N_3, O_3

*Other reference structures may be present. To allow for these, one has to employ the appropriate higher-order group.

these lines are also. An oriented polymer, on the other hand, may exhibit uniformity and macroscopic isotropy over parallel surfaces. The material may be straightened so these are parallel planes. Finally, a material may exhibit macroscopic isotropy in all directions. A glass and an ordinary liquid are examples Symmetry operations that occur form the groups listed in Table 4.2.

4.2
Stress Arrays

The properties that remain after the averaging and smoothing consist of scalars, vectors, dyads, and polyads associated with points, small elementary volumes, and surfaces of such volumes. To be typical, an elementary volume needs to be chosen, with planar faces, to fit the symmetry of the material.

The mechanical influence of the surroundings on such a volume is represented by force vectors acting on the faces and by the corresponding forces per unit area—the stresses. Each possible set of stress vectors acting on a unit of the material is like an array of displacement vectors for a vibrating system. Thus, we can treat it similarly.

In conventional continuum theory, a person considers the stresses acting on an arbitrary small rectangular parallelepiped or cube. Cartesian axes are set up perpendicular to the faces as shown in figure 4.1. The stresses acting on each face are resolved along these axes. The components can then be grouped as figures 4.2 through 4.7

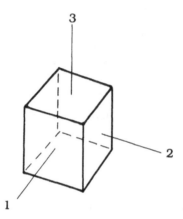

Figure 4.1 Cartesian axes for the rectangular element.

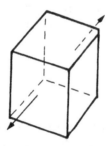

Figure 4.2 The σ_{11} array of stress vectors. The magnitude of each vector equals quantity σ_{11}.

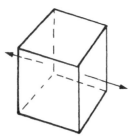

Figure 4.3 The σ_{22} array of stress vectors.

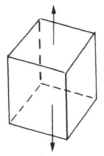

Figure 4.4 The σ_{33} array of stress vectors.

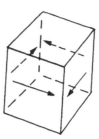

Figure 4.5 The σ_{12} array of stress vectors. The magnitude of each vector equals quantity σ_{12}.

indicate. Each of the constituent sets is then described by the magnitude σ_{jk} of a representative vector. The indices jk identify the face and direction of the representative vector when it is positive.

In the continuum approximation, a cubic element can be the unit for all materials except the trigonal and hexagonal ones. Further-

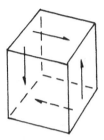

Figure 4.6 The σ_{23} array of stress vectors.

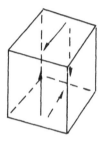

Figure 4.7 The σ_{31} array of stress vectors.

more, the axes perpendicular to the faces (of figure 4.1) can be chosen so that they agree with the choices employed for the symmetry operations. With a trigonal or hexagonal material, a corresponding element is constructed. The usual transformation laws are used to obtain the effects of rotations by ⅓ turn and by ⅔ turn about the third axis.

As in Chapter 3, the primitive symmetry species are given by the common eigenarrays for the class sums. From formula (3.12), we construct

$$\phi_{lm} = N \sum y_k^{(r)} \mathscr{C}_{\bar{k}} \sigma_{lm}. \qquad (4.1)$$

Here N is the normalization factor, $y_k^{(r)}$ the kth component of character vector $\chi^{(r)}$, $\mathscr{C}_{\bar{k}}$ the sum of the inverses of the operators in $\mathscr{C}_k$, and σ_{lm} the lmth stress array. From these results, we proceed as described in Chapter 3.

Results from the groups in tables 4.1 and 4.2 appear in tables 4.3 and 4.4. The species that is completely symmetric with respect to the

Stress Arrays

Table 4.3 Primitive Symmetry Species of Stress Arrays and of the Corresponding Strains for the Crystal Point Groups

	C_1, C_i	
A	$\sigma_{11}, \sigma_{22}, \sigma_{33}, \sigma_{12}, \sigma_{23}, \sigma_{31}$	$\varepsilon_{11}, \varepsilon_{22}, \varepsilon_{33}, \varepsilon_{12}, \varepsilon_{23}, \varepsilon_{31}$

	C_s, C_2, C_{2h}	
A	$\sigma_{11}, \sigma_{22}, \sigma_{33}, \sigma_{12}$	$\varepsilon_{11}, \varepsilon_{22}, \varepsilon_{33}, \varepsilon_{12}$
B	σ_{23}, σ_{31}	$\varepsilon_{23}, \varepsilon_{31}$

	C_{2v}, D_2, D_{2h}	
A	$\sigma_{11}, \sigma_{22}, \sigma_{33}$	$\varepsilon_{11}, \varepsilon_{22}, \varepsilon_{33}$
B_1	σ_{12}	ε_{12}
B_2	σ_{23}	$\begin{pmatrix}\varepsilon_{23}\\\varepsilon_{31}\end{pmatrix}$
B_3	σ_{31}	

	S_4, C_4, C_{4h}	
A	$\sigma_{11} + \sigma_{22}, \sigma_{33}$	$\varepsilon_{11} + \varepsilon_{22}, \varepsilon_{33}$
B	$\sigma_{11} - \sigma_{22}, \sigma_{12}$	$\varepsilon_{11} - \varepsilon_{22}, \varepsilon_{12}$
E	$\begin{pmatrix}\sigma_{23}\\\sigma_{31}\end{pmatrix}$	$\begin{pmatrix}\varepsilon_{23}\\\varepsilon_{31}\end{pmatrix}$

	C_{4v}, D_4, D_{2d}, D_{4h}	
A	$\sigma_{11} + \sigma_{22}, \sigma_{33}$	$\varepsilon_{11} + \varepsilon_{22}, \varepsilon_{33}$
B_1	$\sigma_{11} - \sigma_{22}$	$\varepsilon_{11} - \varepsilon_{22}$
B_2	σ_{12}	ε_{12}
E	$\begin{pmatrix}\sigma_{23}\\\sigma_{31}\end{pmatrix}$	$\begin{pmatrix}\varepsilon_{23}\\\varepsilon_{31}\end{pmatrix}$

	C_3, S_6, C_{3v}, D_3, D_{3d}	
A	$\sigma_{11} + \sigma_{22}, \sigma_{33}$	$\varepsilon_{11} + \varepsilon_{22}, \varepsilon_{33}$
E	$\begin{pmatrix}\sigma_{11} + \omega^2 C_3\sigma_{11} + \omega C_3^2\sigma_{11}\\\sigma_{11} + \omega C_3\sigma_{11} + \omega^2 C_3^2\sigma_{11}\end{pmatrix}$, $\begin{pmatrix}\sigma_{31} + \omega^2 C_3\sigma_{31} + \omega C_3^2\sigma_{31}\\\sigma_{31} + \omega C_3\sigma_{31} + \omega^2 C_3^2\sigma_{31}\end{pmatrix}$	$\begin{pmatrix}\varepsilon_{11} + \omega^2 C_3\varepsilon_{11} + \omega C_3^2\varepsilon_{11}\\\varepsilon_{11} + \omega C_3\varepsilon_{11} + \omega^2 C_3^2\varepsilon_{11}\end{pmatrix}$, $\begin{pmatrix}\varepsilon_{31} + \omega^2 C_3\varepsilon_{31} + \omega C_3^2\varepsilon_{31}\\\varepsilon_{31} + \omega C_3\varepsilon_{31} + \omega^2 C_3^2\varepsilon_{31}\end{pmatrix}$

	C_6, C_{3h}, C_{6h}, C_{6v}, D_6, D_{3h}, D_{6h}	
A	$\sigma_{11} + \sigma_{22}, \sigma_{33}$	$\varepsilon_{11} + \varepsilon_{22}, \varepsilon_{33}$
E_1	$\begin{pmatrix}\sigma_{11} + \omega^2 C_3\sigma_{11} + \omega C_3^2\sigma_{11}\\\sigma_{11} + \omega C_3\sigma_{11} + \omega^2 C_3^2\sigma_{11}\end{pmatrix}$	$\begin{pmatrix}\varepsilon_{11} + \omega^2 C_3\varepsilon_{11} + \omega C_3^{22}\varepsilon_{11}\\\varepsilon_{11} + \omega C_3\varepsilon_{11} + \omega^2 C_3^2\varepsilon_{11}\end{pmatrix}$
E_2	$\begin{pmatrix}\sigma_{31} + \omega^2 C_3\sigma_{31} + \omega C_3^2\sigma_{31}\\\sigma_{31} + \omega C_3\sigma_{31} + \omega^2 C_3^2\sigma_{31}\end{pmatrix}$	$\begin{pmatrix}\varepsilon_{31} + \omega^2 C_3\varepsilon_{31} + \omega C_3^2\varepsilon_{31}\\\varepsilon_{31} + \omega C_3\varepsilon_{31} + \omega^2 C_3^2\varepsilon_{31}\end{pmatrix}$

Table 4.3 *continued*

T, T$_h$, T$_d$, O, O$_h$

A	$\sigma_{11} + \sigma_{22} + \sigma_{33}$	$\varepsilon_{11} + \varepsilon_{22} + \varepsilon_{33}$
E	$\begin{pmatrix} \sigma_{11} + \omega^2\sigma_{22} + \omega\sigma_{33} \\ \sigma_{11} + \omega\sigma_{22} + \omega^2\sigma_{33} \end{pmatrix}$	$\begin{pmatrix} \varepsilon_{11} + \omega^2\varepsilon_{22} + \omega\varepsilon_{33} \\ \varepsilon_{11} + \omega\varepsilon_{22} + \omega^2\varepsilon_{33} \end{pmatrix}$
F	$\begin{pmatrix} \sigma_{12} \\ \sigma_{23} \\ \sigma_{31} \end{pmatrix}$	$\begin{pmatrix} \varepsilon_{12} \\ \varepsilon_{23} \\ \varepsilon_{31} \end{pmatrix}$

$\omega = \exp(2\pi i/3)$

Table 4.4 Species of Stresses and Strains for the Partially and Completely Isotropic Materials of Table 4.2

C$_s$, C$_2$, C$_{2h}$

A	$\sigma_{11}, \sigma_{22}, \sigma_{33}, \sigma_{12}$	$\varepsilon_{11}, \varepsilon_{22}, \varepsilon_{33}, \varepsilon_{12}$
B	σ_{23}, σ_{31}	$\varepsilon_{23}, \varepsilon_{31}$

C$_\infty$, C$_{\infty v}$, D$_{\infty h}$

A	$\sigma_{11} + \sigma_{22}, \sigma_{33}$	$\varepsilon_{11} + \varepsilon_{22}, \varepsilon_{33}$
E$_1$	$\begin{pmatrix} \sigma_{11} + \omega^2 C_3\sigma_{11} + \omega C_3^2\sigma_{11} \\ \sigma_{11} + \omega C_3\sigma_{11} + \omega^2 C_3^2\sigma_{11} \end{pmatrix}$	$\begin{pmatrix} \varepsilon_{11} + \omega^2 C_3\varepsilon_{11} + \omega C_3^2\varepsilon_{11} \\ \varepsilon_{11} + \omega C_3\varepsilon_{11} + \omega^2 C_3^2\varepsilon_{11} \end{pmatrix}$
E$_2$	$\begin{pmatrix} \sigma_{31} + \omega^2 C_3\sigma_{31} + \omega C_3^2\sigma_{31} \\ \sigma_{31} + \varepsilon C_3\sigma_{31} + \omega^2 C_3^2\sigma_{31} \end{pmatrix}$	$\begin{pmatrix} \varepsilon_{31} + \omega^2 C_3\varepsilon_{31} + \omega C_3^2\varepsilon_{31} \\ \varepsilon_{31} + \omega C_3\varepsilon_{31} + \omega^2 C_3^2\varepsilon_{31} \end{pmatrix}$

I, I$_h$, N$_{3p}$, N$_3$, O$_3$

A	$\sigma_{11} + \sigma_{22} + \sigma_{33}$	$\varepsilon_{11} + \varepsilon_{22} + \varepsilon_{33}$
H	$\begin{pmatrix} \sigma_{11} + \omega^2\sigma_{22} + \omega\sigma_{33} \\ \sigma_{11} + \omega\sigma_{22} + \omega^2\sigma_{33} \\ \sigma_{12} \\ \sigma_{23} \\ \sigma_{31} \end{pmatrix}$	$\begin{pmatrix} \varepsilon_{11} + \omega^2\varepsilon_{22} + \omega\varepsilon_{33} \\ \varepsilon_{11} + \omega\varepsilon_{22} + \omega^2\varepsilon_{33} \\ \varepsilon_{12} \\ \varepsilon_{23} \\ \varepsilon_{31} \end{pmatrix}$

$\omega = \exp(2\pi i/3)$

operations in the given group is labeled A. Other nondegenerate species are labeled B, with a distinguishing subscript where necessary. Species with a degeneracy of 2 are labeled E, with a distinguishing subscript where necessary. The species with a degeneracy of 3 is labeled F; that with a degeneracy of 5 is labeled H. Operator C_3 effects rotation by ⅓ turn about the third axis.

4.3
Elements of Strain as Generalized Coordinates

Because materials yield, applying forces to any small unit of a given system causes mechanical work to be done on the unit. If the process is reversible, in the thermodynamic sense, the density W of this work is a function of position in the system. Indeed, if the process occurs at constant temperature, W equals the Helmholtz free energy density increase associated with the strain about the given point. If the process occurs adiabatically, W equals the internal energy density increase for the imposed strain about the point.

The elements of strain $\varepsilon_{11}, \varepsilon_{12}, \ldots, \varepsilon_{33}$ may be defined as the generalized coordinates for which

$$\frac{\partial W}{\partial \varepsilon_{jk}} = \sigma_{jk} \tag{4.2}$$

with the ε_{jk}'s equal to zero when the σ_{jk}'s are all zero. From equation (4.2) we obtain

$$dW = \sum_{j,k} \sigma_{jk} \, d\varepsilon_{jk}. \tag{4.3}$$

The density of strain energy is presumably an analytic function of the generalized coordinates, under the imposed conditions. But W is taken to be zero when the ε_{jk}'s are zero. Also, the strains are measured from a state of zero stress. So the power series for W begins with the quadratic term; we have

$$W = \sum_{i,j,k,l} \tfrac{1}{2} C_{ijkl} \varepsilon_{ij} \varepsilon_{kl} + \sum_{i,j,k,l,m,n} \tfrac{1}{3} C_{ijklmn} \varepsilon_{ij} \varepsilon_{kl} \varepsilon_{mn} + \ldots \tag{4.4}$$

Since

$$\sigma_{jk} = \sigma_{kj}, \tag{4.5}$$

equation (4.2) implies that

$$\varepsilon_{jk} = \varepsilon_{kj}. \tag{4.6}$$

But symmetry (4.6) requires that C_{ijkl} be symmetric in interchanges of i and j and in interchanges of k and l. Because the ε's in each quadratic term of equation (4.4) commute, C_{ijkl} is also symmetric with respect to interchanges of ij with kl. A *reciprocity* is said to prevail. Similar symmetries are present in the higher coefficients.

Under the group operations, the elements of strain behave as the analogous elements of stress. Thus, the primitive symmetry species are related as Tables 4.3 and 4.4 indicate.

4.4
Dependence of Stress on Strain

Applying forces to an object formed of condensed material leads to a distribution of stresses throughout the system. As the stresses act on a given small element, strains and/or rates of strain appear. The work associated with imposition of the stresses has a density W dependent on the material and on the temperature and pressure path that is followed. Under given conditions, this density W can be expressed as a power series in the strain elements. At small strains all except the quadratic terms can be neglected. One can then derive a linear stress-strain relationship.

Differentiating equation (4.4) and introducing equation (4.3) for dW leads to

$$\begin{aligned} dW &= \sum_{i,j,k,l} C_{ijkl}\varepsilon_{ij}d\varepsilon_{kl} + \sum_{i,j,k,l,m,n} C_{ijklmn}\varepsilon_{ij}\varepsilon_{kl}d\varepsilon_{mn} + \ldots \\ &= \sum_{k,l} \sigma_{kl}d\varepsilon_{kl}, \end{aligned} \tag{4.7}$$

whence

$$\sigma_{kl} = \sum_{i,j} C_{ijkl}\varepsilon_{ij} + \sum_{i,j,m,n} C_{mnijkl}\varepsilon_{mn}\varepsilon_{ij} + \ldots$$

$$= \sum_{i,j} C_{klij}\varepsilon_{ij} + \sum_{i,j,m,n} C_{klmnij}\varepsilon_{mn}\varepsilon_{ij} + \ldots \quad (4.8)$$

An arbitrary linear combination of the elements of stress then obeys

$$\sum_{k,l} B_{kl}\sigma_{kl} = \sum_{i,j,k,l} B_{kl} C_{klij}\varepsilon_{ij} + \sum_{i,j,k,l,m,n} B_{kl} C_{klmnij}\varepsilon_{mn}\varepsilon_{ij} + \ldots \quad (4.9)$$

The reorientations that transform a given small unit of a system into an equivalent unit are symmetry operations. Under each of these, both sides of an equation describing the unit must behave in the same manner (or else it could not be physically significant). When the coefficients B_{kl} are chosen to make the left side of equation (4.9) belong to a row of a primitive symmetry species of a pertinent group, then the right side must also belong to this row.

All linear combinations of ε_{ij}'s belonging to the row of the primitive symmetry species, and only these, may than appear in the first sum—each with a separate coefficient. These coefficients depend on the material and on the conditions under which the straining occurs.

The theory of the symmetry properties of products enables one to determine the combinations that may appear in the second and higher sums on the right of equation (4.9). For the sake of simplicity, we will not consider these here. Instead, for the various symmetry species, we will only construct the linear stress-strain relationships

$$\sum_{k,l} B_{kl}\sigma_{kl} = \sum_{i,j,k,l} B_{kl} C_{klij}\varepsilon_{ij}. \quad (4.10)$$

for the various symmetry species.

4.5
Constitutive Relations for the Different Crystal Types

In the continuum approximation, the appropriate volume elements of a given crystal are considered to exhibit the symmetry of a unit cell, as listed in Table 4.1. With respect to such elements, the stress and the strain arrays are classified as in Table 4.3. Now, the left side of equation (4.10) may be constructed so that it belongs to a row of a primitive symmetry species of the pertinent symmetry group; then, the right side must also belong to this row.

The unit elements of *triclinic* material are transformed into equivalent elements only by operation I and, in various cases, by operation i. The corresponding groups are labelled $\mathbf{C}_1$ and $\mathbf{C}_i$. According to table 4.3, the standard components of stress and strain are then of the same type. So here, all ε_{ij}'s can contribute to each σ_{lm}. The only symmetry present in the equations is that due to reciprocity; the contribution per unit magnitude of ε_{11} to σ_{22} equals that of ε_{22} to σ_{11}, and so on. Thus, we obtain the constitutive equations

$$\sigma_{11} = E_{11}\varepsilon_{11} + E_{12}\varepsilon_{22} + E_{13}\varepsilon_{33} + E_{14}\varepsilon_{12} + E_{15}\varepsilon_{23} + E_{16}\varepsilon_{31}, \tag{4.11}$$

$$\sigma_{22} = E_{12}\varepsilon_{11} + E_{22}\varepsilon_{22} + E_{23}\varepsilon_{33} + E_{24}\varepsilon_{12} + E_{25}\varepsilon_{23} + E_{26}\varepsilon_{31}, \tag{4.12}$$

$$\sigma_{33} = E_{13}\varepsilon_{11} + E_{23}\varepsilon_{22} + E_{33}\varepsilon_{33} + E_{34}\varepsilon_{12} + E_{35}\varepsilon_{23} + E_{36}\varepsilon_{31}, \tag{4.13}$$

$$\sigma_{12} = E_{14}\varepsilon_{11} + E_{24}\varepsilon_{22} + E_{34}\varepsilon_{33} + E_{44}\varepsilon_{12} + E_{45}\varepsilon_{23} + E_{46}\varepsilon_{31}, \tag{4.14}$$

$$\sigma_{23} = E_{15}\varepsilon_{11} + E_{25}\varepsilon_{22} + E_{35}\varepsilon_{33} + E_{45}\varepsilon_{12} + E_{55}\varepsilon_{23} + E_{56}\varepsilon_{31}, \tag{4.15}$$

$$\sigma_{31} = E_{16}\varepsilon_{11} + E_{26}\varepsilon_{22} + E_{36}\varepsilon_{33} + E_{46}\varepsilon_{12} + E_{56}\varepsilon_{23} + E_{66}\varepsilon_{31}, \tag{4.16}$$

When properly oriented, the continuum-approximation volume elements of *monoclinic* material belong to $\mathbf{C}_s$, $\mathbf{C}_2$, or $\mathbf{C}_{2h}$ groups. According to table 4.3, the character tables then place σ_{11}, σ_{22}, σ_{33}, σ_{12}, and the corresponding ε_{jk}'s into primitive symmetry species A, while σ_{23}, σ_{31}, and ε_{23}, ε_{31} go into primitive symmetry species B. Since small stresses can only produce strains of the same type, we have

$$\sigma_{11} = E_{11}\varepsilon_{11} + E_{12}\varepsilon_{22} + E_{13}\varepsilon_{33} + E_{14}\varepsilon_{12}, \tag{4.17}$$

$$\sigma_{22} = E_{12}\varepsilon_{11} + E_{22}\varepsilon_{22} + E_{23}\varepsilon_{33} + E_{24}\varepsilon_{12}, \tag{4.18}$$

Constitutive Relations for the Different Crystals Types 117

$$\sigma_{33} = E_{13}\varepsilon_{11} + E_{23}\varepsilon_{22} + E_{33}\varepsilon_{33} + E_{34}\varepsilon_{12}, \quad (4.19)$$

$$\sigma_{12} = E_{14}\varepsilon_{11} + E_{24}\varepsilon_{22} + E_{34}\varepsilon_{33} + E_{44}\varepsilon_{12}, \quad (4.20)$$

$$\sigma_{23} = E_{55}\varepsilon_{23} + E_{56}\varepsilon_{31}, \quad (4.21)$$

$$\sigma_{31} = E_{56}\varepsilon_{23} + E_{66}\varepsilon_{31}. \quad (4.22)$$

The unit elements of *orthorhombic* material form a basis for operations of the C_{2v}, D_2, or D_{2h} groups. The character tables then yield the classification under the heading C_{2v}, D_2, D_{2h} in table 4.3. With the axes properly oriented, stresses σ_{11}, σ_{22}, σ_{33} and strains ε_{11}, ε_{22}, ε_{33} are of type A; σ_{12} and ε_{12} belong to B_1; σ_{23} and ε_{23} belong to B_2; σ_{31} and ε_{31} belong to B_3. Since small stresses can only produce strains of the same symmetry type, we have

$$\sigma_{11} = E_{11}\varepsilon_{11} + E_{12}\varepsilon_{22} + E_{13}\varepsilon_{33}, \quad (4.23)$$

$$\sigma_{22} = E_{12}\varepsilon_{11} + E_{22}\varepsilon_{22} + E_{23}\varepsilon_{33}, \quad (4.24)$$

$$\sigma_{33} = E_{13}\varepsilon_{11} + E_{23}\varepsilon_{22} + E_{33}\varepsilon_{33}, \quad (4.25)$$

$$\sigma_{12} = E_{44}\varepsilon_{12}, \quad (4.26)$$

$$\sigma_{23} = E_{55}\varepsilon_{23}, \quad (4.27)$$

$$\sigma_{31} = E_{66}\varepsilon_{31}. \quad (4.28)$$

To *tetragonal* material whose most symmetric unit cell is transformed into itself by operations of the S_4, C_4, or C_{4h} groups, the fourth classification in table 4.3 applies. Here $\sigma_{11} + \sigma_{22}$, σ_{33} and $\varepsilon_{11} + \varepsilon_{22}$, ε_{33} exhibit symmetry A; $\sigma_{11} - \sigma_{22}$, σ_{12} and $\varepsilon_{11} - \varepsilon_{22}$, ε_{12} exhibit symmetry B; $(\sigma_{23}, \sigma_{31})$ and $(\varepsilon_{23}, \varepsilon_{31})$ belong to the degenerate symmetry type E. Thus, we find that

$$\sigma_{11} + \sigma_{22} = a(\varepsilon_{11} + \varepsilon_{22}) + 2b\varepsilon_{33}, \quad (4.29)$$

$$\sigma_{33} = b(\varepsilon_{11} + \varepsilon_{22}) + c\varepsilon_{33}, \quad (4.30)$$

$$\sigma_{11} - \sigma_{22} = d(\varepsilon_{11} - \varepsilon_{22}) + 2e\varepsilon_{12}, \quad (4.31)$$

$$\sigma_{12} = e(\varepsilon_{11} - \varepsilon_{22}) + f\varepsilon_{12}, \quad (4.32)$$

$$\sigma_{23} = g\varepsilon_{23}, \qquad (4.33)$$

$$\sigma_{31} = g\varepsilon_{31}. \qquad (4.34)$$

The last two equations involve the same coefficient since σ_{23} and σ_{31} are degenerate. Indeed, an operation of the group changes σ_{23} into σ_{31} and ε_{23} into ε_{31}; so it must change equation (4.33) into equation (4.34).

For other *tetragonal* materials, the next classification in table 4.3 applies. Going to it from the preceding set, symmetry species B splits into B_1 and B_2. Since small stresses can only produce strains of the same type, $\varepsilon_{11} - \varepsilon_{22}$ can no longer affect σ_{12}, and ε_{12} can no longer affect $\sigma_{11} - \sigma_{22}$. The bases, however, are the same as before and so equations (4.29) through (4.34) apply with

$$e = 0. \qquad (4.35)$$

The continuum-approximation small units of a *trigonal* material form a basis for operations of the $\mathbf{C}_3$, $\mathbf{S}_6$, $\mathbf{C}_{3v}$, $\mathbf{D}_3$, or $\mathbf{D}_{3d}$ groups. According to the sixth classification in Table 4.3, $\sigma_{11} + \sigma_{22}$, σ_{33}, $\varepsilon_{11} + \varepsilon_{22}$, ε_{33} still exhibit symmetry A. Thus, formulas (4.29) and (4.30) hold. Now, the combinations for species E are given in a symmetrical, complex form. Relating each stress combination in either row linearly to the strain combinations in the same row, rotating the dyads as indicated (C_3 indicates rotation by ⅓ turn), separating the real from the imaginary parts, and reducing, leads to relationships of equations (4.38) through (4.41):

$$\sigma_{11} + \sigma_{22} = a(\varepsilon_{11} + \varepsilon_{22}) + 2b\varepsilon_{33}, \qquad (4.36)$$

$$\sigma_{33} = b(\varepsilon_{11} + \varepsilon_{22}) + c\varepsilon_{33}, \qquad (4.37)$$

$$\sigma_{11} - \sigma_{22} = d(\varepsilon_{11} - \varepsilon_{22}) + e(\varepsilon_{31} - \varepsilon_{23}), \qquad (4.38)$$

$$\sigma_{31} - \sigma_{23} = e(\varepsilon_{11} - \varepsilon_{22}) + f(\varepsilon_{31} - \varepsilon_{23}), \qquad (4.39)$$

$$\sigma_{12} = d\varepsilon_{12} + \frac{e}{\sqrt{2}}(\varepsilon_{31} + \varepsilon_{23}), \qquad (4.40)$$

$$\sigma_{31} + \sigma_{23} = \sqrt{2}e\varepsilon_{12} + f(\varepsilon_{31} + \varepsilon_{23}). \qquad (4.41)$$

In *hexagonal* material, the most symmetric unit cell is transformed into itself only by operations of the C_6, C_{3h}, C_{6h}, C_{6v}, D_6, D_{3h}, or D_{6h} groups. These have the same bases as those constructed for the trigonal materials in table 4.3; however, symmetry species E has split into E_1 and E_2, which do not interact. Equations (4.36) through (4.41) are applicable with

$$e = 0. \tag{4.42}$$

In *cubic* material, the most symmetric elementary cells belong to T, T_h, T_d, O, or O_h groups. In the last section of table 4.3, the stresses and strains for the primitive symmetry species appear in symmetric form, with

$$\omega = \exp(2\pi i/3). \tag{4.43}$$

Since small stresses can only produce strains of the same type, we find that

$$\sigma_{11} + \sigma_{22} + \sigma_{33} = a(\varepsilon_{11} + \varepsilon_{22} + \varepsilon_{33}), \tag{4.44}$$

$$\sigma_{11} + \omega^2\sigma_{22} + \omega\sigma_{33} = b(\varepsilon_{11} + \omega^2\varepsilon_{22} + \omega\varepsilon_{33}), \tag{4.45}$$

$$\sigma_{11} + \omega\sigma_{22} + \omega^2\sigma_{33} = b(\varepsilon_{11} + \omega\varepsilon_{22} + \omega^2\varepsilon_{33}), \tag{4.46}$$

$$\sigma_{12} = c\varepsilon_{12}, \tag{4.47}$$

$$\sigma_{23} = c\varepsilon_{23}, \tag{4.48}$$

$$\sigma_{31} = c\varepsilon_{31}. \tag{4.49}$$

Equations (4.45) and (4.46) combine to yield

$$2\sigma_{11} - \sigma_{22} - \sigma_{33} = b(2\varepsilon_{11} - \varepsilon_{22} - \varepsilon_{33}), \tag{4.50}$$

$$\sigma_{22} - \sigma_{33} = b(\varepsilon_{22} - \varepsilon_{33}). \tag{4.51}$$

In practice, one would probably employ equation (4.44) and equations (4.47) through (4.51).

On increasing the local symmetry from T to I (from tetrahdral to icosahedral) or higher, primitive symmetry species E and F coalesce, forming a 5-dimensional species labeled H. The same bases can still be used. Thus, equations (4.44) and (4.47) through (4.51) apply with

$$c = b. \qquad (4.52)$$

The number of independent parameters has been reduced to 2.

Example 4.1

How may dyads **ii**, **jj**, **ij**, and **ji** be correlated with arrays σ_{11}, σ_{22}, and σ_{12}?

We recall that a planar area is represented by a vector normal to the surface and equal in magnitude to the area. A face of the rectangular element in Figures 4.1 through 4.7 is thus represented by a vector of appropriate size, pointing out.

In figure 4.2, a unit vector perpendicular to a front face of area unity thus represents **ii**; a unit vector perpendicular to a back face of area unity represents $(-\mathbf{i})(-\mathbf{i}) = \mathbf{ii}$. We have two representations of the same dyad. It is similar for **jj**..

But in Figure 4.5, a unit vector lying in the front face of area unity, as shown, represents **ij**; a unit vector lying in the back face of area unity, as shown, represents $(-\mathbf{i})(-\mathbf{j}) = \mathbf{ij}$. The unit vector lying in the right face of area unity, as shown, represents **ji**; the unit vector lying in the left face of area unity, as shown, represents $(-\mathbf{j})(-\mathbf{i}) = \mathbf{ji}$. We thus have two representations of **ij** and two of **ji** in this array.

Array σ_{11} in figure 4.2 behaves as **ii** insofar as operations of the pertinent group are concerned. Similarly, σ_{22} behaves as **jj** and σ_{12} as **ij** and **ji** together; however, this combination has to be normalized. We consider σ_{12} to be represented by $(1/\sqrt{2})(\mathbf{ij} + \mathbf{ji})$.

Example 4.2

Rotate dyad **ii** by angle ϕ about the z axis and determine the resulting components.

When unit vector **i** is rotated by angle ϕ about the z axis, we obtain

$$\mathbf{i}' = \cos\phi\,\mathbf{i} + \sin\phi\,\mathbf{j}.$$

So when dyad $\mathbf{ii}$ is rotated by angle ϕ, we have

$$\mathbf{i'i'} = (\cos\phi\,\mathbf{i} + \sin\phi\,\mathbf{j})(\cos\phi\,\mathbf{i} + \sin\phi\,\mathbf{j})$$
$$= \cos^2\phi\,\mathbf{ii} + \sin^2\phi\,\mathbf{jj} + \sin\phi\cos\phi\,\mathbf{ij} + \sin\phi\cos\phi\,\mathbf{ji}.$$

Example 4.3

Express $C_3\sigma_{11}$ and $C_3^2\sigma_{11}$ in terms of σ_{11}, σ_{22}, and σ_{12}.
For rotation C_3, angle ϕ is $2\pi/3$ and

$$\cos\tfrac{1}{3}(2\pi) = -\tfrac{1}{2}, \qquad \sin\tfrac{1}{3}(2\pi) = \frac{\sqrt{3}}{2}$$

The formula in Example 4.2 then yields

$$C_3\mathbf{ii} = \tfrac{1}{4}\mathbf{ii} + \tfrac{3}{4}\mathbf{jj} - \frac{\sqrt{3}}{4}\mathbf{ij} - \frac{\sqrt{3}}{4}\mathbf{ji}.$$

For rotation C_3^2, we similarly have

$$\cos\tfrac{2}{3}(2\pi) = -\tfrac{1}{2}, \qquad \sin\tfrac{2}{3}(2\pi) = -\frac{\sqrt{3}}{2}$$

whence

$$C_3^2\mathbf{ii} = \tfrac{1}{4}\mathbf{ii} + \tfrac{3}{4}\mathbf{jj} + \frac{\sqrt{3}}{4}\mathbf{ij} + \frac{\sqrt{3}}{4}\mathbf{ji}.$$

From Example 4.1, arrays σ_{11}, σ_{22}, and σ_{12} behave as $\mathbf{ii}$, $\mathbf{jj}$, and $(1/\sqrt{2})(\mathbf{ij} + \mathbf{ji})$, respectively. So these results lead to

$$C_3\sigma_{11} = \tfrac{1}{4}\sigma_{11} + \tfrac{3}{4}\sigma_{22} - \frac{\sqrt{3}}{2\sqrt{2}}\sigma_{12}$$

and

$$C_3^2\sigma_{11} = \tfrac{1}{4}\sigma_{11} + \tfrac{3}{4}\sigma_{22} + \frac{\sqrt{3}}{2\sqrt{2}}\sigma_{12}.$$

Example 4.4

Determine the sum and the difference of $C_3\sigma_{11}$ and $C_3^2\sigma_{11}$.
From the results in example 4.3, we find that

$$C_3\sigma_{11} + C_3^2\sigma_1 = \tfrac{1}{2}\sigma_{11} + \tfrac{3}{2}\sigma_{22}$$

and

$$C_3\sigma_{11} - C_3^2\sigma_{11} = -\frac{\sqrt{3}}{\sqrt{2}}\sigma_{12}$$

This sum and difference, and the analogous ones for $C_3\varepsilon_{11}$ and $C_3^2\varepsilon_{11}$, arise in constructing equations (4.38) and (4.40).

4.6
Constitutive Relations for Other Materials

A homogeneous amorphous solid, glass, or unoriented polymer is macroscopically isotropic. The effective local symmetry would be **I** or higher; the primitive species, A and H. Such a material is governed by equations (4.44) through (4.51), with equation (4.52) imposed. In the linear stress-strain relationships there are two independent parameters.

Now, materials can be oriented by the forces to which they are subjected while they are being formed. The final material may preserve a 2-dimensional isotropy in parallel planes. The effective local symmetry would be $\mathbf{D}_{\infty h}$, $\mathbf{C}_{\infty v}$, or $\mathbf{C}_\infty$; the primitive species, A, E_1, and E_2. Then equations (4.36) through (4.41) hold with equation (4.42) imposed. There are now five independent stress-strain parameters.

If only linear isotropy, in parallel lines, is preserved, the effective local symmetry would be reduced to $\mathbf{C}_{2h}$, $\mathbf{C}_2$, or $\mathbf{C}_s$. Equations (4.17) through (4.22) apply, with thirteen independent parameters.

The foregoing arguments can be applied to fluid media if ε_{jk} is allowed to equal the jkth rate of strain, while the negative pressure, $-p$, is separately added to σ_{11}, to σ_{22}, and to σ_{33}. For isotropic liquids, equations (4.44) through (4.51), with limitation (4.52), would be thus modified. Furthermore, we would have

$$a = 3\kappa \tag{4.53}$$

and

$$b = c = 2\mu \tag{4.54}$$

with κ the bulk, and μ the shear, coefficient of viscosity.

In some liquids, below a transition temperature, interaction among the molecules leads to the formation of anisotropic regions. If the effective symmetry were merely reduced to that of a cube, limitation (4.52) would no longer apply. Then two independent shear coefficients would exist

$$b = 2\mu_1, \quad c = 2\mu_2. \tag{4.55}$$

If the fluid molecules tended to stack as logs in a neat pile, the effective local symmetry would be hexagonal. Then equations (4.36) through (4.41), with condition (4.42) imposed, modified as indicated above, would apply. There are now five independent coefficients. The μ_2 in equation (4.55) has split into a longitudinal and a transverse coefficient. The κ in equation (4.53) has similarly split. The remaining coefficient is a mixed one.

4.7
Basic Ideas

For relating macroscopic properties, a person can impose the continuum approximation on any given solid or fluid material. This approximation does not however, eliminate the nontranslational geometric symmetries that are present. The remaining symmetry operations make up one of the standard point groups.

As a consequence, the canonical stress arrays σ_{lm} can be combined in the form

$$\phi_{lm} = N \sum_k y_k^{(r)} \mathscr{C}_{\bar{k}} \sigma_{lm} \tag{4.56}$$

where $y_k^{(r)}$ is the kth component of the rth primitive character vector and $\mathscr{C}_{\bar{k}}$ is the sum of the inverses of the operators in the kth class sum.

The result is an array belonging to the rth primitive symmetry species of the pertinent group.

The strain arrays have similar forms. Furthermore, in the linear approximation stresses of a given species produce strains of the same species and no others.

Beside the symmetries imposed by the geometric group, we have those from the relations

$$\sigma_{jk} = \sigma_{kj}, \tag{4.57}$$

$$\sigma_{ij}\sigma_{kl} = \sigma_{kl}\sigma_{ij}, \tag{4.58}$$

and from the corresponding relations among the strain elements. The coefficients in the resulting constitutive relations depend on the particular material and on the thermodynamic conditions imposed. Thus, they depend on the temperature and pressure initially. They also depend on how the stresses are applied, on whether the process is isothermal or adiabatic, reversible or irreversible. In a fluid, the jkth rate of strain plays a role similar to the jkth strain itself in a solid.

Discussion Questions

4.1 In what ways are liquids and crystals similar? In what ways do they differ?

4.2 For each kind of lattice in Table 4.1, cite an example.

4.3 How may a system possess more symmetry than that of a crystal? Cite examples.

4.4 In conventional continuum theory, how are the stresses defined?

4.5 Why do we describe the stress elements as arrays?

4.6 How are the symmetry-adapted stress arrays found for a given system?

4.7 How are the elements of strain defined geometrically?

4.8 How do the stresses perform work on a small element of a system?

4.9 When does this work increase (a) the Helmholz free energy, (b) the internal energy, of the element?

4.10 How can the elements of strain be taken as generalized coordinates for each small element of the system?

4.11 Why may we start the power series expansion of W with the quadratic term?

4.12 Explain the symmetries that the C coefficients exhibit.

4.13 Why is the symmetry of $d\varepsilon_{jk}$ the same as that of σ_{jk}?

4.14 How does one obtain the dependence of stress on strain from the expression for the strain energy density W?

4.15 How do symmetry conditions limit the strain combinations on which a given stress combination depends?

4.16 What form do the linear constitutive equations assume when the only symmetries are those necessarily present because

$$\sigma_{jk} = \sigma_{kj}, \quad \varepsilon_{jk} = \varepsilon_{kj},$$

$$\sigma_{ij}\sigma_{kl} = \sigma_{kl}\sigma_{ij}, \quad \varepsilon_{ij}\varepsilon_{kl} = \varepsilon_{kl}\varepsilon_{ij}?$$

4.17 (a) How many independent coefficients may there be in the linear constitutive equations? How are these reduced in going to (b) a monoclinic structure, (c) an orthorhombic structure?

4.18 Show that for a tetragonal material the interactions within the A species and within the B species are correctly represented by equations (4.29), (4.30), (4.31), and (4.32).

4.19 Show that equations (4.36) through (4.41) are consistent with the reciprocity principle.

4.20 Show how equations (4.50) and (4.51) are obtained from equations (4.45) and (4.46).

4.21 (a) How many independent linear stress-strain coefficients are needed for a cubic crystal? (b) How may these be classified? (c) How are these reduced on going to an amorphous structure?

4.22 How can the constitutive equations for solids be altered so they apply to the flow of a material?

Problems

4.1 The unit cell of an orthorhombic crystal exhibits $\mathbf{D}_2$ symmetry. Determine the effect of each group operation on each element of stress.

4.2 Employ the results from Problem 4.1 in constructing and identifying the primitive species for the stresses in the orthorhombic crystal.

4.3 Consider a $\mathbf{C}_{4v}$ tetragonal crystal in the continuum approximation and depict the elements of stress. Then determine the effect of each group operation on each of these.

4.4 Apply formula (4.1) to the results in Problem 4.3 to generate the corresponding symmetry adapted stress arrays.

4.5 Explain how the results in Problem 4.4 are altered if the symmetry of the unit cell in the crystal is reduced to that of the $\mathbf{C}_4$ group.

4.6 Rotate dyad $(1/\sqrt{2})(\mathbf{ij} + \mathbf{ji})$ by angle ϕ about the z axis and determine the resulting components.

4.7 Express $C_3\sigma_{12}$ and $C_3^2\sigma_{12}$ in terms of σ_{11} σ_{22}, and σ_{12}.

4.8 Rotate dyads $\mathbf{jk}$ and $\mathbf{ki}$ by angle ϕ about the z axis and determine the resulting components.

4.9 Express $C_3\sigma_{23}$, $C_3^2\sigma_{23}$, $C_3\sigma_{31}$, $C_3^2\sigma_{31}$ in terms of σ_{23} and σ_{31}.

4.10 The unit cell of a trigonal crystal exhibits $\mathbf{C}_3$ symmetry. Determine the effect of each group operation on the standard elements of stress. You may use the results obtained in Problems 4.7 and 4.9.

4.11 In the $\mathbf{C}_3$ group, the characters for the D symmetry species are the complex conjugates of those for the C symmetry species. How may formulas (4.1) for these two species be combined to yield real expressions?

4.12 Employ the results from Problems 4.10 and 4.11 to construct real expressions representing the symmetry species for the stresses in the trigonal crystal.

4.13 Consider a $\mathbf{T}$ cubic crystal in the continuum approximation and determine the effect of each group operation on the elements of stress.

4.14 Apply formula (4.1) to the results in Problem 4.13 to generate the corresponding symmetry adapted stress arrays.

References

Books

Bisplinghoff, R. L., Mar, J. W., and Pian, T. H. H.: 1965, *Statics of Deformable Solids*, Addison-Wesley, Reading, Mass., pp. 168–205.

In Chapter 7, the authors present the simple basic ideas governing elastic stress-strain relationships.

Spencer, A. J. M.: 1971, "Theory of Invariants," in Eringen, A. C. (editor), *Continuum Physics*, Academic Press, New York, pp. 239-353.
Spencer discusses the invariants of vectors and tensors for three dimensions in a rigorous mathematical manner.

Articles

Callen, H.: 1968, "Crystal Symmetry and Macroscopic Laws," *Am. J. Phys.* **36**, 735-748.

Callen, H., Callen, E., and Kalva, Z.: 1970, "Crystal Symmetry and Macroscopic Laws II," *Am. J. Phys.* **38**, 1278-1284.

CHAPTER 5

Matrix Representations

5.1
Explicit Forms for the Elements of a Group

Symmetry in a physical system is manifested by the existence of reorientations, permutations, and/or conversions that transform the system into an equivalent system. The complete set of symmetry operations and their closed subsets constitute groups.

Now, a group is characterized by how its elements combine. Explicit mathematical operators that combine as the elements do can be constructed for various physically significant groups. Such operators *represent* the elements.

Certain attributes or properties are associated with equivalent positions or parts of a symmetric system. Others are associated with an invariant structure. The corresponding expressions may be chosen so that they are mixed linearly by each operation of the covering group.

The coefficients generated by any operation acting singly on the expressions can be arranged in a matrix. With n independent expressions, one obtains g matrices $n \times n$ in size, where g is the number of elements in the group. Since such matrices multiply as the group elements, they represent the group.

A set of standard expressions that behave thus are said to make up a *basis* for the representation. The expressions may be numbers, coordinates of equivalent positions, employed as in Section 1.3. They may be vectors, such as the mutually perpendicular **i, j, k** erected on a center of symmetry. They may be arrays of vectors, or dyads, such as those in Chapters 3 and 4. They may be particular generalized coordinates. The radius vectors drawn to equivalent positions may serve, as in Section 1.5. Products or functions of coordinates or vectors may also be employed. The standard expressions may furthermore be mutually independent kets, associated with equivalent aspects, orientations, or parts of the given system, as in Chapter 6.

5.2
Generating Representations from Bases

A representation of a group is a set of mathematical expressions that combine as the group elements do. Most common and useful are the matrix representations that various bases generate. We will here employ vectors as bases. But if other forms were used, they would replace the base vectors in the formulas.

Let us symbolize certain bases associated with the given system by the linearly independent vectors

$$\mathbf{u}_1, \mathbf{u}_2, \ldots, \mathbf{u}_n. \tag{5.1}$$

We will suppose that these make up a complete set so that any operation of the covering group acting on any one of them merely yields a linear combination of them.

If C and D are operations in the group, we then have

$$C\mathbf{u}_k = \sum_j \mathbf{u}_j \Gamma_{jk}(C) \tag{5.2}$$

and

$$D\mathbf{u}_j = \sum_i \mathbf{u}_i \Gamma_{ij}(D). \tag{5.3}$$

Here $\Gamma_{jk}(C)$ is the coefficient of $\mathbf{u}_j$ in the sum produced when C acts on $\mathbf{u}_k$; $\Gamma_{ij}(D)$ is the number multiplying $\mathbf{u}_i$ in the sum produced when D acts on $\mathbf{u}_j$. Similarly,

$$DC\mathbf{u}_k = \sum_i \mathbf{u}_i \Gamma_{ik}(DC). \tag{5.4}$$

But combining equations (5.2) and (5.3) yields

$$DC\mathbf{u}_k = \sum_{i,j} \mathbf{u}_i \Gamma_{ij}(D) \Gamma_{jk}(C). \tag{5.5}$$

Since the left sides of equations (5.4) and (5.5) are the same, the right sides are equal identically and the coefficients of $\mathbf{u}_i$ must be equal:

$$\Gamma_{ik}(DC) = \sum_j \Gamma_{ij}(D)\Gamma_{jk}(C). \tag{5.6}$$

Thus, we have the matrix relationship

$$\mathbf{\Gamma}(DC) = \mathbf{\Gamma}(D)\mathbf{\Gamma}(C). \tag{5.7}$$

The $n \times n$ matrices constructed from the coefficients combine as the corresponding elements of the group do. Indeed, these matrices represent the group.

The *dimensionality* of a representation is given by the number of linearly independent expressions needed for a basis. The dimensionality of set (5.1) is n.

To describe a mode of degeneracy d' requires the use of d' independent generalized coordinates or vectors. Likewise, describing a symmetry species of dimensionality d requires the use of d linearly independent bases. A set (5.1) that one may construct, however, generally involves a superposition of a_1 different sets of the first symmetry species, a_2 different sets of the second symmetry species, and so on.

The set describing a single species cannot be reduced further without violating the symmetry properties of the physical system; the corresponding representation is said to be *irreducible*. When the set (5.1) involves more than a single species once—that is, when sum $a_1 + a_2 + \ldots$ is greater than 1—the corresponding representation is said to be *reducible*.

Example 5.1

What base vectors and groups may a person employ in studying properties of an equilateral-triangle molecule of equivalent atoms?

This array is depicted in Figure 5.1. Unit vectors are erected on each equilibrium position as shown. Here $\mathbf{r}_j$ is directed inward toward the center of mass, $\mathbf{t}_j$ is directed perpendicular to $\mathbf{r}_j$ counterclockwise in the plane of the equilibrium positions, while $\mathbf{s}_j$ is directed perpendicular to the plane of $\mathbf{r}_j$ and $\mathbf{t}_j$ in the right-handed sense, so $\mathbf{t}_j \times \mathbf{r}_j = \mathbf{s}_j$.

Going down the flow chart of Figure 1.2, we find that the complete group for the system is $\mathbf{D}_{3h}$. In many applications, however, use of a

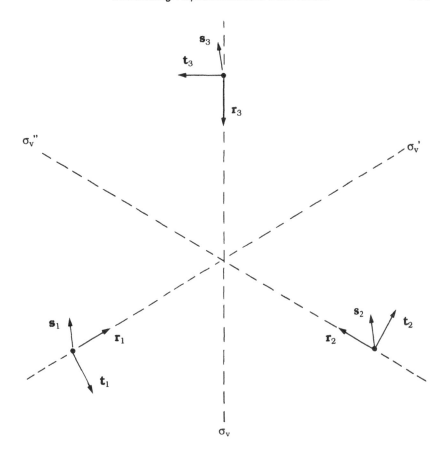

Figure 5.1 Three equivalent interacting particles at their equilibrium positions together with a suitable set of base vectors.

subgroup may be adequate and more convenient. Advantageous possibilities include C_{3h}, D_3, C_{3v}, C_3.

Example 5.2

Show that the combination

$$\mathbf{e}_1 = \frac{1}{\sqrt{3}} (\mathbf{r}_1 + \mathbf{r}_2 + \mathbf{r}_3)$$

of vectors from Figure 5.1 is a complete basis for a representation of the C_3 group.

Let each operation of C_3 act on the jth inwardly directed unit vector $\mathbf{r}_j$ in turn. We find that

$$I\mathbf{r}_j = \mathbf{r}_j,$$
$$C_3\mathbf{r}_j = \mathbf{r}_{j+1},$$
$$C_3{}^2\mathbf{r}_j = \mathbf{r}_{j+2},$$

if indicial numbers 1, 2, 3 make up a cyclic system, with 1 following 3. Then apply these formulas to the given combination and identify the results:

$$I\mathbf{e}_1 = I \frac{1}{\sqrt{3}}(\mathbf{r}_1 + \mathbf{r}_2 + \mathbf{r}_3) = \frac{1}{\sqrt{3}}(\mathbf{r}_1 + \mathbf{r}_2 + \mathbf{r}_3) = \mathbf{e}_1,$$

$$C_3\mathbf{e}_1 = C_3 \frac{1}{\sqrt{3}}(\mathbf{r}_1 + \mathbf{r}_2 + \mathbf{r}_3) = \frac{1}{\sqrt{3}}(\mathbf{r}_2 + \mathbf{r}_3 + \mathbf{r}_1) = \mathbf{e}_1,$$

$$C_3{}^2\mathbf{e}_1 = C_3{}^2 \frac{1}{\sqrt{3}}(\mathbf{r}_1 + \mathbf{r}_2 + \mathbf{r}_3) = \frac{1}{\sqrt{3}}(\mathbf{r}_3 + \mathbf{r}_1 + \mathbf{r}_2) = \mathbf{e}_1.$$

We see that each operation acting on $\mathbf{e}_1$ gives an expression equivalent to $\mathbf{e}_1$. So $\mathbf{e}_1$ is a complete basis for a representation. The resulting correspondence of group elements to matrices is

$$I \doteq (1), \quad C_3 \doteq (1), \quad C_3{}^2 \doteq (1).$$

We have here, in the ones, a completely symmetric representation of the group.

Example 5.3

Prove that the combination

$$\mathbf{e}_2 = \frac{1}{\sqrt{3}}(\mathbf{r}_1 + \omega^2\mathbf{r}_2 + \omega\mathbf{r}_3)$$

in which

$$\omega = \exp(2\pi i/3)$$

is a complete basis for a representation of the C_3 group.

Applying the formulas from Example 5.2 for the action of the group elements on $\mathbf{r}_j$ leads to

$$I\mathbf{e}_2 = \frac{1}{\sqrt{3}} (\mathbf{r}_1 + \omega^2 \mathbf{r}_2 + \omega \mathbf{r}_3) = \mathbf{e}_2,$$

$$C_3 \mathbf{e}_2 = \frac{1}{\sqrt{3}} (\mathbf{r}_2 + \omega^2 \mathbf{r}_3 + \omega \mathbf{r}_1) = \omega \frac{1}{\sqrt{3}} (\omega^2 \mathbf{r}_2 + \omega \mathbf{r}_3 + \mathbf{r}_1) = \omega \mathbf{e}_2,$$

$$C_3^2 \mathbf{e}_2 = \frac{1}{\sqrt{3}} (\mathbf{r}_3 + \omega^2 \mathbf{r}_1 + \omega \mathbf{r}_2)$$

$$= \omega^2 \frac{1}{\sqrt{3}} (\omega \mathbf{r}_3 + \mathbf{r}_1 + \omega^2 \mathbf{r}_2) = \omega^2 \mathbf{e}_2.$$

Each operation acting on $\mathbf{e}_2$ yields a number times $\mathbf{e}_2$. So $\mathbf{e}_2$ is a complete basis for a representation. The correspondence of group element to matrix is now

$$I \doteq (1), \quad C_3 \doteq (\omega), \quad C_3^2 \doteq (\omega^2).$$

In like manner, we find that

$$\mathbf{e}_3 = \frac{1}{\sqrt{3}} (\mathbf{r}_1 + \omega \mathbf{r}_2 + \omega^2 \mathbf{r}_3)$$

is a complete basis, with the correspondence

$$I \doteq (1), \quad C_3 \doteq (\omega^2), \quad C_3^2 \doteq (\omega).$$

Example 5.4

Show that the combinations

$$\mathbf{e}_4 = \frac{\sqrt{2}}{\sqrt{3}} (\mathbf{r}_1 - \tfrac{1}{2}\mathbf{r}_2 - \tfrac{1}{2}\mathbf{r}_3),$$

$$\mathbf{e}_5 = \frac{1}{\sqrt{2}} (\mathbf{r}_3 - \mathbf{r}_2)$$

make up a complete basis for a representation of the $\mathbf{C}_3$ group. Applying the key formulas from Example 5.2 leads to

$$I\mathbf{e}_4 = \mathbf{e}_4,$$

$$I\mathbf{e}_5 = \mathbf{e}_5,$$

$$\mathbf{C}_3 \mathbf{e}_4 = \frac{\sqrt{2}}{\sqrt{3}} (\mathbf{r}_2 - \tfrac{1}{2}\mathbf{r}_3 - \tfrac{1}{2}\mathbf{r}_1)$$

$$= -\frac{1}{2} \frac{\sqrt{2}}{\sqrt{3}} (\mathbf{r}_1 - \tfrac{1}{2}\mathbf{r}_2 - \tfrac{1}{2}\mathbf{r}_3) - \frac{\sqrt{3}}{2} \frac{1}{\sqrt{2}} (\mathbf{r}_3 - \mathbf{r}_2)$$

$$= -\frac{1}{2} \mathbf{e}_4 - \frac{\sqrt{3}}{2} \mathbf{e}_5,$$

$$\mathbf{C}_3 \mathbf{e}_5 = \frac{1}{\sqrt{2}} (\mathbf{r}_1 - \mathbf{r}_3)$$

$$= \frac{\sqrt{3}}{2} \frac{\sqrt{2}}{\sqrt{3}} (\mathbf{r}_1 - \tfrac{1}{2}\mathbf{r}_2 - \tfrac{1}{2}\mathbf{r}_3) - \frac{1}{2} \frac{1}{\sqrt{2}} (\mathbf{r}_3 - \mathbf{r}_2)$$

$$= \frac{\sqrt{3}}{2} \mathbf{e}_4 - \frac{1}{2} \mathbf{e}_5,$$

$$\mathbf{C}_3^2 \mathbf{e}_4 = \frac{\sqrt{2}}{\sqrt{3}} (\mathbf{r}_3 - \tfrac{1}{2}\mathbf{r}_1 - \tfrac{1}{2}\mathbf{r}_2)$$

$$= -\frac{1}{2} \mathbf{e}_4 + \frac{\sqrt{3}}{2} \mathbf{e}_5,$$

$$C_3{}^2 e_5 = \frac{1}{\sqrt{2}}(r_2 - r_1)$$

$$= -\frac{\sqrt{3}}{2} e_4 - \frac{1}{2} e_5.$$

Each altering operation of C_3 converts bases e_4 and e_5 into a linear combination of the two. Thus, the two together make up a complete basis for a representation. The results may be summarized with the matrix equations:

$$I(e_4 \;\; e_5) = (e_4 \;\; e_5)\begin{pmatrix} 1 & 0 \\ 0 & 1 \end{pmatrix},$$

$$C_3(e_4 \;\; e_5) = (e_4 \;\; e_5)\begin{pmatrix} -\frac{1}{2} & \frac{\sqrt{3}}{2} \\ -\frac{\sqrt{3}}{2} & -\frac{1}{2} \end{pmatrix},$$

$$C_3{}^2(e_4 \;\; e_5) = (e_4 \;\; e_5)\begin{pmatrix} -\frac{1}{2} & -\frac{\sqrt{3}}{2} \\ \frac{\sqrt{3}}{2} & -\frac{1}{2} \end{pmatrix}.$$

The 2 × 2 matrices constructed here multiply as the corresponding group elements. Thus, they form a matrix representation for the group C_3.

Example 5.5

Show that the representation in Example 5.4 is composed of the two distinct representations listed in Example 5.3.

First, we recall Euler's formula and apply it to numbers ω and ω^2:

$$\omega = \cos(2\pi/3) + i\sin(2\pi/3) = -\frac{1}{2} + \frac{\sqrt{3}}{2}i,$$

$$\omega^2 = \cos(4\pi/3) + i\sin(4\pi/3) = -\frac{1}{2} - \frac{\sqrt{3}}{2}i.$$

The sum and the difference of these are

$$\omega + \omega^2 = -1, \qquad \omega - \omega^2 = \sqrt{3}\,i.$$

These results are then employed in simplifying the normalized sum and the normalized difference of the two bases:

$$\frac{1}{\sqrt{2}}(\mathbf{e}_2 + \mathbf{e}_3) = \frac{1}{\sqrt{2}\sqrt{3}}(2\mathbf{r}_1 - \mathbf{r}_2 - \mathbf{r}_3)$$

$$= \frac{\sqrt{2}}{\sqrt{3}}\left(\mathbf{r}_1 - \frac{1}{2}\mathbf{r}_2 - \frac{1}{2}\mathbf{r}_3\right) = \mathbf{e}_4$$

and

$$\frac{1}{\sqrt{2}\,i}(\mathbf{e}_2 - \mathbf{e}_3) = \frac{1}{\sqrt{2}\,i\sqrt{3}}(-\sqrt{3}\,i\mathbf{r}_2 + \sqrt{3}\,i\mathbf{r}_3)$$

$$= \frac{1}{\sqrt{2}}(\mathbf{r}_3 - \mathbf{r}_2) = \mathbf{e}_5.$$

Thus, $\mathbf{e}_4$ and $\mathbf{e}_5$ are linear combinations of $\mathbf{e}_2$ and $\mathbf{e}_3$. But $\mathbf{e}_2$ and $\mathbf{e}_3$ are, individually, complete bases for the 1-dimensional representations in Example 5.3, while set $(\mathbf{e}_4, \mathbf{e}_5)$ generates the 2 × 2 matrices in Example 5.4. So the 2-dimensional representation is reducible to the two 1-dimensional representations.

Example 5.6

Show how the set of vectors $(\mathbf{r}_1, \mathbf{r}_2, \mathbf{r}_3)$ generates a 3-dimensional representation of the $\mathbf{C}_3$ group.

From the formulas for the action of the group elements on $\mathbf{r}_j$ in Example 5.2, we obtain

Generating Representations from Bases

$$I(\mathbf{r}_1\ \mathbf{r}_2\ \mathbf{r}_3) = (\mathbf{r}_1\ \mathbf{r}_2\ \mathbf{r}_3)\begin{pmatrix} 1 & 0 & 0 \\ 0 & 1 & 0 \\ 0 & 0 & 1 \end{pmatrix},$$

$$C_3(\mathbf{r}_1\ \mathbf{r}_2\ \mathbf{r}_3) = (\mathbf{r}_1\ \mathbf{r}_2\ \mathbf{r}_3)\begin{pmatrix} 0 & 0 & 1 \\ 1 & 0 & 0 \\ 0 & 1 & 0 \end{pmatrix},$$

$$C_3^2(\mathbf{r}_1\ \mathbf{r}_2\ \mathbf{r}_3) = (\mathbf{r}_1\ \mathbf{r}_2\ \mathbf{r}_3)\begin{pmatrix} 0 & 1 & 0 \\ 0 & 0 & 1 \\ 1 & 0 & 0 \end{pmatrix}.$$

The correspondence of group element to matrix is

$$I \rightleftharpoons \begin{pmatrix} 1 & 0 & 0 \\ 0 & 1 & 0 \\ 0 & 0 & 1 \end{pmatrix},$$

$$C_3 \rightleftharpoons \begin{pmatrix} 0 & 0 & 1 \\ 1 & 0 & 0 \\ 0 & 1 & 0 \end{pmatrix},$$

$$C_3^2 \rightleftharpoons \begin{pmatrix} 0 & 1 & 0 \\ 0 & 0 & 1 \\ 1 & 0 & 0 \end{pmatrix}.$$

Thus, the set of vectors $(\mathbf{r}_1, \mathbf{r}_2, \mathbf{r}_3)$ generates a 3-dimensional representation. But by Examples 5.2 and 5.3, this representation is reducible to the three irreducible representations.

Example 5.7

How is the 3-dimensional matrix representation affected when the set $(\mathbf{r}_1, \mathbf{r}_2, \mathbf{r}_3)$ is rearranged as set $(\mathbf{e}_1, \mathbf{e}_2, \mathbf{e}_3)$?

From the discussions in Example 5.2 and 5.3, we obtain

$$I(e_1 \ e_2 \ e_3) = (e_1 \ e_2 \ e_3) \begin{pmatrix} 1 & 0 & 0 \\ 0 & 1 & 0 \\ 1 & 0 & 0 \end{pmatrix},$$

$$C_3(e_1 \ e_2 \ e_3) = (e_1 \ e_2 \ e_3) \begin{pmatrix} 1 & 0 & 0 \\ 0 & \omega & 0 \\ 0 & 0 & \omega^2 \end{pmatrix},$$

$$C_3^2(e_1 \ e_2 \ e_3) = (e_1 \ e_2 \ e_3) \begin{pmatrix} 1 & 0 & 0 \\ 0 & \omega^2 & 0 \\ 0 & 0 & \omega \end{pmatrix}.$$

The correspondence of group element to matrix is now

$$I \rightleftharpoons \begin{pmatrix} 1 & 0 & 0 \\ 0 & 1 & 0 \\ 0 & 0 & 1 \end{pmatrix},$$

$$C_3 \rightleftharpoons \begin{pmatrix} 1 & 0 & 0 \\ 0 & \omega & 0 \\ 0 & 0 & \omega^2 \end{pmatrix},$$

$$C_3^2 \rightleftharpoons \begin{pmatrix} 1 & 0 & 0 \\ 0 & \omega^2 & 0 \\ 0 & 0 & \omega \end{pmatrix}.$$

Note that the only nonzero matrix elements occur on the principal diagonal.

One generalizes this result as follows. Whenever a basis set is rearranged to a form made up of bases for irreducible representations, the matrices for the reducible representation are altered to the form

$$\begin{pmatrix} (\ldots) & (0) & \ldots & (0) \\ (0) & (\ldots) & \ldots & (0) \\ \cdot & \cdot & \cdot & \cdot \\ \cdot & \cdot & \cdot & \cdot \\ \cdot & \cdot & \cdot & \cdot \\ (0) & (0) & \ldots & (\ldots) \end{pmatrix}$$

Here each submatrix on the principal diagonal is a matrix of the pertinent irreducible representation; each of the off-diagonal submatrices consists of zeroes.

5.3
Traces of the Matrices in a Representation

The expansions obtained when a symmetry operation acts on base expressions from a given complete set depend not only on the operation and on the symmetry species contributing to the basis set, but also on how the set is arranged or partitioned into these expressions. Nevertheless, as long as the species composition is unaltered, the sum of numerical elements on the principal diagonal of each resulting representation matrix is independent of the particular arrangement, or division, chosen. The set of such sums therefore typifies the composition of a representation.

Let us consider a physical system covered by a group of symmetry operations. As in Section 5.2, let us associate with aspects, orientations, or parts of the system, expressions that the symmetry operations mix linearly. The coefficients in the linear transformation equations form matrices that combine as the corresponding operations of the group. Thus, these matrices represent the corresponding operations.

Let us choose two operations of the group, C and D, belonging to the same class. From the definition of class, these are linked by a similarity transformation,

$$D = PCP^{-1}, \tag{5.8}$$

with some other operation of the group, P. With a particular choice of base expressions, the corresponding matrices are found to be **C, D, P, P**$^{-1}$.

Let us call the sum of elements along the principal diagonal of each matrix its *trace*. Thus,

$$\operatorname{tr} \mathbf{C} = \sum_k C_{kk}, \quad \operatorname{tr} \mathbf{D} = \sum_j D_{jj}, \tag{5.9}$$

where tr **C** and tr **D** stand for the trace of **C** and the trace of **D**.

Corresponding to equation (5.8), we have the matrix equation

$$\mathbf{D} = \mathbf{PCP}^{-1}. \tag{5.10}$$

Let us take the trace of each side, rearrange factors in each term, and simplify:

$$\operatorname{tr} \mathbf{D} = \sum_j (\mathbf{PCP}^{-1})_{jj} = \sum_{j,k,l} P_{jk} C_{kl} P^{-1}{}_{lj} = \sum_{j,k,l} P^{-1}{}_{lj} P_{jk} C_{kl}$$

$$= \sum_{k,l} \delta_{lk} C_{kl} = \sum_k C_{kk} = \operatorname{tr} \mathbf{C}. \tag{5.11}$$

In a representation of a group, matrices belonging to the same class have the same trace.

Following equations (2.24) and (2.25), a person can add the matrices in each class to construct a matrix representation of its class sum:

$$\mathscr{C}_1 = \mathbf{I}, \tag{5.12}$$

$$\mathscr{C}_j = \mathbf{C} + \mathbf{D} + \ldots + \mathbf{F}. \tag{5.13}$$

Taking the trace of each side of equation (5.12) yields

$$\operatorname{tr} \mathscr{C}_1 = \operatorname{tr} \mathbf{I} = n \tag{5.14}$$

where n is the number of elements along the principal diagonal, the dimesionality of the representation. From equation (5.13) and the fact that all matrices in the same class have the same trace, we obtain

$$\begin{aligned}\operatorname{tr}\mathscr{C}_j &= \operatorname{tr}\mathbf{C} + \operatorname{tr}\mathbf{D} + \ldots + \operatorname{tr}\mathbf{F}\\ &= \operatorname{tr}\mathbf{C} + \operatorname{tr}\mathbf{C} + \ldots + \operatorname{tr}\mathbf{C}\\ &= h_j \operatorname{tr}\mathbf{C},\end{aligned} \qquad (5.15)$$

where h_j is the number of operations in the jth class.

For each class-sum matrix, one can construct the eigenvalue equation

$$\mathscr{C}_j \mathbf{u} = \lambda_j \mathbf{u}. \qquad (5.16)$$

Here expression $\mathbf{u}$ is a characteristic arrangement of bases, in an n-element column matrix. Possible eigenvalues can be found by solving the corresponding secular equation. But since equation (5.16) involves the same acting operator as equation (2.37), the eigenvalues must be the same.

Corresponding to eigenvalues $\lambda_j^{(r)}, \lambda_j^{(r)}, \ldots, \lambda_j^{(s)}, \ldots$, can be found n orthogonal eigenmatrices. Let us choose a normalization and phasing, label the results $\mathbf{u}_1, \mathbf{u}_2, \ldots, \mathbf{u}_n$, and order them in the square matrix

$$\mathbf{U} = (\mathbf{u}_1 \quad \mathbf{u}_2 \quad \ldots \quad \mathbf{u}_n), \qquad (5.17)$$

With the eigenvalues, we also construct

$$\Lambda = \begin{pmatrix} \lambda_j^{(r)} & 0 & \cdot & \cdot & \cdot & \cdot & \cdot \\ 0 & \lambda_j^{(r)} & \cdot & \cdot & \cdot & \cdot & \cdot \\ \cdot & \cdot & \cdot & \cdot & \cdot & \cdot & \cdot \\ \cdot & \cdot & \cdot & \cdot & \cdot & \cdot & \cdot \\ \cdot & \cdot & \cdot & \cdot & \lambda_j^{(s)} & \cdot & \cdot \\ \cdot & \cdot & \cdot & \cdot & \cdot & \cdot & \cdot \\ \cdot & \cdot & \cdot & \cdot & \cdot & \cdot & \cdot \end{pmatrix}. \qquad (5.18)$$

The set of satisfied eigenvalue equations can then be written in the form

$$\mathscr{C}_j \mathbf{U} = \mathbf{U}\lambda \qquad (5.19)$$

whence

$$\mathbf{U}^{-1}\mathscr{C}_j\mathbf{U} = \boldsymbol{\lambda}. \qquad (5.20)$$

Equation (5.20) has the same form as equation (5.10); so the argument in equation (5.11) applies. A similarity transformation does not alter the trace. Indeed, the trace of $\mathscr{C}_j$ equals the sum of the diagonal elements in $\boldsymbol{\lambda}$. If we let the number of times that $\lambda_j^{(k)}$ appears be $a^{(k)}y_1^{(k)}$ and introduce relationship (2.47), we obtain

$$\operatorname{tr}\mathscr{C}_j = \lambda_j^{(r)} + \lambda_j^{(r)} + \ldots + \lambda_j^{(s)} + \ldots$$

$$= \sum_k a^{(k)}y_1^{(k)}\lambda_j^{(k)} = \sum_k a^{(k)}h_j y_j^{(k)}. \qquad (5.21)$$

Combining equation (5.21) with equation (5.15) and canceling h_j gives

$$\operatorname{tr}\mathbf{C}_j = \sum_k a^{(k)}y_j^{(k)}, \qquad (5.22)$$

if we add subscript j to symbol $\mathbf{C}$ to indicate the class.

We have noted how the bases for a representation are transformed by operations of the group. A complete set of bases for a reducible representation can be reduced in number if they are combined linearly (rearranged) in an appropriate manner. When the representation is irreducible, however, no linear rearrangement will yield a smaller complete set.

For equation (5.16), an n-dimensional basis yields n eigenvalues. Each irreducible representation of dimensionality $y_1^{(k)}$ yields $y_1^{(k)}$ identical roots. So if a $\lambda_j^{(k)}$ appears $a^{(k)}y_1^{(k)}$ times, coefficient $a^{(k)}$ is the number of times the kth irreducible representation occurs.

5.4
Character Vectors for Representations

The $y_j^{(k)}$ of equation (5.22) was identified (in Chapter 2) as the jth component of the kth character vector for the given group. This vector is important because it pairs with the group operators in determining properties of the kth primitive symmetry species of the group. (Recall Section 3.3, for instance.)

Now when matrix $\mathbf{C}_j$ is the identity matrix $\mathbf{I}$, index j is 1 and equation (5.22) becomes

$$\operatorname{tr} \mathbf{I} = \sum_k a^{(k)} y_1^{(k)}. \tag{5.23}$$

Since $\mathbf{I}$ has only 1's on its principal diagonal, the tr $\mathbf{I}$ equals the number of rows, and the number of columns, in each representation matrix. Consequently, it equals the number of independent bases for the representation. But this equals the dimensionality of the representation.

In equation (5.22) as it stands, component $y_j^{(k)}$ is multiplied by $a^{(k)}$ and added to similar products for the other symmetry species (irreducible representations). The result is the jth component of a composite vector χ, which obeys the equation

$$\chi = \sum_k a^{(k)} \chi^{(k)}. \tag{5.24}$$

Expression χ is called the *character vector* for the given reducible representation. According to equation (5.22), the trace of matrix $\mathbf{C}_j$ is the jth component of χ:

$$\chi_j = \operatorname{tr} \mathbf{C}_j. \tag{5.25}$$

In calculating tr $\mathbf{C}_j$, one need only determine the elements on the principal diagonal and add these. But equation (5.2) may be put in the form

$$\mathbf{C}_j \mathbf{u}_k = b_k \mathbf{u}_k + c_k \mathbf{u}_{k+1} + \ldots \tag{5.26}$$

Thus when an operation belonging to the jth class acts on a base, the result equals a linear combination of the bases. Coefficient b_k is the kth element on the principal diagonal of $\mathbf{C}_j$. We add these to get the jth character component:

$$\chi_j = \sum_k b_k. \tag{5.27}$$

Example 5.8

From the representation matrices, determine the primitive symmetry species that contribute to the bases in Examples 5.6, 5.4, and 5.3.

Adding the elements along each principal diagonal of the matrices representing I, C_3, and C_3^2 in Example 5.6 yields the numbers:

$$\text{tr } \mathbf{I} = 1 + 1 + 1 = 3,$$

$$\text{tr } \mathbf{C}_3 = 0 + 0 + 0 = 0,$$

$$\text{tr } \mathbf{C}_3^2 = 0 + 0 + 0 = 0.$$

From our discussions of equations (5.22) and (5.24), these are the components of a character vector. Let us label the corresponding symmetry species Γ_1 and list the components in Table 5.1.

Euler's formula for an exponential tells us that

$$\exp(2\pi i/3) = \cos(2\pi/3) + i\sin(2\pi/3) = -\frac{1}{2} + \frac{\sqrt{3}}{2}i,$$

$$\exp(4\pi i/3) = \cos(4\pi/3) + i\sin(4\pi/3) = -\frac{1}{2} - \frac{\sqrt{3}}{2}i.$$

With

$$\exp(2\pi i/3) = \omega,$$

we have

$$1 + \omega + \omega^2 = 0.$$

Table 5.1 Components of Character Vectors for the C_3 Group

	I	C_3	C_3^2
A ($k=1$)	1	1	1
C ($k=2$)	1	ω	ω^2
D ($k=3$)	1	ω^2	ω
Γ_1	3	0	0
Γ_2	2	-1	-1
Γ_3	1	ω	ω^2
Γ_4	1	ω^2	ω
	$\omega = \exp(2\pi i/3)$		

Now in Table 5.1, the components for one A, one C, and one D character vector add to produce those for the Γ_1 character vector. So $a^{(k)}$ in equation (5.24) equals 1 for each k and we write

$$\Gamma_1 = A + C + D.$$

Similarly, the traces of the matrices representing I, C_3, and C_3^2 in Example 5.4 are

$$\text{tr } \mathbf{I} = 1 + 1 = 2,$$

$$\text{tr } \mathbf{C}_3 = -\frac{1}{2} - \frac{1}{2} = -1,$$

$$\text{tr } \mathbf{C}_3^2 = -\frac{1}{2} - \frac{1}{2} = -1.$$

These make up the row labeled Γ_2 in Table 5.1. Since

$$\omega + \omega^2 = -1,$$

the character components for one C and one D add to give those for Γ_2. Coefficient $a^{(k)}$ in (5.24) equals 0 for $k = 1$, 1 for $k = 2$, and 1 for $k = 3$. We say that

$$\Gamma_2 = C + D.$$

For basis $\mathbf{e}_2$ in Example 5.3, the traces are

$$\text{tr } \mathbf{I} = 1, \quad \text{tr } \mathbf{C}_3 = \omega, \quad \text{tr } \mathbf{C}_3^2 = \omega^2.$$

These form the character components in the Γ_3 row of Table 5.1. For basis $\mathbf{e}_3$ in Example 5.3, the traces are

$$\text{tr } \mathbf{I} = 1, \quad \text{tr } \mathbf{C}_3 = \omega^2, \quad \text{tr } \mathbf{C}_3^2 = \omega.$$

These are in the Γ_4 row of Table 5.1. On comparing components, we see that the Γ_3 character vector is the same as the C character vector, the Γ_4 character vector is the same as the D character vector. In summary,

$$\Gamma_3 = C, \quad \Gamma_4 = D.$$

Example 5.9

From the bases that are left unchanged (or merely changed in phase) by a typical operation of each class, determine the character vector for the representation that $\mathbf{r}_1$, $\mathbf{r}_2$, and $\mathbf{r}_3$ generate.

For the arrangement of base vectors, refer again to Figure 5.1. Under operation I, each of these remains unchanged. Then equation (5.27) yields the character component

$$\chi_I = 1 + 1 + 1 = 3.$$

But under operations C_3 and C_3^2, all the bases are changed to different bases. So the first term in equation (5.26) is zero in each instance and equation (5.27) yields

$$\chi_{C_3} = 0 + 0 + 0 = 0,$$

$$\chi_{C_3^2} = 0 + 0 + 0 = 0.$$

These are the components for Γ_1 in Table 5.1.

Example 5.10

Use equation (5.26) in determining the character vector for the representation that $\mathbf{e}_4$ and $\mathbf{e}_5$ in Example 5.4 generate.

For operation I, we note that $\mathbf{e}_4$ and $\mathbf{e}_5$ are unchanged. So

$$\chi_I = 1 + 1 = 2.$$

On comparing the results for $C_3 \mathbf{e}_4$ and $C_3 \mathbf{e}_5$ with equation (5.26), we find that equation (5.27) yields

$$\chi_{C_3} = -\tfrac{1}{2} - \tfrac{1}{2} = -1.$$

On comparing the results for $C_3^2 \mathbf{e}_4$ and $C_3^2 \mathbf{e}_5$ with equation (5.26), we find that equation (5.27) gives us

$$\chi_{C_3^2} = -\tfrac{1}{2} - \tfrac{1}{2} = -1.$$

These are the components for Γ_2 in Table 5.1.

5.5
Symmetry Species in a Representation

When the order of a group is finite, the orthogonality relationship can be used to determine the number of times a symmetry species occurs in a representation.

Consider a group whose class sums are

$$\mathscr{C}_1, \mathscr{C}_2, \ldots, \mathscr{C}_n, \qquad (5.28)$$

as before. Let a representation be constructed and let the corresponding sums be formed

$$\pmb{\mathscr{C}}_1, \pmb{\mathscr{C}}_2, \ldots, \pmb{\mathscr{C}}_n. \qquad (5.29)$$

Equation (5.21) now tells us that

$$\operatorname{tr} \pmb{\mathscr{C}}_j = \sum_k a^{(k)} h_j y_j^{(k)}. \qquad (5.30)$$

Multiplying both sides by $y_{\bar{j}}^{(s)}$ and summing over j leads to

$$\sum_j (\operatorname{tr} \pmb{\mathscr{C}}_j) \, y_{\bar{j}}^{(s)} = \sum_{j,k} a^{(k)} h_j y_j^{(k)} y_{\bar{j}}^{(s)} = \sum_k a^{(k)} g \delta_{ks}$$

$$= g a^{(s)}. \qquad (5.31)$$

In the second equality, equation (2.64) has been introduced; in the last step, the effect of the Kronecker delta has been brought in. Solving for the number of times the sth symmetry species occurs yields

$$a^{(s)} = \frac{1}{g} \sum_j (\operatorname{tr} \pmb{\mathscr{C}}_j) \, y_{\bar{j}}^{(s)}. \qquad (5.32)$$

Here g is the order of the group, $y_{\bar{j}(s)}$ the component of the sth character vector for the jth inverse class sum, and $\operatorname{tr} \pmb{\mathscr{C}}_j$ the trace of the matrix for the jth class sum.

Example 5.11

Using formula (5.32), determine the composition of the reducible representation of the $\mathbf{C}_3$ group in Example 5.4.

The C_3 group consists of I, C_3, and C_3^2, each in a class by itself, with C_3^2 the inverse of C_3 and C_3 the inverse of C_3^2. So the traces of the class sum matrices are the same as the traces of the representation matrices. For the reducible representation of Example 5.4, these are

$$\text{tr } I = 2, \quad \text{tr } \mathbf{C}_3 = -1, \quad \text{tr } \mathbf{C}_3^2 = -1.$$

Since the group contains 3 elements, its order is 3.

Inserting these numbers and the character components from Table 5.1 into equation (5.32) yields

$$a_A = \frac{1}{3}[2 \cdot 1 - 1 \cdot 1 - 1 \cdot 1] = 0$$

$$a_C = \frac{1}{3}[2 \cdot 1 - 1 \cdot \omega^2 - 1 \cdot \omega]$$

$$= \frac{1}{3}[2 - 1(\omega^2 + \omega)] = 1$$

$$a_D = \frac{1}{3}[2 \cdot 1 - 1 \cdot \omega - 1 \cdot \omega^2] = 1.$$

The same result is obtained as in Example 5.8.

5.6
Freedom in Representing a Group

A group of order g contains g independent elements. Associated with these may be g linearly independent vectors in a g-dimensional abstract space. But how may these be related to representations?

Consider a matrix representation

$$\mathbf{A}, \mathbf{B}, \ldots, \mathbf{P}, \mathbf{Q} \tag{5.33}$$

of a group of order g constructed from the quantities

$$A_{jk}, B_{jk}, \ldots, P_{jk}, Q_{jk} \tag{5.34}$$

with

$$j = 1, 2, \ldots, d \tag{5.35}$$

and

$$k = 1, 2, \ldots, d. \qquad (5.36)$$

Here d is the dimensionality of the representation.

Each number in line (5.34) is associated with a different group element and there are g elements. So we may let these be the components of a g-dimensional vector. From equations (5.35) and (5.36), the total number of such vectors is $d \times d$.

Now, a representation may be constructed such that, in equation (5.23), coefficient $a^{(k)}$ is 1 when $k = r$ and zero otherwise. Equation (5.23) then tells us that the dimensionality of the representation is $y_1^{(r)}$. And the representation yields $y_1^{(r)}$ times $y_1^{(r)}$ of the g-dimensional vectors. Reducing $a^{(r)}$ to a possible fraction would keep this species from developing its potentialities; indeed, this symmetry species is irreducible. Also, each time the contribution from a symmetry species is increased, it should be by an integral amount, so that all potentialities of the addition are allowed.

A representation made up of more than a single irreducible representation yields a number of linearly independent vectors equal to $y_1^{(r)}$ times $y_1^{(r)}$ for each time the rth representation is used. Coefficient $a^{(r)}$ gives the number of times the rth symmetry species contributes to the bases being employed. It is the number of times the rth irreducible representation appears in the reducible one, as we have noted.

If each symmetry species is employed once, then orthogonality relationship (2.67) yields

$$\sum_r y_1^{(r)} y_1^{(r)} = g. \qquad (5.37)$$

But the total number of linearly independent vectors that a person can construct in a g-dimensional space is g. So this representation, in which each symmetry species is used once, spans the possibilities. No additional freedom exists in representing the group.

5.7
Schur's Lemma

Conditions that distinguish irreducible representations from reducible ones are provided by two theorems of Schur.

Again consider a complete set of linearly independent operands

$$\mathbf{u}_1, \mathbf{u}_2, \ldots, \mathbf{u}_n \tag{5.38}$$

that each element of the group transforms linearly:

$$A\mathbf{u}_k = \sum_{j=1}^{n} \mathbf{u}_j \Gamma_{jk}(A). \tag{5.39}$$

If the representation is reducible, there are m operands

$$\mathbf{e}_i = \sum \mathbf{u}_k c_{ki}, \tag{5.40}$$

with $m < n$ such that

$$A\mathbf{e}_i = \sum_{h=1}^{m} \mathbf{e}_h \Gamma'_{hi}(A) \tag{5.41}$$

for each element in the group.

Operate on equation (5.40) with A and introduce equation (5.39) into the result:

$$A\mathbf{e}_i = \sum_k A\mathbf{u}_k c_{ki} = \sum_j \sum_k \mathbf{u}_j \Gamma_{jk}(A) c_{ki}. \tag{5.42}$$

Also expand equation (5.41) with equation (5.40):

$$A\mathbf{e}_i = \sum_k \sum_h \mathbf{u}_k c_{kh} \Gamma'_{hi}(A). \tag{5.43}$$

Combining equations (5.42) and (5.43) yields

$$\sum_j \sum_k \mathbf{u}_j \Gamma_{jk}(A) c_{ki} = \sum_j \sum_k \mathbf{u}_j c_{jk} \Gamma'_{ki}(A). \tag{5.44}$$

Since the $\mathbf{u}_j$'s are linearly independent, their coefficients must be equal:

$$\sum_{k=1}^{n} \Gamma_{jk}(A) c_{ki} = \sum_{k=1}^{m} c_{jk} \Gamma'_{ki}(A), \tag{5.45}$$

whence

$$\mathbf{\Gamma}(A)\mathbf{C} = \mathbf{C}\,\mathbf{\Gamma}'(A). \tag{5.46}$$

Matrix $\mathbf{C}$ is constant; it does not depend on operation A in the group.

This argument can be reversed. Suppose that each matrix $\Gamma(A)$ of an n-dimensional representation multiplied by one $n \times m$ matrix $\mathbf{C}$ equals $\mathbf{C}$ times the corresponding matrix $\Gamma'(A)$ of an m-dimensional representation

$$\Gamma(A)\mathbf{C} = \mathbf{C}\Gamma'(A). \tag{5.47}$$

Then

$$\sum_{k=1}^{n} \Gamma_{jk}(A)c_{ki} = \sum_{k=1}^{m} c_{jk}\Gamma'_{ki}(A). \tag{5.48}$$

Multiply both sides by the jth base for the $\Gamma(A)$'s and sum from 1 to n:

$$\sum_{j}\sum_{k} \mathbf{u}_j \Gamma_{jk}(A) c_{ki} = \sum_{j}\sum_{k} \mathbf{u}_j c_{jk} \Gamma'_{ki}(A). \tag{5.49}$$

Reduce with equation (5.39),

$$\sum_{k} A\mathbf{u}_k c_{ki} = \sum_{j}\sum_{k} \mathbf{u}_j c_{jk} \Gamma'_{ki}(A); \tag{5.50}$$

then let

$$\sum_{k} \mathbf{u}_k c_{ki} = \mathbf{e}_i, \quad \sum_{j} \mathbf{u}_j c_{jk} = \mathbf{e}_k \tag{5.51}$$

to get

$$A\mathbf{e}_i = \sum_{k=1}^{m} \mathbf{e}_k \Gamma'_{ki}(A), \tag{5.52}$$

for any A in the group. Thus if $m < n$, the unprimed representation $\Gamma(A)$, $\Gamma(B)$, ... is reducible. *Schur's first theorem* can now be formulated.

Part 1. If the Γ and Γ' representations are irreducible and $m < n$, the only matrix $\mathbf{C}$ satisfying equation (5.47) is the null matrix.

When $m = n$, the Γ and Γ' matrices have the same dimensionality. And if the Γ matrix is irreducible, the m $\mathbf{e}_h$'s must be linearly independent. As long as they all exist, relationship (5.40) can be inverted and $\mathbf{C}^{-1}$ exists; however equation (5.46) may be constructed with a singular matrix, one for which $\mathbf{C}^{-1}$ does not exist.

Part 2. If the Γ representation is irreducible and $m = n$, either equation (5.47) can be rewritten in the form

$$\Gamma(A) = \mathbf{C}\Gamma'(A)\mathbf{C}^{-1} \tag{5.53}$$

or $\mathbf{C}$ is singular. When equation (5.53) holds, the two representations are equivalent.

What can one deduce concerning a matrix that commutes with all the matrices of an irreducible representation? If $\mathbf{C}$ is a nonzero matrix satisfying

$$\Gamma(A)\mathbf{C} = \mathbf{C}\Gamma(A), \tag{5.54}$$

where $\Gamma(A)$ may be any $n \times n$ matrix in an irreducible representation of the given group, what is the nature of $\mathbf{C}$?

The square matrix $\mathbf{C}$ may act in an n-dimensional vector space. It will not alter the directions of some vectors $\mathbf{v}$; thus

$$\mathbf{C}\mathbf{v} = \lambda\mathbf{v}. \tag{5.55}$$

Now act on this eigenvalue equation with $\Gamma(A)$:

$$\Gamma(A)\,\mathbf{C}\mathbf{v} = \Gamma(A)\lambda\mathbf{v}. \tag{5.56}$$

By equation (5.54), the square matrices on the left can be commuted. Furthermore, λ commutes with $\Gamma(A)$. Thus,

$$\mathbf{C}\Gamma(A)\,\mathbf{v} = \lambda\Gamma(A)\mathbf{v}, \tag{5.57}$$

expression $\Gamma(A)\mathbf{v}$ is an eigenvector of $\mathbf{C}$.

Since $\mathbf{C}$ is an $n \times n$ matrix, it has n independent eigenvectors. If all of them are obtained with form $\Gamma(A)\mathbf{v}$, then only one eigenvalue exists. But if all are not obtainable in this way, operations of the group, through $\Gamma(A)$ acting on $\mathbf{v}$, would mix only m vectors, where $m < n$. Then Γ would be reducible, contrary to assumption.

Thus, a nonzero matrix that commutes with all matrices in an irreducible representation has n equal eigenvalues. And there is a similarity transformation that would diagonalize the matrix, with the eigenvalues on the resulting diagonal:

Schur's Lemma

$$\mathbf{UCU}^{-1} = \begin{pmatrix} \lambda & 0 & \cdots & 0 \\ 0 & \lambda & \cdots & 0 \\ \cdot & \cdot & \cdot & \cdot \\ \cdot & \cdot & \cdot & \cdot \\ \cdot & \cdot & \cdot & \cdot \\ 0 & 0 & \cdots & \lambda \end{pmatrix} = \lambda \mathbf{E}. \quad (5.58)$$

Solving for **C** gives us

$$\mathbf{C} = \mathbf{U}^{-1} \lambda \mathbf{E} \mathbf{U} = \lambda \mathbf{E} \mathbf{U}^{-1} \mathbf{U} = \lambda \mathbf{E}. \quad (5.59)$$

Schur's second theorem summarizes this result. When $\boldsymbol{\Gamma}$ and $\boldsymbol{\Gamma}'$ are identical irreducible representations in equation (5.47), as in equation (5.54), we have

$$\mathbf{C} = \lambda \mathbf{E}. \quad (5.60)$$

Here **C** equals a number times the pertinent unit matrix.

Example 5.12

The permutation matrices

$$\mathbf{I} = \begin{pmatrix} 1 & 0 & 0 \\ 0 & 1 & 0 \\ 0 & 0 & 1 \end{pmatrix}, \mathbf{A} = \begin{pmatrix} 0 & 0 & 1 \\ 1 & 0 & 0 \\ 0 & 1 & 0 \end{pmatrix}, \mathbf{B} = \begin{pmatrix} 0 & 1 & 0 \\ 0 & 0 & 1 \\ 1 & 0 & 0 \end{pmatrix},$$

$$\mathbf{C} = \begin{pmatrix} 1 & 0 & 0 \\ 0 & 0 & 1 \\ 0 & 1 & 0 \end{pmatrix}, \mathbf{D} = \begin{pmatrix} 0 & 0 & 1 \\ 0 & 1 & 0 \\ 1 & 0 & 0 \end{pmatrix}, \mathbf{E} = \begin{pmatrix} 0 & 1 & 0 \\ 1 & 0 & 0 \\ 0 & 0 & 1 \end{pmatrix}$$

form a group. Show that this representation of $\mathscr{S}_3$ is reducible.

A matrix that commutes with all these matrices is

$$\mathbf{H} = \begin{pmatrix} 0 & 1 & 1 \\ 1 & 0 & 1 \\ 1 & 1 & 0 \end{pmatrix}.$$

But the secular equation for **H** is

$$\begin{vmatrix} -\lambda & 1 & 1 \\ 1 & -\lambda & 1 \\ 1 & 1 & -\lambda \end{vmatrix} = 0$$

or

$$\lambda^3 - 3\lambda - 2 = 0.$$

The roots of this equation are

$$\lambda_1 = -1, \quad \lambda_2 = -1, \quad \lambda_3 = 2.$$

Since the eigenvalues are not all equal, the representation is reducible.

5.8
An Overview

The symmetries of a physical system are embodied in the various operations that change it to an equivalent system. Each such operation transforms a given local physical attribute or property. In general, the result can be expressed as a *linear combination* of independent standard expressions associated with different parts of the physical system. The expressions may appear as magnitudes, vectors, dyads, polyads, functions of any of these, or kets. The standard expressions in a given complete set are called bases. Usually, these are normalized in a uniform way.

The coefficients obtained with a given complete set of bases make up matrix representations of the group elements. The matrices are characterized by their traces. A person can combine the bases linearly and normalize the results. The new set is complete when all operations of the group merely mix members of the new set linearly.

When a complete set containing fewer members than the original set can be constructed, the original set is said to be reducible. When this is not possible, the set is said to be irreducible. The corresponding representations are also labeled reducible and irreducible, respectively.

An irreducible complete set of basis expressions, and linear combinations of these, are manifestations of a primitive symmetry species. Character vectors for each such species of a group are described in character tables.

Now, the trace of each matrix in a given representation equals a linear combination of the corresponding components of the contributing primitive character vectors:

$$\text{tr } \mathbf{C}_j = \sum_k a^{(k)} y_j^{(k)}. \tag{5.61}$$

Here j identifies the class to which the matrix belongs while k identifies the symmetry species that contribute. The set of traces constitute the components of a character vector χ. From equation (5.61),

$$\chi = \sum_k a^{(k)} \chi^{(k)}. \tag{5.62}$$

The number of times the sth symmetry species contributes to a representation is given by

$$a^{(s)} = \frac{1}{g} \sum_j (\text{tr } \mathscr{C}_j) y_j^{(s)}. \tag{5.63}$$

Here g is the order of the group, tr $\mathscr{C}_j$ the trace of the matrix for the jth class sum, and $y_j^{(s)}$ the component of the sth character vector for the jth inverse class sum.

Discussion Questions

5.1 What may represent elements of a group?
5.2 What properties must expressions possess to serve as bases for such representations?
5.3 How do bases generate representations of groups?
5.4 When is a representation (a) reducible, (b) irreducible?

5.5 What is the trace of a matrix? Why is the trace for each matrix in a class the same?

5.6 How is the trace for a class sum related to the trace of a matrix in the class?

5.7 How is the trace for a class sum related to eigenvalues for the class-sum operator?

5.8 How is the trace for a class sum related to the character components for the primitive symmetry species?

5.9 How is the trace of a representation matrix related to standard character components?

5.10 Why is a representation of dimensionality equal to its $y_1^{(r)}$ irreducible? What is the significance of coefficient $a^{(k)}$ in

$$\mathrm{tr}\, \mathbf{C}_j = \sum_k a^{(k)} y_j^{(k)}?$$

5.11 How does the orthogonality relationship enable one to calculate the number of times a symmetry species contributes to a representation?

5.12 Why does the formula for $a^{(s)}$ fail when the order of the group is infinite? How might one then determine $a^{(s)}$?

5.13 When does a set of linearly independent basis vectors span a space? What is the dimensionality of the space?

5.14 What representations are needed to span the group-element space?

Problems

5.1 From the appropriate reorientation matrices, deduce the representation

$$\begin{pmatrix} 1 & 0 \\ 0 & 1 \end{pmatrix}, \begin{pmatrix} 0 & -1 \\ 1 & 0 \end{pmatrix}, \begin{pmatrix} -1 & 0 \\ 0 & -1 \end{pmatrix}, \begin{pmatrix} 0 & 1 \\ -1 & 0 \end{pmatrix},$$

of the C_4 group.

5.2 Four identical particles are bound together so that they oscillate about the four corners of a square. Construct a possible set of equilibrium positions. At each of these, erect three mutually perpendicular unit vectors as in Figure 5.1. Number the origins consecutively in the counterclockwise direction. At the jth one, let unit vector $\mathbf{r}_j$ point directly toward the center, unit vector $\mathbf{t}_j$ point perpendicular to $\mathbf{r}_j$ in the plane of the square in the counterclockwise direction, while $\mathbf{s}_j$ is oriented perpendicular to the plane of the square upwards. Thus, we have $\mathbf{r}_j \times \mathbf{s}_j = \mathbf{t}_j$.

5.3 Consider the C_4 subgroup for the system in Problem 5.2, with the base vectors as described. Show that the linear combinations

$$\mathbf{e}_1 = \frac{1}{\sqrt{2}} (\mathbf{r}_1 - \mathbf{r}_3),$$

$$\mathbf{e}_2 = \frac{1}{\sqrt{2}} (\mathbf{r}_2 - \mathbf{r}_4)$$

form a complete basis for a reducible representation.

5.4 Show how $\mathbf{e}_1$ and $\mathbf{e}_2$ in Problem 5.3 can be combined linearly to form bases for irreducible representations.

5.5 For the system in Problem 5.2, construct the function

$$f = \mathbf{r}_1 \cdot \mathbf{r}_1 - \mathbf{r}_2 \cdot \mathbf{r}_2 + \mathbf{r}_3 \cdot \mathbf{r}_3 - \mathbf{r}_4 \cdot \mathbf{r}_4.$$

Determine the representation of the C_4 subgroup for which this function is a basis.

5.6 Prove that, for each irreducible representation but one in any given group, the sum of the characters over all operations is zero.

5.7 For the system in Problem 5.2, determine the characters of the reducible representation of D_{4h} for which we have as bases (a) the $\mathbf{r}_j$ vectors, (b) the $\mathbf{s}_j$ vectors, (c) the $\mathbf{t}_j$ vectors. Let the planes for σ_v, σ_v' and the axes for $(C_2')_a$, $(C_2')_b$ pass through the chosen equilibrium positions. The classes in the group include I, $2C_4$, C_2, $2\sigma_v$, $2\sigma_d$, σ_h, $2S_4$, i, $2C_2'$, $2C_2''$.

5.8 For the square system treated in Problem 5.7, identify the irreducible representations (primitive symmetry species) that can be formed from (a) the $\mathbf{r}_j$ vectors, (b) the $\mathbf{s}_j$ vectors, (c) the $\mathbf{t}_j$ vectors.

5.9 For the square system of Problem 5.7, tabulate the effects of each symmetry operation on $\mathbf{r}_1$, $\mathbf{s}_1$, $\mathbf{t}_1$, and of I, C_2, i, σ_h on $\mathbf{r}_2$, $\mathbf{s}_2$, $\mathbf{t}_2$.

5.10 Use the results from Problem 5.9 to construct form $\sum_k y_k^{(r)} \mathscr{C}_k \mathbf{u}_l$ as in equation (3.12). Then choose normalization multiplier N so that the forms are normalized to one. Are these forms bases for irreducible representations of the $\mathbf{D}_{4h}$ group?

5.11 Out of the doubly occurring E_u bases in Problem 5.10, separate the translational modes from the vibrational modes.

5.12 From the appropriate reorientation matrices, deduce the representation

$$\begin{pmatrix} 1 & 0 & 0 \\ 0 & 1 & 0 \\ 0 & 0 & 1 \end{pmatrix}, \begin{pmatrix} -1 & 0 & 0 \\ 0 & -1 & 0 \\ 0 & 0 & 1 \end{pmatrix}, \begin{pmatrix} 1 & 0 & 0 \\ 0 & -1 & 0 \\ 0 & 0 & -1 \end{pmatrix}, \begin{pmatrix} -1 & 0 & 0 \\ 0 & 1 & 0 \\ 0 & 0 & -1 \end{pmatrix}$$

of the $\mathbf{D}_2$ group. From these matrices, construct three 1-dimensional representations. Also, write down the completely symmetric representation.

5.13 Show that C_n can be represented as multiplication by

$$\omega = e^{2\pi i(m/n)},$$

where m and n are integers. Under what circumstances can C_n be represented as multiplication by -1?

5.14 Six identical particles form a vibrating system with the symmetry of a regular hexagon. Pick out a set of equilibrium positions. Let these be numbered consecutively in the counterclockwise direction. On the jth one, erect a unit vector $\mathbf{r}_j$ pointing inward toward the center of the system. Then show that the vector

$$\mathbf{e} = \frac{1}{\sqrt{6}} (\mathbf{r}_1 + \omega \mathbf{r}_2 + \omega^2 \mathbf{r}_3 + \omega^3 \mathbf{r}_4 + \omega^4 \mathbf{r}_5 + \omega^5 \mathbf{r}_6),$$

where $\omega = e^{2\pi i/6}$, is a complete basis for an irreducible representation of the $\mathbf{C}_6$ subgroup. Also obtain the components of the character vector for this representation.

5.15 Four equivalent particles oscillate about alternate corners of a cube, the equilibrium positions forming a regular tetrahedron. Let the three cube edges from the back lower (occupied) corner be the x, y, z axes, with corresponding unit vectors $\mathbf{u}_1$, $\mathbf{u}_2$, $\mathbf{u}_3$. Erect similar unit vectors $\mathbf{u}_4$, $\mathbf{u}_5$, $\mathbf{u}_6$ along the cube edges drawn from the opposite corner of the base, with $\mathbf{u}_4$ pointing in the $-x$ direction, $\mathbf{u}_5$ in the $-y$ direction, and $\mathbf{u}_6$ in the z direction. Go to the top corner from which the edges proceed in the $-x$, y, $-z$ directions and label the corresponding unit vectors $\mathbf{u}_7$, $\mathbf{u}_8$, $\mathbf{u}_9$. From the opposite upper corner construct unit vectors $\mathbf{u}_{10}$, $\mathbf{u}_{11}$, $\mathbf{u}_{12}$ pointing in the x, $-y$, $-z$ directions. Locate axes for the C_2 symmetry operations on the tetrahedron (by dots on the cube) and label them a, b, c. Locate axes for the C_3 symmetry operations on the tetrahedron (by dots on the cube) and label them d, e, f, g.

5.16   Show that the vectors

$$\mathbf{e}_1 = \tfrac{1}{2}(\mathbf{u}_1 + \mathbf{u}_4 + \mathbf{u}_7 + \mathbf{u}_{10}),$$
$$\mathbf{e}_2 = \tfrac{1}{2}(\mathbf{u}_2 + \mathbf{u}_5 + \mathbf{u}_8 + \mathbf{u}_{11}),$$
$$\mathbf{e}_3 = \tfrac{1}{2}(\mathbf{u}_3 + \mathbf{u}_6 + \mathbf{u}_9 + \mathbf{u}_{12})$$

form a complete basis for a representation of the **T** subgroup for the system in Problem 5.15. Note that the classes of the group include I, $3C_2$, $4C_3$, $4C_3^2$.

5.17   Determine the characters of the reducible representation of **T** for which the $\mathbf{u}_j$ vectors in Problem 5.15 are bases.

5.18   Identify the irreducible representations that are based on linear combinations of the $\mathbf{u}_j$ vectors of Problem 5.15.

5.19   Generate bases for these irreducible representations of **T** from the $\mathbf{u}_j$ vectors of Problem 5.15.

5.20   Deduce the characters and the composition of the representation of which vectors $\mathbf{e}_1$, $\mathbf{e}_2$, $\mathbf{e}_3$ of Problem 5.16 are the basis. Then show how these vectors need to be combined to form bases for the constituent irreducible representations.

5.21   Show that, if we rewrote equations (5.2) and (5.3) as

$$C\mathbf{u}_k = \sum_j \Gamma_{kj}(C)\mathbf{u}_j,$$
$$D\mathbf{u}_j = \sum_i \Gamma_{ji}(D)\mathbf{u}_i,$$

the resulting $n \times n$ matrices would not combine in the same order as the group elements. Instead, we would have

$$\Gamma(DC) = \Gamma(C)\Gamma(D).$$

References

Books

Cornwell, J. F.: 1984, *Group Theory in Physics*, vol. 1, Academic Press, London, pp. 68–129. Modern mathematicians employ a style that outsiders find very difficult. Cornwell has tempered this style without relinquishing detail or rigor. In Chapters 4 and 5, he develops the theory of matrix representations, covering the foundations and bases for applications.

Duffey, G. H.: 1980, *Theoretical Physics: Classical and Modern Views*, Krieger, Melbourne, Fla., pp. 237-264. The emphasis here is on how matrix representations are generated from sets of operands.

Wigner, E. P. (translated by Griffin, J. J.): 1959, *Group Theory and Its Application to the Quantum Mechanics of Atomic Spectra*, Academic Press, London, pp. 58-66, 72-87, 102-123. This is the classic presentation of matrix representation theory for physicists.

Articles

Martinez-Torres, E., Lopez-Gonzalez, J. J., Fernandez-Gomes, M., and Cardenete-Espinosa, A.: 1989, "Icosahedral Matrix Representations as a Function of Eulerian Angles," *J. Chem. Educ.* **66**, 706-709.

Wolbarst, A. B.: 1977, "An Intuitive Approach to Group Representation Theory," *Am. J. Phys.* **45**, 803-810.

CHAPTER 6

Symmetry-Adapted Kets and Bras

6.1
Quantum Mechanical States

Each definite state of a quantum mechanical system is represented by a normalized vector in a Hilbert space generated by the independent variables. This space may be constructed so that the vector is a ket or, alternatively, a bra. The interactions, internal and external, to which the system is subjected are embodied in a Hamiltonian operator. This acts in the pertinent Hilbert space linearly. And each ket describing a possible energy state is an eigenket of the governing Hamiltonian operator.

Constructing the Hamiltonian operator is generally not too much of a problem. Solving the resulting eigenvalue equation generally is. As we have seen in other contexts, however, any symmetry present imposes very useful constraints.

Now, a quantum mechanical system possesses symmetry whenever there are distinct operations (such as reorientations or permutations) that do not alter its Hamiltonian operator. These operations together constitute a group. Subsets that are closed form smaller groups called subgroups.

The operations can be sorted into classes, as we have seen. Class sums can be formed and the corresponding eigenoperators defined. The eigenoperators common to the different class sums may act on any suitable set of basis kets (or bras). The corresponding eigenkets (or eigenbras) are then defined.

These eigenkets may also be formulated so that they are eigenkets of the Hamiltonian operator, since the Hamiltonian operator commutes with the class sums. Then they represent possible quantum mechanical states.

6.2
Symmetry Properties of the Eigenstates

Solving many quantum mechanical problems is aided by the fact that the energy eigenstates belong to primitive symmetry species of each covering group for the given system.

We consider a physical system with energy states determined by a Hamiltonian operator

$$H. \qquad (6.1)$$

We also suppose that the class sums for a group of symmetry operations for the system are

$$\mathscr{C}_1, \mathscr{C}_2, \ldots, \mathscr{C}_n. \qquad (6.2)$$

From the way classes are defined (recall Section 2.3), the class sums commute:

$$\mathscr{C}_j \mathscr{C}_k = \mathscr{C}_k \mathscr{C}_j. \qquad (6.3)$$

Because symmetry operations do not alter the Hamitonian, the Hamiltonian commutes with each class sum:

$$\mathscr{C}_j H = H \mathscr{C}_j. \qquad (6.4)$$

Any ket that has a representation in the space of the physical system is acted on by the linear operators in lines (6.1) and (6.2). Consequently, it is acted on by each linear combination

$$\mathscr{E}^{(r)} \qquad (6.5)$$

that is a common eigenoperator for the class sums. Each of the common eigenoperators transforms a given ket to either nothing or a ket that is a common unnormalized eigenket for the class sums. If $|f>$ is the original ket, we have

Symmetry Properties of the Eigenstates

$$\mathscr{C}_j \mathscr{E}^{(r)}|f\rangle = \lambda_j^{(r)} \mathscr{E}^{(r)}|f\rangle. \tag{6.6}$$

Formally, equation (6.6) is obtained by allowing each side of equation (3.8) to act on ket $|f\rangle$.

From equation (3.10), we find that

$$\mathscr{E}^{(r)}|f\rangle = \sum_k y_k^{(r)} \mathscr{C}_{\bar{k}}|f\rangle, \tag{6.7}$$

where $y_k^{(k)}$ is the kth component of character $\chi^{(r)}$ and $\mathscr{C}_{\bar{k}}$ is the sum of the inverses of the operators in $\mathscr{C}_k$. Ket (6.7) can be normalized when necessary.

Relationship (6.4) implies that H and the $\mathscr{C}_j$'s have a common set of eigenkets. Consequently, the kets satisfying

$$H|j\rangle = E|j\rangle \tag{6.8}$$

for any allowed energy E can be constructed so that they are common eigenkets for the class sums. Each of these common eigenkets, however, belongs to a definite eigenvalue

$$\lambda_1^{(r)}, \lambda_2^{(r)}, \ldots, \lambda_n^{(r)} \tag{6.9}$$

and so to a definite character vector

$$\chi^{(r)}. \tag{6.10}$$

Each eigenstate of a physical system correlates with one such vector; the eigenket belongs to the corresponding primitive symmetry species of the group.

The symmetry of a system may require the degeneracy of a given level to be d. Then d mutually orthogonal normalized eigenkets

$$|1\rangle, |2\rangle, \ldots, |d\rangle \tag{6.11}$$

can be constructed for the level. Each of these is said to define a *row* of the symmetry species and d is called the dimensionality of the species.

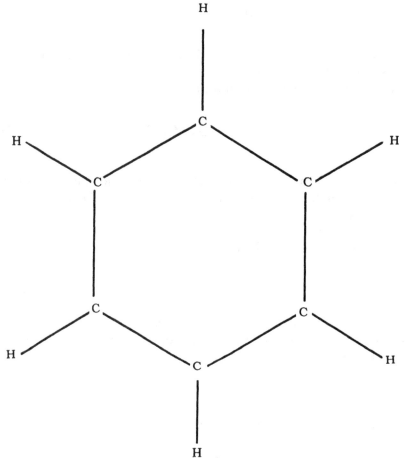

Figure 6.1 Arrangement of atoms and bonds in the benzene molecule.

Example 6.1

What group should a person employ in studying the benzene molecule?

The benzene molecule consists of 6 carbon atoms vibrating about the corners of a regular hexagon with 6 hydrogen atoms bound radially, on the average, as Figure 6.1 shows. Noting what bases for symmetry operations are present and using them in going down the flow chart of Figure 1.2, we find that the pertinent group is $\mathbf{D}_{6h}$. Describing fewer aspects of symmetry of the molecule are subgroups $\mathbf{C}_{6h}$, $\mathbf{D}_6$, $\mathbf{C}_{6v}$, $\mathbf{C}_6$, and so on.

On changing from a large group to a subgroup, the distinctions between some primitive symmetry species disappear. In a system with the lower symmetry, the eigenkets from combining species interact and shift apart.

On changing from the full group to a subgroup, on the other hand, some degenerate species become split into smaller species; eigenkets

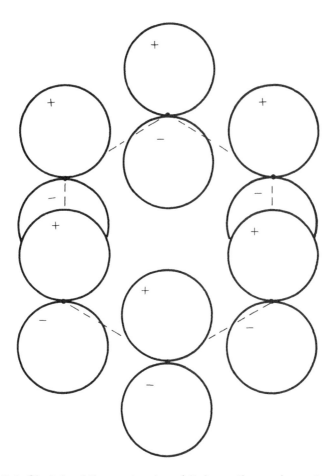

Figure 6.2 Sketch of the π atomic orbitals on the carbon atoms in the benzene molecule.

for the split species may then be employed for the degenerate one. Thus, eigenkets for different rows can be obtained without difficulty. This situation prevails with benzene; so we will construct species of its subgroup $\mathbf{C}_6$, rather than of the full group $\mathbf{D}_{6h}$.

Example 6.2

Construct a complete set of symmetry-adapted combinations of pi kets for the benzene molecule.

The atomic orbitals that combine to form π bonds in the benzene molecule are the real 2p carbon orbitals directed perpendicular to the plane of the ring, hybridized with higher atomic orbitals having the same pertinent symmetry properties. A sketch of these orbitals appears in Figure 6.2.

Let the ket for the resulting atomic orbital on the jth atom be $|j>$. Number the atoms as in Figure 6.3. Let each common eigenoperator $\mathscr{E}^{(r)}$ act on the first ket. Equation (6.7) then yields the common eigenket

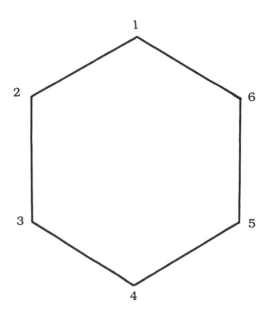

Figure 6.3 Numbering of positions on the benzene ring.

Symmetry Properties of the Eigenstates

$$\mathscr{E}^{(r)}|1\rangle = \sum_k y_k^{(r)} \mathscr{C}_k |1\rangle = \sum_k y_{\bar{k}}^{(r)} \mathscr{C}_{\bar{k}} |1\rangle.$$

The characters for the C_6 group appear in Table 6.1; the results of each operation acting on $|1\rangle$ appear in Table 6.2. Since each of the class sums contains only one operation, the entries in Table 6.2 are also the results of each class sum acting on $|1\rangle$.

Table 6.1 Components of the Character Vectors for the C_6 Group

	I	C_6	C_3	C_2	C_3^{-1}	C_6^{-1}
A	1	1	1	1	1	1
B	1	-1	1	-1	1	-1
C_1	1	$-\omega^2$	ω	-1	ω^2	$-\omega$
D_1	1	$-\omega$	ω^2	-1	ω	$-\omega^2$
C_2	1	ω	ω^2	1	ω	ω^2
D_2	1	ω^2	ω	1	ω^2	ω

($\omega = \exp(2\pi i/3)$)

Table 6.2 Effect of Each Operation of the C_6 Group on the Ket for the First π Orbital of Benzene

I	C_6	C_3	C_2	C_3^{-1}	C_6^{-1}
$|1\rangle$	$|2\rangle$	$|3\rangle$	$|4\rangle$	$|5\rangle$	$|6\rangle$

Substituting these results into the formula leads to the expression

$$|f\rangle = |1\rangle + |2\rangle + |3\rangle + |4\rangle + |5\rangle + |6\rangle$$

for symmetry species A. For symmetry species C_1, we obtain

$$|f\rangle = |1\rangle - \omega|2\rangle + \omega^2|3\rangle - |4\rangle + \omega|5\rangle - \omega^2|6\rangle.$$

For symmetry species D_1,

$$|f\rangle = |1\rangle - \omega^2|2\rangle + \omega|3\rangle - |4\rangle + \omega^2|5\rangle - \omega|6\rangle.$$

For symmetry species C_2,

$$|f\rangle = |1\rangle + \omega^2|2\rangle + \omega|3\rangle + |4\rangle + \omega^2|5\rangle + \omega|6\rangle.$$

For symmetry species D_2,

$$|f\rangle = |1\rangle + \omega|2\rangle + \omega^2|3\rangle + |4\rangle + \omega|5\rangle + \omega^2|6\rangle.$$

For symmetry species B,

$$|f\rangle = |1\rangle - |2\rangle + |3\rangle - |4\rangle + |5\rangle - |6\rangle.$$

The procedure could be repeated with $|2\rangle$, $|3\rangle$, $|4\rangle$, $|5\rangle$, and $|6\rangle$ replacing $|1\rangle$ in equation (6.7). But since we have already obtained six independent combinations, the maximum possible from six independent basis kets, no new results would be obtained. In many problems, however, more than one choice of $|f\rangle$ in equation (6.7) would be needed. Because the combinations we have constructed belong to distinct symmetry species, they describe completely distinct states and are mutually orthogonal.

Example 6.3

From the symmetry-adapted combinations that we have just found, construct expressions for the energy levels of the pi electrons in the benzene molecule.

Any acceptable solution of a Schrödinger equation for a physical system belongs to a single primitive symmetry species of each pertinent symmetry group. The solution cannot contain contributions from other species of the group. Consequently, the linear variation function for a particular secular equation need contain contributions from a single species alone.

Each of the kets that we have assembled belongs to a different species. Therefore, it need not be combined with any of the other kets; and the secular equation for it assumes the form

$$|H_{11} - S_{11}E| = 0$$

whence

Symmetry Properties of the Eigenstates

$$E = \frac{H_{11}}{S_{11}}$$

where

$$H_{11} = <f|H|f>$$

and

$$S_{11} = <f|f>.$$

Multiplying H_{11} and S_{11} out gives rise to sums of various products of atomic bras with atomic kets. Let us represent these products by lower case symbols:

$$h_{jk} = <j|H|k>$$

and

$$s_{jk} = <j|k>$$

where

$$j = 1, 2, 3, 4, 5, \text{ or } 6$$

and

$$k = 1, 2, 3, 4, 5, \text{ or } 6.$$

Because all pi kets in the benzene molecule are equivalent, different products occur only for different spacings between the atoms involved. Thus,

$$h_{12} = h_{23} = h_{34} = h_{45} = h_{56} = h_{61},$$
$$h_{13} = h_{24} = h_{35} = h_{46} = h_{51} = h_{62},$$
$$h_{14} = h_{25} = h_{36} = h_{41} = h_{52} = h_{63},$$

and similarly for the s_{jk}'s.

Substituting the first combination of Example 6.2 into

$$S_{11} = \langle f | f \rangle$$

yields the terms in Table 6.3. These add to give

$$S_{11} = 6(1 + 2s_{12} + 2s_{13} + s_{14}).$$

Substituting the first ket $|f\rangle$ and the corresponding bra $\langle f|$ into

$$H_{11} = |f|H|f\rangle$$

yields the terms in Table 6.4. These add to give

$$H_1 = 6(h_1 + 2h_{12} + 2h_{13} + h_{14}).$$

Dividing this H_{11} by the normalization integral S_{11}, as we noted before, leads to the energy expression

$$E = \frac{h_{11} + 2h_{12} + 2h_{13} + h_{14}}{1 + 2s_{12} + 2s_{13} + s_{14}}$$

for symmetry species A.

The four succeeding combinations in Example 6.2 contain complex coefficients. The corresponding bras must contain the complex conjugates of these coefficients. For symmetry species C_1, we thus obtain the multiplication Tables 6.5 and 6.6, from which

Table 6.3 Products of Individual Terms in the A Bra with Those in the A Ket

	$|1\rangle$	$|2\rangle$	$|3\rangle$	$|4\rangle$	$|5\rangle$	$|6\rangle$
$\langle 1|$	1	s_{12}	s_{13}	s_{14}	s_{13}	s_{12}
$\langle 2|$	s_{12}	1	s_{12}	s_{13}	s_{14}	s_{13}
$\langle 3|$	s_{13}	s_{12}	1	s_{12}	s_{13}	s_{14}
$\langle 4|$	s_{14}	s_{13}	s_{12}	1	s_{12}	s_{13}
$\langle 5|$	s_{13}	s_{14}	s_{13}	s_{12}	1	s_{12}
$\langle 6|$	s_{12}	s_{13}	s_{14}	s_{13}	s_{12}	1

Symmetry Properties of the Eigenstates

Table 6.4 Products of Individual Terms in the A Bra with H Times Those in the A Ket

	$H\|1>$	$H\|2>$	$H\|3>$	$H\|4>$	$H\|5>$	$H\|6>$
$<1\|$	h_{11}	h_{12}	h_{13}	h_{14}	h_{13}	h_{12}
$<2\|$	h_{12}	h_{11}	h_{12}	h_{13}	h_{14}	h_{13}
$<3\|$	h_{13}	h_{12}	h_{11}	h_{12}	h_{13}	h_{14}
$<4\|$	h_{14}	h_{13}	h_{12}	h_{11}	h_{12}	h_{13}
$<5\|$	h_{13}	h_{14}	h_{13}	h_{12}	h_{11}	h_{12}
$<6\|$	h_{12}	h_{13}	h_{14}	h_{13}	h_{12}	h_{11}

Table 6.5 Products of Individual Terms in the C_1 Bra with Those in the C_1 Ket

	$\|1>$	$-\omega\|2>$	$\omega^2\|3>$	$-\|4>$	$\omega\|5>$	$-\omega^2\|6>$
$<1\|$	1	$-\omega s_{12}$	$\omega^2 s_{13}$	$-s_{14}$	ωs_{13}	$-\omega^2 s_{12}$
$-\omega^2<2\|$	$-\omega^2 s_{12}$	1	$-\omega s_{12}$	$\omega^2 s_{13}$	$-s_{14}$	ωs_{13}
$\omega<3\|$	ωs_{13}	$-\omega^2 s_{12}$	1	$-\omega s_{12}$	$\omega^2 s_{13}$	$-s_{14}$
$-<4\|$	$-s_{14}$	ωs_{13}	$-\omega^2 s_{12}$	1	$-\omega s_{12}$	$\omega^2 s_{13}$
$\omega^2<5\|$	$\omega^2 s_{13}$	$-s_{14}$	ωs_{13}	$-\omega^2 s_{12}$	1	$-\omega s_{12}$
$-\omega<6\|$	$-\omega s_{12}$	$\omega^2 s_{13}$	$-s_{14}$	ωs_{13}	$-\omega^2 s_{12}$	1

Table 6.6 Products of Individual Terms in the C_1 Bra with H Times Those in the C_1 Ket

	$H\|1>$	$-\omega H\|2>$	$\omega^2 H\|3>$	$-H\|4>$	$\omega H\|5>$	$-\omega^2 H\|6>$
$<1\|$	h_{11}	$-\omega h_{12}$	$\omega^2 h_{13}$	$-h_{14}$	ωh_{13}	$-\omega^2 h_{12}$
$-\omega^2<2\|$	$-\omega^2 h_{12}$	h_{11}	$-\omega h_{12}$	$\omega^2 h_{13}$	$-h_{14}$	ωh_{13}
$\omega<3\|$	ωh_{13}	$-\omega^2 h_{12}$	h_{11}	$-\omega h_{12}$	$\omega^2 h_{13}$	$-h_{14}$
$-<4\|$	$-h_{14}$	ωh_{13}	$-\omega^2 h_{12}$	h_{11}	$-\omega h_{12}$	$\omega^2 h_{13}$
$\omega^2<5\|$	$\omega^2 h_{13}$	$-h_{14}$	ωh_{13}	$-\omega^2 h_{12}$	h_{11}	$-\omega h_{12}$
$-\omega<6\|$	$-\omega h_{12}$	$\omega^2 h_{13}$	$-h_{14}$	ωh_{13}	$-\omega^2 h_{12}$	h_{11}

$$S_{11} = 6(1 + s_{12} - s_{13} - s_{14})$$

and

$$H_{11} = 6(h_{11} + h_{12} - h_{13} - h_{14}).$$

The resulting energy expression is

$$E = \frac{h_{11} + h_{12} - h_{13} - h_{14}}{1 + s_{12} - s_{13} - s_{14}},$$

for symmetry species C_1. Since $|f>$ and $<f|$ for D_1 are the complex conjugates of $|f>$ and $<f|$ for C_1, symmetry species D_1 yields the same terms in the multiplication tables and the same expression for energy E.

On constructing the pertinent multiplication tables and adding up the results, we similarly obtain

$$E = \frac{h_{11} - h_{12} - h_{13} + h_{14}}{1 - s_{12} - s_{13} + s_{14}}$$

for symmetry species C_2 and D_2. For symmetry species B, we obtain

$$E = \frac{h_{11} - 2h_{12} + 2h_{13} - h_{14}}{1 - 2s_{12} + 2s_{13} - s_{14}}.$$

6.3
Matrix Elements for Scalar Physical Operators

Products of bras with transformed kets appear in secular equations and in perturbation formulas. Whenever a system possesses some symmetry, however, bases can be chosen so that some of these products vanish. The equations and formulas are thereby simplified—in many cases, tremendously.

For discussion purposes, let us continue with the system and group described in Section 6.2. Suppose that a physical operator for which matrix elements are needed is

$$A. \qquad (6.12)$$

Let us suppose that A is not affected by operations of the group. Then the sum in the rth common eigenoperator has no effect:

$$\mathscr{E}^{(r)} A = A \mathscr{E}^{(r)}. \qquad (6.13)$$

Matrix Elements for Scalar Physical Operators 173

Let a ket describing certain aspects of the system be

$$|k\rangle. \tag{6.14}$$

To $|k\rangle$, we add mutually orthogonal similar kets until a complete basis is constructed. Operator A acting on $|k\rangle$ then produces a linear combination of kets from this complete set:

$$A|k\rangle = \sum_{l} |l\rangle A_{lk}. \tag{6.15}$$

Multiplying equation (6.15) by bra $\langle j|$, with the vectors adjusted so that

$$\langle j|k\rangle = \delta_{jk}, \tag{6.16}$$

makes

$$\langle j|A|k\rangle = A_{jk}. \tag{6.17}$$

Expressions $\langle j|A|k\rangle$ may be arranged in a square array that combines with other such arrays as a matrix; so A_{jk} is called the jkth matrix element of A.

Operating on equation (6.15) with the eigenoperator $\mathscr{E}^{(r)}$ leads to

$$A\mathscr{E}^{(r)}|k\rangle = \sum_{j} \mathscr{E}^{(r)}|j\rangle A_{jk} \tag{6.18}$$

if A is commuted with $\mathscr{E}^{(r)}$ as in equation (6.13). From the argument in Section 6.2, eigenket $\mathscr{E}^{(r)}|k\rangle$ and all nonvanishing $\mathscr{E}^{(r)}|j\rangle$'s in equation (6.18) belong to a definite primitive symmetry species. The action of A on a ket of a certain kind, as in equation (6.15), has produced a linear combination of kets of the same kind. There are *no* contributions from other species. So when $|k\rangle$ in equation (6.15) belongs to a definite species, A_{lk} must vanish for each $|l\rangle$ of a different species. We have

$$A_{jl} = 0 \tag{6.19}$$

whenever $|j\rangle$ and $|l\rangle$ belong to different primitive symmetry species, as long as operator A is not altered by operations of the group.

The kets can also be classified into rows, as indicated at the end of Section 6.2. If $|k\rangle$ defines a row of a definite symmetry species, then

$$A|k\rangle \qquad (6.20)$$

belongs to the same row, as long as A is not affected by operations of the group. Then since kets in different rows are mutually orthogonal, we have

$$A_{jl} = 0 \qquad (6.21)$$

when $|j\rangle$ and $|l\rangle$ belong to different rows of the same symmetry species.

Example 6.4

If the symmetry of a system requires zeros to appear in the matrix for its Hamiltonian H only in the indicated positions,

$$\begin{pmatrix} H_{11} & 0 & H_{13} & 0 & 0 & H_{16} \\ 0 & H_{22} & 0 & H_{24} & H_{25} & 0 \\ H_{31} & 0 & H_{33} & 0 & 0 & H_{36} \\ 0 & H_{42} & 0 & H_{44} & H_{45} & H_{46} \\ 0 & H_{52} & 0 & H_{54} & H_{55} & 0 \\ H_{61} & 0 & H_{63} & H_{64} & 0 & H_{66} \end{pmatrix}$$

when basis kets $|1\rangle, |2\rangle, |3\rangle, |4\rangle, |5\rangle$, and $|6\rangle$ are employed, what can a person say about the symmetry species present in the basis kets?

Because the Hamiltonian operator H is not affected by any operations of a covering group, ket $H|j\rangle$ belongs to the same symmetry species and row as $|j\rangle$. And since the products of $\langle 1|$ with $H|3\rangle$ and $H|6\rangle$ may differ from zero, then $|1\rangle, |3\rangle$, and $|6\rangle$ contain contributions from the same symmetry species and row. Let us label this A.

Since the products of $\langle 2|$ with $H|4\rangle$ and $H|5\rangle$ may differ from zero, then $|2\rangle, |4\rangle$, and $|5\rangle$ contain contributions from the same symmetry species and row. Let us label this C. But since the products of $\langle 1|$ with $H|2\rangle, H|4\rangle$, and $H|5\rangle$ are zero, then C must differ in species or row from A. Because $\langle 4|H|6\rangle$ differs from zero, then $|4\rangle$ and $|6\rangle$

both contain contributions from a symmetry species or row different from A and C. Let us label this B.

6.4
Generating Representations of Groups with Kets

In Section 5.2, the bases for generating matrix representations of groups were taken to be generalized vectors. These bases, however, may be functions of the coordinates for a given system. Since such functions are representations of kets, appropriate kets (or bras) may serve as bases.

Consider a physical system for which certain aspects are described by the linearly independent kets

$$|1>, |2>, \ldots, |n>. \tag{6.22}$$

Furthermore, suppose that these form a complete set in the sense that each operation in the covering group to be employed transforms any one of them into a linear combination of members of the set.

If C and D are operations in the group under consideration, we then have

$$C|k> = \sum_j |j> \Gamma_{jk}(C) \tag{6.23}$$

and

$$D|j> = \sum_i |i> \Gamma_{ij}(D). \tag{6.24}$$

Here $\Gamma_{jk}(C)$ is the coefficient of $|j>$ in the superposition produced when C acts on $|k>$; $\Gamma_{ij}(D)$ is the number multiplying $|i>$ in the expansion generated when operation D acts on $|j>$. Similarly,

$$DC|k> = \sum_i |i> \Gamma_{ik}(DC). \tag{6.25}$$

But combining equations (6.23) and (6.24) yields

$$DC|k> = \sum_{i,j} |i> \Gamma_{ij}(D) \Gamma_{jk}(C). \tag{6.26}$$

Since the left sides of equations (6.25) and (6.26) are the same, the right sides are identically equal. As a consequence, the coefficients of $|i\rangle$ must be equal:

$$\Gamma_{ik}(DC) = \sum_j \Gamma_{ij}(D)\Gamma_{jk}(C), \tag{6.27}$$

whence

$$\mathbf{\Gamma}(DC) = \mathbf{\Gamma}(D)\mathbf{\Gamma}(C). \tag{6.28}$$

The $n \times n$ matrices constructed from the coefficients combine as the corresponding elements of the group do and so they form an explicit representation of the group.

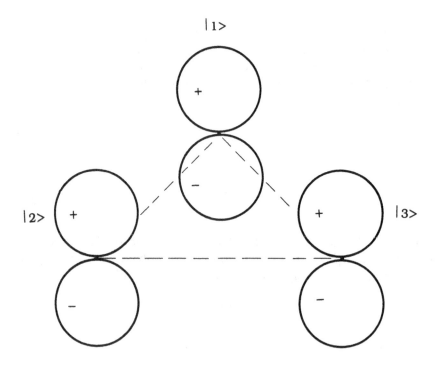

Figure 6.4 Pi orbitals from three equivalent atoms arranged symmetrically in a ring.

Example 6.5

Show how three equivalent pi atomic orbitals at the corners of an equilateral triangle generate a representation for the **C**$_3$ group.

First label the ket for each orbital as in Figure 6.4. Then, determine the effect of each operation of the group on each ket:

$$I|j\rangle = |j\rangle,$$
$$C_3|j\rangle = |j + 1\rangle,$$
$$C_3{}^2|j\rangle = |j + 2\rangle.$$

Here numbers 1, 2, 3 appear in cyclic order, with 2 following 1, 3 following 2, and 1 following 3.

These equations are special cases of equations (6.23) and (6.24). The corresponding matrix relationship for the identity operation is

$$I(|1\rangle \ |2\rangle \ |3\rangle) = (|1\rangle \ |2\rangle \ |3\rangle)\begin{pmatrix} 1 & 0 & 0 \\ 0 & 1 & 0 \\ 0 & 0 & 1 \end{pmatrix}.$$

From the effects of C_3, we similarly construct the matrix equation

$$C_3 (|1\rangle \ |2\rangle \ |3\rangle) = (|1\rangle \ |2\rangle \ |3\rangle)\begin{pmatrix} 0 & 0 & 1 \\ 1 & 0 & 0 \\ 0 & 1 & 0 \end{pmatrix},$$

and from the effects of $C_3{}^2$, we obtain

$$C_3{}^2(|1\rangle \ |2\rangle \ |3\rangle) = (|1\rangle \ |2\rangle \ |3\rangle)\begin{pmatrix} 0 & 1 & 0 \\ 0 & 0 & 1 \\ 1 & 0 & 0 \end{pmatrix}.$$

Each matrix of coefficients behaves algebraically as the corresponding operation of the group. Thus, we find that the expressions

$$\begin{pmatrix} 1 & 0 & 0 \\ 0 & 1 & 0 \\ 0 & 0 & 1 \end{pmatrix}, \begin{pmatrix} 0 & 0 & 1 \\ 1 & 0 & 0 \\ 0 & 1 & 0 \end{pmatrix}, \begin{pmatrix} 0 & 1 & 0 \\ 0 & 0 & 1 \\ 1 & 0 & 0 \end{pmatrix}$$

combine in multiplication as do I, C_3, C_3^2. Consequently, these matrices represent the group $\mathbf{C}_3$. The kets $|1>$, $|2>$, and $|3>$ form a basis for the representation.

Example 6.6

Show how the combinations

$$|\mathrm{I}> = \frac{\sqrt{2}}{\sqrt{3}}(|1> - \tfrac{1}{2}|2> - \tfrac{1}{2}|3>),$$

$$|\mathrm{II}> = \frac{1}{\sqrt{2}}(|2> - |3>)$$

of π orbitals from Figure 6.4 generate a representation of the $\mathbf{C}_3$ group.

Since the identity operator leaves each pi orbital unchanged, we have

$$I|\mathrm{I}> = |\mathrm{I}>,$$
$$I|\mathrm{II}> = |\mathrm{II}>.$$

The second formula in Example 6.5, together with the definitions of $|\mathrm{I}>$ and $|\mathrm{II}>$, tells us that

$$C_3|\mathrm{I}> = C_3 \frac{\sqrt{2}}{\sqrt{3}}(|1> - \tfrac{1}{2}|2> - \tfrac{1}{2}|3>) = \frac{\sqrt{2}}{\sqrt{3}}(|2> - \tfrac{1}{2}|3> - \tfrac{1}{2}|1>)$$

$$= -\frac{1}{2}\frac{\sqrt{2}}{\sqrt{3}}(|1> - \tfrac{1}{2}|2> - \tfrac{1}{2}|3>) + \frac{\sqrt{3}}{2}\frac{1}{\sqrt{2}}(|2> - |3>)$$

$$= -\frac{1}{2}|\mathrm{I}> + \frac{\sqrt{3}}{2}|\mathrm{II}>,$$

Generating Representations of Groups with Kets

$$C_3|II\rangle = C_3 \frac{1}{\sqrt{2}}(|2\rangle - |3\rangle) = \frac{1}{\sqrt{2}}(|3\rangle - |1\rangle)$$

$$= -\frac{\sqrt{3}}{2}\frac{\sqrt{2}}{\sqrt{3}}(|1\rangle - \tfrac{1}{2}|2\rangle - \tfrac{1}{2}|3\rangle) - \frac{1}{2}\frac{1}{\sqrt{2}}(|2\rangle - |3\rangle)$$

$$= -\frac{\sqrt{3}}{2}|I\rangle - \frac{1}{2}|II\rangle.$$

From the third formula in Example 6.5, we similarly find that

$$C_3^2|I\rangle = -\frac{1}{2}|I\rangle - \frac{\sqrt{3}}{2}|II\rangle,$$

$$C_3^2|II\rangle = \frac{\sqrt{3}}{2}|I\rangle - \frac{1}{2}|II\rangle.$$

These equations are special instances of equations (6.23) and (6.24). The corresponding matrix relationships are

$$I(|I\rangle \ |II\rangle) = (|I\rangle \ |II\rangle)\begin{pmatrix} 1 & 0 \\ 0 & 1 \end{pmatrix},$$

$$C_3(|I\rangle \ |II\rangle) = (|I\rangle \ |II\rangle)\begin{pmatrix} -\frac{1}{2} & -\frac{\sqrt{3}}{2} \\ \frac{\sqrt{3}}{2} & -\frac{1}{2} \end{pmatrix},$$

$$C_3^2(|I\rangle \ |II\rangle) = (|I\rangle \ |II\rangle)\begin{pmatrix} -\frac{1}{2} & \frac{\sqrt{3}}{2} \\ -\frac{\sqrt{3}}{2} & -\frac{1}{2} \end{pmatrix}.$$

Each of the square matrices behaves as the corresponding operation of the group. Indeed, the expressions

$$\begin{pmatrix} 1 & 0 \\ 0 & 1 \end{pmatrix}, \begin{pmatrix} -\frac{1}{2} & -\frac{\sqrt{3}}{2} \\ \frac{\sqrt{3}}{2} & -\frac{1}{2} \end{pmatrix}, \begin{pmatrix} -\frac{1}{2} & \frac{\sqrt{3}}{2} \\ -\frac{\sqrt{3}}{2} & -\frac{1}{2} \end{pmatrix}$$

combine in multiplication as do I, C_3, C_3^2 and so represent the group C_3. Kets $|I\rangle$ and $|II\rangle$ form the basis for this representation.

Example 6.7

Show how the combination

$$|IV\rangle = \frac{1}{\sqrt{3}}(|1\rangle + \omega|2\rangle + \omega^2|3\rangle)$$

of π orbitals from Figure 6.4 generates a representation of the C_3 group.

Applying the first three formulas in Example 6.5 leads to

$$I|IV\rangle = |IV\rangle,$$

$$C_3|IV\rangle = \frac{1}{\sqrt{3}}(|2\rangle + \omega|3\rangle + \omega^2|1\rangle)$$

$$= \omega^2 \frac{1}{\sqrt{3}}(|1\rangle + \omega|2\rangle + \omega^2|3\rangle)$$

$$= |IV\rangle\, \omega^2,$$

$$C_3^2|IV\rangle = \frac{1}{\sqrt{3}}(|3\rangle + \omega|1\rangle + \omega^2|2\rangle)$$

$$= \omega \frac{1}{\sqrt{3}}(|1\rangle + \omega|2\rangle + \omega^2|3\rangle)$$

$$= |IV\rangle\, \omega.$$

Each of these equations by itself fits equation (6.23), with just one element in the corresponding Γ. The transformation matrices (the Γ's) consist of

$$(1), \quad (\omega^2), \quad (\omega).$$

Since these combine in multiplication as do

$$I, \ C_3, \ C_3^2,$$

they represent the group $\mathbf{C}_3$. Ket $|IV\rangle$ is the basis for this representation. Because this representation is 1-dimensional, it is irreducible.

Similarly, the ket

$$|V\rangle = \frac{1}{\sqrt{3}} (|1\rangle + \omega^2|2\rangle + \omega|3\rangle)$$

yields the irreducible representation

$$(1), \quad (\omega), \quad (\omega^2).$$

And the ket

$$|VI\rangle = \frac{1}{\sqrt{3}} (|1\rangle + |2\rangle + |3\rangle)$$

is the basis for the irreducible representation

$$(1), \quad (1), \quad (1).$$

6.5
Reduction of Symmetry Induced by Degeneracy

In a system in which a degenerate level is only partially occupied, stability tends to be increased by introducing a distortion (or interaction) that removes the degeneracy and splits the level. This is because the split is symmetric to a first approximation and the particle, or particles, may shun the resulting higher levels, occupying only the resulting lower ones. The gain in stability here tends to more than cancel the loss of stability in other occupied levels, as long as the distortion is not too large.

As an example, consider the triangular cyclopropenyl radical. The most symmetric structure that it may assume forms a basis for the $\mathbf{D}_{3h}$ group. Its pi atomic orbitals would combine as Example 6.7 indicates.

The lowest pi molecular orbital, which would be occupied by a pair of electrons, is then

$$|\text{I}\rangle = \frac{1}{\sqrt{3}} (|1\rangle + |2\rangle + |3\rangle). \tag{6.29}$$

The two complex orbitals in Example 6.7 would be occupied by a single electron. These may be combined to yield the orthogonal real orbitals.

$$|\text{II}\rangle = \frac{1}{\sqrt{6}} (2|1\rangle - |2\rangle - |3\rangle) \tag{6.30}$$

and

$$|\text{III}\rangle = \frac{1}{\sqrt{2}} (|2\rangle - |3\rangle). \tag{6.31}$$

Since the complex orbitals are degenerate, this rearrangement does not affect the calculated energy, as long as the $\mathbf{D}_{3h}$ structure is maintained. But the degeneracy may be removed by distorting the radical to an isosceles triangle form. Thus, the distance between atoms 2 and 3 may be decreased somewhat, while the distances between atoms 1 and 2 and between atoms 1 and 3 may be increased by the same amount.

The resulting structure is covered by operations of the $\mathbf{C}_{2v}$ group. Orbital (6.29) now belongs to primitive species B_2. The complex orbitals, on the other hand, contain contributions from both species B_2 and A_2. But combination (6.30) is pure B_2, while combination (6.31) is pure A_2. And, eigenstates of the Hamiltonian operator must belong to primitive symmetry species.

The relative phasing of the π orbitals is depicted in Figure 6.5. From this, we see that the splitting of the degenerate level causes the level for expression (6.30) to move down and that for expression (6.31) to move up. Furthermore, the distortion causes the center for all the levels to move up somewhat. Also, the interaction between the two B_2 levels causes them to move apart further. But because the originally degenerate level was not fully occupied, distortion by a moderate amount leads to a lower total energy.

This effect was first discussed by H. A. Jahn and Edward Teller. It

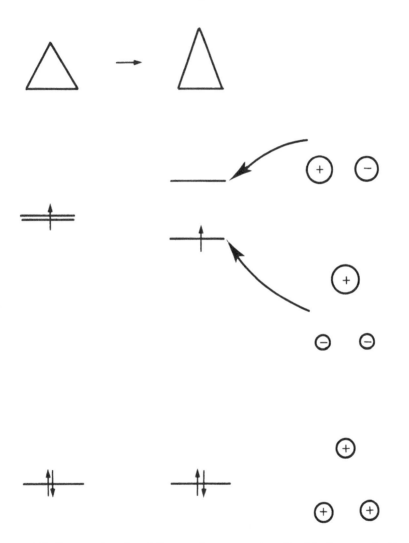

Figure 6.5 Energy levels of the cyclopropenyl radical before and after distortion from the most symmetric configuration, from a $\mathbf{D}_{3h}$ to a $\mathbf{C}_{2v}$ arrangement. The relative phasing of the π orbitals is indicated on the right.

is summarized in the statement: If in a certain nonlinear configuration, the given state for a system is degenerate, then a distortion that destroys the symmetry causing the degeneracy will generally yield a more stable state. The resulting change is called a *Jahn-Teller distortion*.

6.6
Summary

A coherent state of a quantum mechanical system is described by a ket $|g\rangle$. A property of the state is obtainable from the ket through action of an appropriate operator A in the expression

$$\frac{\langle g|A|g\rangle}{\langle g|g\rangle}. \qquad (6.32)$$

Factor $\langle g|$ is the bra corresponding to the ket $|g\rangle$, while $|g\rangle$ itself arises as a linear combination of orthogonal eigenkets satisfying the equation

$$A|j\rangle = a|j\rangle. \qquad (6.33)$$

Any group whose elements act on these eigenkets while leaving operator A unaltered has only class sums $\mathscr{C}_1, \mathscr{C}_2, \ldots, \mathscr{C}_n$ that commute with the operator. The class sums also commute with each other. So the $\mathscr{C}_k$'s and A have a common set of eigenkets and the kets satisfying equation (6.33) fall into sets characterized by the tabulated character vectors. Indeed, each eigenket $|j\rangle$ belongs to a single primitive symmetry species of the group.

Any ket $|f\rangle$ that determines a function in the given physical space is presumably acted on by operator A and by the class sums. Consequently, it is acted on by each common eigenoperator $\mathscr{E}^{(r)}$ for the class sums:

$$\mathscr{E}^{(r)}|f\rangle = \sum_k y_k^{(r)} \mathscr{C}_{\bar{k}}|f\rangle. \qquad (6.34)$$

Here $y_k^{(r)}$ is the kth component of the character vector for the rth primitive symmetry species and $\mathscr{C}_{\bar{k}}$ is the class sum containing the inverses of the operations in $\mathscr{C}_k$.

Ket (6.34) and the independent kets obtained by sustituting each $G|f>$ for $|f>$ as operand form a set to which the rth primitive symmetry species may contribute more than once. (Here G is any operation of the group.) From the way operations of the group transform members of the set among themselves, a person may determine how to combine members linearly to get independent orthogonal sets belonging to the symmetry species.

In a linear-variation calculation, a person need combine only ket vectors from the same row of a symmetry species. Possible vectors with other symmetry properties cannot contribute to the eigenket that is being approximated.

Destroying or breaking a symmetry that caused a level to be multiple splits the level. To a first approximation, the center of the level shifts but a small amount. If the level was partially filled, the constituent particles need occupy only the lower levels. Then energy evolution would accompany the symmetry breaking.

Discussion Questions

6.1 What attributes must a ket possess to be acted on (a) by the Hamiltonian operator for a given system, (b) by the class sums for a given group?

6.2 Into what does a common eigenoperator of the class sums transform a ket?

6.3 Why do the eigenkets of the Hamiltonian operator belong to the primitive symmetry species of the relevant groups?

6.4 How may the row of a symmetry species be defined?

6.5 Under what conditions may it be expedient not to employ the full symmetry group for a system?

6.6 When does

$$\sum_k y_{\bar{k}}^{(r)} \mathscr{C}_k | f >$$

not yield a ket belonging to the rth primitive symmetry species? How does one then construct such a ket?

6.7 Why does a linear variation function only need to contain contributions from one primitive symmetry species? Why does it need to contain contributions from only one row of a degenerate species?

6.8 How can a person systematize calculations of H_{jk} and S_{jk} when the pertinent kets are symmetry adapted combinations of certain bases?

6.9 What behavior characterizes the atomic orbitals that combine to form the π bonds in the benzene molecule?

6.10 In what approximation does the interaction between the pi atomic orbitals of the benzene molecule split the level of the atomic orbitals symmetrically, so that if all of them were doubly occupied there would be no net shift of energy?

6.11 When does the operation of A on a ket belonging to a given symmetry species produce a ket belonging to the same species? Is the Hamiltonian such an operator?

6.12 When is matrix element A_{jl} (a) equal to zero, (b) different from zero?

6.13 Under what circumstances does a secular equation factor?

6.14 What properties must a set of kets have to serve as a basis for a representation of a group?

6.15 How does such a set generate the representation?

6.16 When is such a representation reducible?

6.17 What can cause a multidimensional species to become split?

6.18 How can reduction in the symmetry of a system be a spontaneous process?

Problems

6.1 Consider the benzene π orbitals obtained in Example 6.2. Combine the complex ones for C_1 and D_1 to form two orthogonal real molecular orbitals. Where are the pertinent nodal planes in these orbitals?

6.2 Calculate the energy levels for the real molecular orbitals constructed in Problem 6.1.

6.3 An H atom is placed in a field that is transformed into itself by all operations of the D_2 group. Determine the primitive symmetry species of D_2 to which the atom's ns, np, and nd orbitals belong. Do any of the orbitals need to remain degenerate in the field?

6.4 A field that is transformed into itself only by the operations of C_4 is imposed on an atom. What np and nd orbitals remain as eigenfunctions of the Hamiltonian H? To what symmetry species do these belong?

6.5 Construct a complete set of symmetry-adapted combinations of the pi kets in cyclobutadiene

$$\begin{array}{c} H\diagdown C = C \diagup H \\ | \bigcirc | \\ \diagup C = C \diagdown \\ H H \end{array}$$

6.6 From the combinations found in Problem 6.5, construct expressions for the energy levels of the pi electrons in the cyclobutadiene molecule.

6.7 Show how the original pi kets in Problem 6.5 generate a representation of the C_4 group.

6.8 Using the pertinent character components, calculate the number of times each primitive symmetry species of C_4 contributes to the representation in Problem 6.7.

6.9 From the combinations given in Example 6.7, construct expressions for the energy levels of the pi electrons in the cyclopropenyl radical.

6.10 Show how the complex cyclopropenyl radical orbitals of Example 6.7 are combined to yield the orthogonal real orbitals of equations (6.30) and (6.31).

6.11 Construct expressions for the energy levels of the distorted cyclopropenyl radical in the approximation that the two A_1 orbitals do not interact.

6.12 The standard angular factors for the s, p, and d orbitals of a hydrogenlike atom are

Y_0^0	Y_1^0	$Y_1^{\pm 1}$	Y_2^0	$Y_2^{\pm 1}$	$Y_2^{\pm 2}$
1	$\dfrac{z}{r}$	$\dfrac{x \pm iy}{r}$	$\dfrac{3z^2 - r^2}{r^2}$	$\dfrac{(x \pm iy)z}{r^2}$	$\dfrac{(x \pm iy)^2}{r^2}$

Find the effects of each operation of D_3 on each of these factors. Then determine the primitive symmetry species of D_3 to which an atom's s, p, and d orbitals belong.

6.13 Show how the arguments in Problem 6.12 are affected when the field to which the atom is exposed is reduced from $\mathbf{D}_3$ to $\mathbf{C}_3$ symmetry.

6.14 Construct all symmetry-adapted combinations of the pi kets in the cyclopentadienyl radical

6.15 From the combinations found in Problem 6.14, construct expressions for the energy levels of the pi electrons in the cyclopentadienyl radical.

6.16 Show how the original pi kets in Problem 6.14 generate a matrix representation of the $\mathbf{C}_5$ group.

6.17 Using the pertinent character components, calculate the number of times each primitive symmetry species of $\mathbf{C}_5$ contributes to the representation in Problem 6.16.

6.18 To the three equivalent pi orbitals placed at the corners of an equilateral triangle, a fourth pi orbital can be added at the center. An example is provided by the trimethylenemethane molecule (or diradical)

From these orbitals, construct a complete set of symmetry-adapted combinations of pi kets.

6.19 Construct and solve the secular equation for the two A combinations found in Problem 6.18. For simplicity, consider that $s_{jk} = 0$ when $j \neq k$ and that $h_{12} = h_{23} = h_{31} = 0$.

References

Books

Douglas, B. E., and Hollingsworth, C. A.: 1985, *Symmetry in Bonding and Spectra*, Academic Press, Orlando, pp. 109-253.
After reviewing the pertinent group theory, Douglas and Hollingsworth turn to applications. Symmetry properties of the Hamiltonian, atomic orbitals, spectral terms, sigma bonding, pi bonding, multicenter bonding, ligand field theory are all considered in detail. Numerous helpful figures are included.

Duffey, G. H.: 1984, *A Development of Quantum Mechanics Based on Symmetry Considerations*, D. Reidel, Dordrecht, pp. 63-72.
Simple symmetry arguments are used to get the Θ (θ) function from the $\Phi(\phi)$ function for an atomic orbital and for a rotational eigenfunction.

Nussbaum, A.: 1971, *Applied Group Theory for Chemists, Physicists and Engineers*, Prentice-Hall, Englewood Cliffs, N.J., pp. 191-232.
In Chapter 4, Nussbaum applies group theory to atomic and molecular structures. Molecules of water, methane, and singly-ionized hydrogen are considered in some detail.

Articles

Hsu, C.-Y., and Orchin, M.: 1973, "A Simple Method for Generating Sets of Orthonormal Hybrid Atomic Orbitals," *J. Chem. Educ.* **50**, 114-118.

Wolbarst, A. B.: 1979, "An Intuitive Approach to Group Representation Theory II," *Am. J. Phys.* **47**, 103-112.

CHAPTER 7 / *Combinations of Products of Bases*

7.1
Significant Combinations

As they stand, the Cartesian coordinates x, y, z are basis functions for a representation of any given subgroup of the full rotation group about the origin. These coordinates can be combined to form homogeneous polynomials of any integral degree l. Particular combinations are basis functions for representations of the chosen subgroup of the full rotation group. Coefficients in such polynomials can be adjusted so that the corresponding representation is irreducible. Dividing the result by r^l yields an angular factor $Y(\theta, \phi)$ for an atomic orbital.

Possible orbitals for single particles, and similar geometric or physical structures, are represented by kets. Since probabilities combine multiplicatively, multiparticle systems in which the constituent particles are in definite states are described by products of kets. The general situation is represented by a superposition of the allowable products.

The symmetry conditions imposed on the system apply to this superposition. When the combination belongs to a particular symmetry species, the symmetry species of the constituent kets are related in a manner to be investigated here.

A submicroscopic particle generally exhibits a spin angular momentum with respect to any chosen axis. The particle may also exhibit an orbital angular momentum about an axis of symmetry. In both single and multiple particle systems, we need to consider how these angular momenta may combine.

Products of bras with transformed kets appear in matrix elements. The operator transforming the ket need not be completely symmetric, as a scalar is. Instead, it may be a vector or a tensor operator. The behavior of pertinent combinations can be linked to how angular momenta combine.

7.2
Product Representations

The operations of a given group transform the bases in suitable sets linearly. As a consequence, products of single bases in one such set with single bases in another such set are also transformed linearly. The coefficients governing the combined transformations form a representation whose character is derivable from the characters of the representations for the sets being combined.

Suppose that a complete set of bases for one representation of a given group is

$$|1,1>, |1,2>, \ldots, |1,r>, \tag{7.1}$$

while a complete set of bases for a second representation of the group consists of

$$|2,1>, |2,2>, \ldots, |2,s>. \tag{7.2}$$

Since each operation G of the group transforms each basis ket linearly, from the way bases are constructed, we have

$$G|1,l> = \sum_j |1,j> \Gamma_{jl}^{(1)}(G), \tag{7.3}$$

$$G|2,m> = \sum_k |2,k> \Gamma_{km}^{(2)}(G). \tag{7.4}$$

When G acts on the product $|1,l>|2,m>$, the result is the product of the right sides of equations (7.3) and (7.4); thus

$$\begin{aligned}
G|1, l>|2, m> &= \sum_j |1,j> \Gamma_{jl}^{(1)}(G) \sum_k |2, k> \Gamma_{km}^{(2)}(G) \\
&= \sum_{j,k} |1,j>|2,k> \Gamma_{jl}^{(1)}(G) \Gamma_{km}^{(2)}(G) \\
&= \sum_{j,k} |1,j>|2,k> \Gamma_{jk,lm}(G).
\end{aligned} \tag{7.5}$$

In the second step, the factors have been rearranged; in the third step, product $\Gamma_{jl}^{(1)}(G)\Gamma_{km}^{(2)}(G)$ has been designated

$$\Gamma_{jk,lm}(G). \tag{7.6}$$

Let us arrange each pair of indices in the order

$$11, 12, \ldots, 1s, 21, 22, \ldots, 2s, \ldots, r1, r2, \ldots, rs \tag{7.7}$$

and let jk specify the row, lm the column, of matrix $\Gamma(G)$. We call $\Gamma(G)$ the *direct product* of $\Gamma^{(1)}(G)$ and $\Gamma^{(2)}(G)$, and write

$$\mathbf{\Gamma}(G) = \mathbf{\Gamma}^{(1)}(G) \times \mathbf{\Gamma}^{(2)}(G). \tag{7.8}$$

The indices appear on the elements along the principal diagonal of $\Gamma(G)$ in the order

$$11,11; \; 12,12; \ldots; \; 1s,1s; \; 21,21; \; 22,22; \ldots; \; 2s,2s; \ldots;$$
$$r1,r1; \; r2,r2; \ldots; \; rs,rs. \tag{7.9}$$

Consequently, the trace of the product matrix is given by

$$\text{tr } \mathbf{\Gamma}(G) = \sum_{j,k} \Gamma_{jk,jk}(G) = \sum_j \Gamma_{jj}^{(1)}(G) \sum_k \Gamma_{kk}^{(2)}(G). \tag{7.10}$$

Sums $\sum_j \Gamma_{jj}^{(1)}(G)$ and $\sum_k \Gamma_{kk}^{(2)}(G)$ are components of the characters for the first and the second representations; $\text{tr }\Gamma(G)$ is the corresponding component of the character for the product representation. We thus have

$$\chi_i = \chi_i^{(1)} \chi_i^{(2)}. \tag{7.11}$$

The ith component of the character for the product representation equals the ith component of the character for the first representation times the ith component of the character for the second one.

Example 7.1

Compare the matrix product of **A** and **B** with the direct product of **A** and **B**.

A matrix, **A** or **B**, consists of elements, A_{jl} or B_{km}, which can be arranged in a rectangular array. By convention, the first index labels the row, the second index labels the column, in which the element appears. By definition, the *jm*th element of the matrix product is the sum

$$\sum_{k=1}^{n} A_{jk} B_{km}.$$

Here n is the number of columns in **A** and the number of rows in **B**. These two numbers have to be the same in order for **AB** to exist.

To form the array representing the direct product, we have to arrange the number pairs in the order

$$11, 12, \ldots, 1s, 21, 22, \ldots, 2s, \ldots, r1, r2, \ldots rs.$$

Then the element that appears in the *jk*th row and *lm*th column of the direct product is the term

$$A_{jl} B_{km}.$$

Here there is no restriction on the sizes of the matrices that are combined.

Example 7.2

Construct the direct product of

$$\begin{pmatrix} A_{11} & A_{12} \\ A_{21} & A_{22} \end{pmatrix} \quad \text{and} \quad \begin{pmatrix} B_{11} & B_{12} \\ B_{21} & B_{22} \end{pmatrix}$$

The formula for the direct product is

$$C_{jk,lm} = A_{jl} B_{km}$$

in which each number pair on C is arranged in order (7.7). Applying this formula to the given matrices leads to the result

$$= \begin{pmatrix} C_{11,11} & C_{11,12} & C_{11,21} & C_{11,22} \\ C_{12,11} & C_{12,12} & C_{12,21} & C_{12,22} \\ C_{21,11} & C_{21,12} & C_{21,21} & C_{21,22} \\ C_{22,11} & C_{22,12} & C_{22,21} & C_{22,22} \end{pmatrix}$$

$$= \begin{pmatrix} A_{11}B_{11} & A_{11}B_{12} & A_{12}B_{11} & A_{12}B_{12} \\ A_{11}B_{21} & A_{11}B_{22} & A_{12}B_{21} & A_{12}B_{22} \\ A_{21}B_{11} & A_{21}B_{12} & A_{22}B_{11} & A_{22}B_{12} \\ A_{21}B_{21} & A_{21}B_{22} & A_{22}B_{21} & A_{22}B_{22} \end{pmatrix}$$

$$= \begin{pmatrix} A_{11}\mathbf{B} & A_{12}\mathbf{B} \\ A_{21}\mathbf{B} & A_{22}\mathbf{B} \end{pmatrix}.$$

Example 7.3

What representation of the $\mathbf{C}_3$ group is the direct product of the C representation with the D representation thereof?

The character vectors for the irreducible representations of the $\mathbf{C}_3$ group are described in Table 5.1. Take the components for C and D, multiply them, and tabulate the results:

Table 7.1 Character components for $C \times D$ of $\mathbf{C}_3$.

	I	C_3	C_3^2
C	1	ω	ω^2
D	1	ω^2	ω
$C \times D$	1	$\omega^3 = 1$	$\omega^3 = 1$

We see that the products agree with the components of the character for the A representation. Consequently,

$$C \times D = A.$$

7.3
Product Groups

Each operation of a group may be combined with each operation of another group. The result is a group—as long as the elements of the first group commute with those of the second one. Characters for the product group equal products of characters of the constituent groups.

Consider two groups with the operations

$$A_1, A_2, \ldots, A_g; \qquad (7.12)$$

$$B_1, B_2, \ldots, B_h. \qquad (7.13)$$

Typical combinations of elements in the first group with those in the second group are

$$A_j B_k \text{ and } A_l B_m. \qquad (7.14)$$

When each element in the first group commutes with each element in the second group, combining elements (7.14) yields an element of the same kind:

$$A_j B_k A_l B_m = A_j A_l B_k B_m = A_r B_s. \qquad (7.15)$$

Now, the product of the identity element in the first group with the identity element in the second group is the identity element in the set of products. In products (7.14), furthermore, A_l may be the inverse of A_j and B_m the inverse of B_k. Following equation (7.15), then $A_l B_m$ is the inverse of $A_j B_k$. Also because of equation (7.15), each combination of elements is an element in the set. As elements (7.12) and (7.13) both obey the associative law, elements (7.14) obey the associative law. Thus, the product elements form a group.

If

$$|1,1>, |1,2>, \ldots, |1,v> \qquad (7.16)$$

and

$$|2,1>, |2,2>, \ldots, |2,w> \qquad (7.17)$$

are bases for representations of the first and the second groups, respectively, then we have

$$A_j|1,s\rangle = \sum_r |1,r\rangle \Gamma_{rs}^{(1)}(A_j) \tag{7.18}$$

and

$$B_k|2,u\rangle = \sum_t |2,t\rangle \Gamma_{tu}^{(2)}(B_k). \tag{7.19}$$

Let us also suppose that A_j has no effect on $|2,u\rangle$ and that B_k has no effect on $|1,s\rangle$. Then a typical element of the product group acting on the product of a ket from set (7.16) with a ket from set (7.17) gives a linear combination of product kets:

$$\begin{aligned}
A_j B_k |1,s\rangle|2,u\rangle &= \sum_r |1,r\rangle \Gamma_{rs}^{(1)}(A_j) \sum_t |2,t\rangle \Gamma_{tu}^{(2)}(B_k) \\
&= \sum_{r,t} |1,r\rangle|2,t\rangle \Gamma_{rs}^{(1)}(A_j) \Gamma_{tu}^{(2)}(B_k) \\
&= \sum_{r,t} |1,r\rangle|2,t\rangle \Gamma_{rt,su}(A_j B_k).
\end{aligned} \tag{7.20}$$

Thus, the product kets

$$|1,2\rangle|2,1\rangle, |1,1\rangle|2,2\rangle, \ldots, |1,v\rangle|2,w\rangle \tag{7.21}$$

form a basis for a representation that is the direct product

$$\Gamma(A_j B_k) = \Gamma^{(1)}(A_j) \times \Gamma^{(2)}(B_k). \tag{7.22}$$

As in Section 7.2, we find that

$$\operatorname{tr} \Gamma(A_j B_k) = \sum_{r,t} \Gamma_{rt,rt}(A_j B_k) = \sum_r \Gamma_{rr}^{(1)}(A_j) \sum_t \Gamma_{tt}^{(2)}(B_k) \tag{7.23}$$

and

$$\chi_{jk} = \chi_j^{(1)} \chi_k^{(2)}. \tag{7.24}$$

Each component of the character of the product representation equals the component of the character for the chosen class and representation in the first group multiplied by the component of the character for the chosen class and representation in the second group.

Example 7.4

Show that C_{2h} is the direct product of C_2 and S_2.
Group C_2 contains the operations

$$I \text{ and } C_2,$$

while group S_2 contains

$$I \text{ and } i.$$

From Section 1.3, the reorientation matrices for these operations are

$$I = \begin{pmatrix} 1 & 0 & 0 \\ 0 & 1 & 0 \\ 0 & 0 & 1 \end{pmatrix}, \quad C_2 = \begin{pmatrix} -1 & 0 & 0 \\ 0 & -1 & 0 \\ 0 & 0 & 1 \end{pmatrix},$$

$$i = \begin{pmatrix} -1 & 0 & 0 \\ 0 & -1 & 0 \\ 0 & 0 & -1 \end{pmatrix}.$$

Multiplying the matrices for rotation by $\frac{1}{2}$ turn and for inversion yields the matrix for reflection in the xy plane:

$$C_2 i = \begin{pmatrix} -1 & 0 & 0 \\ 0 & -1 & 0 \\ 0 & 0 & 1 \end{pmatrix} \begin{pmatrix} -1 & 0 & 0 \\ 0 & -1 & 0 \\ 0 & 0 & -1 \end{pmatrix} = \begin{pmatrix} 1 & 0 & 0 \\ 0 & 1 & 0 \\ 0 & 0 & -1 \end{pmatrix} = \sigma_h.$$

Similarly,

$$II = I, \quad C_2 I = C_2, \quad Ii = i.$$

The results are summarized in Table 7.2. Note that the products include all elements of the C_{2h} group. Thus

$$C_{2h} = C_2 \times S_2.$$

Table 7.2 Products of Elements of C_2 with Elements of S_2

	I	i
I	I	i
C_2	C_2	σ_h

Example 7.5

Obtain characters for the irreducible representations of the C_{2h} group from those for the C_2 and S_2 groups.

Since C_{2h} is the direct product of C_2 and S_2, equation (7.24) applies. The characters for the C_2 and S_2 groups are described in Table 7.3. Multiplying the components, following equation (7.24), leads to the results in Table 7.4.

Table 7.3 Components of Character Vectors for the C_2 and S_2 Groups

C_2	I	C_2
A	1	1
B	1	-1

S_2	I	i
A_g	1	1
A_u	1	-1

Table 7.4 Components of Character Vectors for the Product Group $C_2 \times S_2$

C_{2h}	II	$C_2 I$	Ii	$C_2 i$
$A \times A_g$	1	1	1	1
$B \times A_g$	1	-1	1	-1
$A \times A_u$	1	1	-1	-1
$B \times A_u$	1	-1	-1	1

7.4
Quantum Angular Momenta

From quantum mechanics, we know that an angular momentum of a particle or small system involves two quantum numbers. The number of units of $\hbar$ exhibited along a chosen axis is m, the magnetic quantum number. The magnitude of the limit on m, for a given angular-momentum energy, is j, the angular momentum quantum number.

Let the operators for angular momentum around the x, y, and z axes be J_x, J_y, and J_z, respectively. Formally, we have

$$J_0 = J_z, \tag{7.25}$$

$$J_\pm = J_x \pm iJ_y, \tag{7.26}$$

$$J^2 = J_x^2 + J_y^2 + J_z^2. \tag{7.27}$$

Furthermore, let units be chosen so that

$$\hbar = 1. \tag{7.28}$$

The eigenkets for the angular momenta then satisfy the equations

$$J_0|j,m\rangle = m|j,m\rangle, \tag{7.29}$$

$$J_\pm|j,m\rangle = [j(j+1) - m(m \pm 1)]^{\frac{1}{2}}|j,m \pm 1\rangle, \tag{7.30}$$

$$J^2|j,m\rangle = j(j+1)|j,m\rangle. \tag{7.31}$$

For orbital motion of particles about an axis, j is an integer. For the spin of a single fermion, j equals $\frac{1}{2}$. For combined behavior, j may be either integral or half integral, depending on the constituents.

In considering how motions combine, we first treat binary combinations. Transbinary combinations are then built up from the rules for binary combinations.

To satisfy the condition for combining probabilities, the kets for independent behaviors combine muliplicatively. If the angular momentum identified by quantum numbers j_1 and m_1 were to combine with that identified by j_2 and m_2, without interaction, the overall ket would be related to the constituent kets by

$$|j,m\rangle = |j_1,m_1\rangle|j_2,m_2\rangle. \tag{7.32}$$

But in general, particles contain electric charges and magnetic moments. The resulting interactions cause the m_1 and m_2 to vary over all the allowed possibilities. Equation (7.32) is replaced by

$$|j,m\rangle = \sum_{m_1,m_2} C(j,m|j_1,m_1; j_2, m_2)|j_1,m_1\rangle|j_2,m_2\rangle. \tag{7.33}$$

Multiplying equation (7.33) by a particular $\langle j_1,m_1|\langle j_2,m_2|$ yields

$$\langle j_1,m_1|\langle j_2,m_2||j,m\rangle = C(j,m|j_1,m_1; j_2,m_2). \tag{7.34}$$

We choose phases so coefficient C is real. Then we can rewrite equation (7.34) as

$$\langle j,m|j_1, m_1; j_2,m_2\rangle = C(j,m|j_1,m_1; j_2,m_2). \tag{7.35}$$

We also make $C(j,j|j_1, j_1; j_2,j-j_1)$ positive and impose the normalization condition

$$\sum_{m_1,m_2} |C(j,m|j_1,m_1; j_2,m_2)|^2 = 1. \tag{7.36}$$

Coefficient C in equations (7.33) and (7.34) or (7.35)—subject to the reality, sign, and normalization conditions—is called the *Clebsch-Gordon coefficient* for the coupling

$$j_1 \otimes j_2 \rightarrow j. \tag{7.37}$$

Example 7.6

How is the system with $j = \frac{1}{2}$, $m = \frac{1}{2}$ constructed from a particle with orbital j equal to 1 and spin j equal to $\frac{1}{2}$?

The possible magnetic quantum numbers for $j_1 = 1$ and $j_2 = \frac{1}{2}$ are

$$m_1 = -1, 0, 1 \quad \text{and} \quad m_2 = -\tfrac{1}{2}, \tfrac{1}{2}.$$

Now, the only combinations that make $m = \frac{1}{2}$ are

$$m_1 = 0, \quad m_2 = \tfrac{1}{2}$$

and

$$m_1 = 1, \quad m_2 = -\tfrac{1}{2}.$$

So sum (7.33) has the form

$$|\tfrac{1}{2}, \tfrac{1}{2}\rangle = c_1 |1,0\rangle |\tfrac{1}{2}, \tfrac{1}{2}\rangle + c_2 |1,1\rangle |\tfrac{1}{2}, -\tfrac{1}{2}\rangle.$$

Apply J_+ to each ket, following equation (7.30):

$$J_+ |\tfrac{1}{2}, \tfrac{1}{2}\rangle = 0,$$

$$J_+ |1,0\rangle = \sqrt{2}\,|1,1\rangle, \; J_+ |1,1\rangle = 0,$$

$$J_+ |\tfrac{1}{2}, -\tfrac{1}{2}\rangle = |\tfrac{1}{2}, \tfrac{1}{2}\rangle.$$

But since J_+ is a differentiating operator, we also have

$$J_+ |a\rangle |b\rangle = (J_+ |a\rangle)|b\rangle + |a\rangle J_+ |b\rangle.$$

So action of J_+ on the superposition from equation (7.33) yields

$$0 = c_1 \sqrt{2}\, |1,1\rangle |\tfrac{1}{2}, \tfrac{1}{2}\rangle + c_2 |1,1\rangle |\tfrac{1}{2}, \tfrac{1}{2}\rangle,$$

whence

$$c_2 = -\sqrt{2}\, c_1.$$

On the other hand, the normalization condition on the sum imposes the restriction

$$c_1^2 + c_2^2 = 1.$$

Eliminating c_2 from the last two equations then gives us

$$c_1^2 + 2c_1^2 = 1,$$

whence

$$c_1^2 = \frac{1}{3}.$$

According to our sign convention, c_2 is positive. So we take the negative root to get

$$c_1 = -\frac{1}{\sqrt{3}}.$$

Then

$$c_2 = -\sqrt{2}\,\frac{-1}{\sqrt{3}} = \frac{\sqrt{2}}{\sqrt{3}}.$$

7.5
Clebsch-Gordon Coefficients

The eigenket shift equation enables one to determine how an angular momentum of given j_1 can couple with an angular momentum of given j_2 to produce a combination of given j

We recall that the basis kets $|j,m\rangle$ and $|j_1,m_1\rangle$, $|j_2,m_2\rangle$ separately satisfy equations (7.29), (7.30), and (7.31). Furthermore, J_x, J_y, J_z, and J^2 are Hermitian operators.

With equation (7.29), we find that

$$\langle j,m|J_0|j_1,m_1;j_2,m_2\rangle = \langle J_0 j,m||j_1,m_1;j_2,m_2\rangle$$
$$= m\langle j,m|j_1,m_1;j_2,m_2\rangle \tag{7.38}$$

and

$$\langle j,m|J_0|j_1,m_1;j_2,m_2\rangle = \langle j,m|(J_0|j_1,m_1\rangle)|j_2,m_2\rangle$$
$$+ \langle j,m||j_1,m_1\rangle J_0|j_2,m_2\rangle = (m_1+m_2)\langle j,m|j_1,m_1;j_2,m_2\rangle. \tag{7.39}$$

In equation (7.38), the Hermiticity of J_0 and the reality of m have been employed. On comparing equation (7.38) with equation (7.39), we see that

$$m = m_1 + m_2. \tag{7.40}$$

Equation (7.40) agrees with the result from the naive vector model.

Similarly employing $J_\pm$ gives us

Clebsch-Gordon Coefficients

$$\langle j,m|J_\pm|j_1,m_1;j_2,m_2\rangle = \langle J_\pm j,m||j_1,m_1;j_2,m_2\rangle$$
$$= [j(j+1) - m(m\pm 1)]^{\frac{1}{2}} \langle j,m\pm 1|j_1,m_1;j_2,m_2\rangle \quad (7.41)$$

and

$$\langle j,m|J_\pm|j_1,m_1;j_2,m_2\rangle$$
$$= [j_1(j_1+1) - m_1(m_1\pm 1)]^{\frac{1}{2}} \langle j,m|j_1,m_1\pm 1;j_2,m_2\rangle$$
$$+ [j_2(j_2+1) - m_2(m_2\pm 1)]^{\frac{1}{2}} \langle j,m|j_1,m_1;j_2,m_2\pm 1\rangle. \quad (7.42)$$

Combining equations (7.41) and (7.42) with equation (7.35) leads to the recursion relation

$$C(j,m\mp 1|j_1,m_1;j_2,m_2)[j(j+1) - m(m\mp 1)]^{\frac{1}{2}}$$
$$= C(j,m|j_1,m_1\pm 1;j_2,m_2)[j_1(j_1+1) - m_1(m_1\pm 1)]^{\frac{1}{2}}$$
$$+ C(j,m|j_1,m_1;j_2,m_2\pm 1)[j_2(j_2+1) - m_2(m_2\pm 1)]^{\frac{1}{2}}. \quad (7.43)$$

This, together with the normalization and phase conditions, enables one to calculate Clebsch-Gordon coefficients.

Example 7.7

Apply equation (7.43) to the system in Example 7.6.
Choose the parameters

$$j = \tfrac{1}{2}, \qquad j_1 = 1, \qquad j_2 = \tfrac{1}{2},$$
$$m = \tfrac{1}{2}, \qquad m_1 = 1, \qquad m_2 = \tfrac{1}{2}.$$

Also, employ the lower signs in equation (7.43) to get

$$C(j,3/2|1,1;\tfrac{1}{2},\tfrac{1}{2})(0) = C(\tfrac{1}{2},\tfrac{1}{2}|1,0;\tfrac{1}{2},\tfrac{1}{2})\sqrt{2} + C(\tfrac{1}{2},\tfrac{1}{2}|1,1;\tfrac{1}{2},-\tfrac{1}{2})(1).$$

But

$$C(\tfrac{1}{2},\tfrac{1}{2}|1,0;\tfrac{1}{2},\tfrac{1}{2}) = c_1$$

and

$$C(\tfrac{1}{2}, \tfrac{1}{2} | 1, 1; \tfrac{1}{2}, -\tfrac{1}{2}) = c_2.$$

So the preceding equation reduces to

$$0 = c_1\sqrt{2} + c_2$$

or

$$c_2 = -\sqrt{2}\, c_1.$$

This agrees with the relationship found in Example 7.6.

Example 7.8

Calculate the first two Clebsch-Gordon coefficients for the coupling

$$j \otimes 1 \to j.$$

Into equation (7.43), insert the parameters

$$j = j, \qquad j_1 = j, \qquad j_2 = 1,$$

$$m = j, \qquad m_1 = j, \qquad m_2 = 1.$$

Also, choose the lower signs to construct

$$C(j, j+1 | j, j; 1, 1)(0) = C(j,j|j,j-1; 1,1)\sqrt{2j} + C(j,j|j,j; 1,0)\sqrt{2}.$$

Rewrite this result in the form

$$0 = c_1\sqrt{j} + c_2,$$

whence

$$c_2 = -\sqrt{j}\, c_1.$$

Now, the normalization condition imposes the restriction

$$c_1^2 + c_2^2 = 1.$$

Eliminate c_2 from the last two equations,

$$c_1^2 + jc_1^2 = 1,$$

and solve for c_1, with our sign convention, to get

$$C(j, j|j, j-1; 1, 1) = c_1 = -\frac{1}{\sqrt{j+1}}$$

and

$$C(j, j|j, j; 1, 0) = c_2 = \frac{\sqrt{j}}{\sqrt{j+1}}.$$

7.6
Relating Tensorial-Operator Matrix Elements

Scalar physical operators behave as we saw in Section 6.3. But more complicated physical operators are transformed by operations of the covering group for the given physical system. When the pertinent group is the full rotation group, or one of its subgroups, such an operator can be expressed as a sum of irreducible ones. Each of these behaves as a ket for a given j and m. For a particular physical observable with label j, one has a family of matrix elements. These are linked by the Wigner-Eckart theorem.

First, consider the system in Section 7.4, where j and m are the angular momentum and magnetic quantum numbers of the system, while j_1, m_1 and j_2, m_2 are these quantum numbers for subsystems. Note that equation (7.35) can be rewritten in the form

$$<j,m||j_1, m_1>|j_2, m_2> = C(j,m|j_1, m_1; j_2, m_2). \qquad (7.44)$$

Letting

$$j_1 = \omega, \quad j_2 = j',$$
$$m_1 = \mu, \quad m_2 = m' \qquad (7.45)$$

changes equation (7.44) to

$$<j,m||\omega, \mu>|j', m'> = C(j,m|\omega, \mu; j', m'). \quad (7.46)$$

In this relationship, ket $|\omega, \mu>$ behaves as an operator acting on $|j', m'>$.

In the setup where ket $|j_1, m_1>$ represents a state of the first subsystem, a general state of this subsystem is represented by a superposition of these kets with all allowable j_1's and m_1's. Likewise, a general operator transforming $|j', m'>$ above may be considered a linear combination of the standard operators.

In the following, we will consider $|j', m'>$ as an alternate ket for the whole system. Also, we will let T_μ^ω be the standard operator that transforms as $|\omega, \mu>$ does. Note, we cannot equate these but we can relate the matrix elements for operators with the same ω.

In Section 7.5, we made the right sides of equations (7.41) and (7.42) equal. The terms of the result contain $|j_1, m_1>, |j_1, m_1 \pm 1>$, and $|j_1, m_1>$ as factors. We can represent how the standard operators must transform by replacing these kets with the corresponding operators, getting

$$<j,m \mp 1|T_\mu^\omega|j', m'>[j(j + 1) - m(m \mp 1)]^{\frac{1}{2}}$$
$$= <j,m|T_{\mu\pm 1}^\omega|j', m'>[\omega(\omega + 1) - \mu(\mu + 1)]^{\frac{1}{2}}$$
$$+ <j,m|T_\mu^\omega|j', m'>[j'(j' + 1) - m'(m' \pm 1)]^{\frac{1}{2}}. \quad (7.47)$$

This recursion relation is like that for the Clebsch-Gordon coefficients, equation (7.43). Consequently, the matrix elements for given j and j' are proportional to these coefficients. Furthermore, the constant of proportionality depends on j and j'. So we have the result

$$<j,m|T_\mu^\omega|j', m'> = C(j,m|\omega, \mu; j', m')<j||T^\omega||j'>. \quad (7.48)$$

Equation (7.48) is a form of the *Wigner-Eckart theorem*.

The constant $<j||T^\omega||j'>$ is called the *reduced matrix element* of T^ω. It is determined by calculating one matrix element for the given ω. A simple choice is to take $\mu = 0, m' = m$.

Remember that ket $|\omega, \mu>$ transforms as a standard tensor of rank ω. So T_μ^ω is called the *irreducible tensor operator* of rank ω.

Example 7.9

Determine the reduced matrix element for the vector angular momentum operator **J**.

The standard components of **J** with respect to the z axis are

$$J_1 = -\frac{1}{\sqrt{2}} J_+,$$

$$J_0 = J_z,$$

$$J_{-1} = \frac{1}{\sqrt{2}} J_-.$$

Since these transform as $|1, \mu\rangle$ does, with

$$\mu = 1, 0, -1,$$

the rank ω equals 1. And equation (7.48) becomes

$$\langle j,m|J_\mu|j', m'\rangle = C(j,m|1, \mu; j', m') \langle j||J||j'\rangle.$$

When J_μ acts on $|j', m'\rangle$, the result is a number times a ket with j' unaltered, following equations (7.29) and (7.30). Furthermore, the standard kets are mutually orthogonal; thus

$$\langle j,m|j', m'\rangle = 0 \quad \text{when} \quad j \neq j'.$$

But for the allowed combinations, we have

$$C(j,m|1, \mu; j', m') \neq 0$$

Substituting into the above form of equation (7.48) now leads to

$$\langle j||J||j'\rangle = 0 \quad \text{when} \quad j \neq j'.$$

For $j = j'$, we choose $\mu = 0$ and $m = j$, $m' = j$. From the last formula in Example 7.8, we have

$$C(j,j|1, 0; j, j) = C(j, j|j, j; 1, 0) = \frac{\sqrt{j}}{\sqrt{j+1}}.$$

Also, employing equation (7.29) gives us

$$< j,j|J_0|j,j> = j<j,j|j,j> = j.$$

Substituting these results into the reduced form of equation (7.48), as before, yields

$$<j||J||j> = \sqrt{j(j+1)}.$$

Consequently, the nonzero matrix elements of J_μ are

$$<j,m|J_\mu|j,m'> = C(j,m|1,\mu;j,m')\sqrt{j(j+1)}.$$

7.7
Synopsis

The bases for representations of a given group can be arranged into symmetry species. The representations themselves yield components of the corresponding character vectors. The products of the character components from two given species make up the character components for a product species:

$$\chi_i^{(1)}\chi_i^{(2)} = \chi_i. \qquad (7.49)$$

Correspondingly, the products of individual bases (in a set for the first representation) with individual bases (in a set for the second representation) produce a set of bases for the product representation. Similar relationships apply among the species in one group, the species in a second group, and the species in a product group.

A matrix in the product representation is the direct product of corresponding matrices in the factor representations:

$$\mathbf{\Gamma}(G) = \mathbf{\Gamma}^{(1)}(G) \times \mathbf{\Gamma}^{(2)}(G). \qquad (7.50)$$

In the direct product matrix, the element in the jkth row and lmth column is

$$C_{jk,lm} = A_{jl}B_{km}. \qquad (7.51)$$

The two digit numbers are arranged in the order

$$11, 12, \ldots, 1s, 21, 22, \ldots, 2s, \ldots, r1, r2, \ldots, rs. \quad (7.52)$$

Each aspect that can be considered separate in a given system is described by an individual ket. Because of the way probabilities combine, the kets governing the different aspects combine multiplicatively. If the same overall state can be formed in more than one way from the constituents, the overall ket would equal a superposition of the possible products.

One may, for instance, have an angular momentum described by the quantum numbers j and m resulting from a combination of angular momenta described by quantum numbers j_1, m_1 and j_2, m_2. When m_1 and m_2 may vary over their various possibilities, we have

$$|j, m> = \sum_{m_1, m_2} C(j, m\, j_1, m_1; j_2, m_2) |j_1, m_1> |j_2, m_2>. \quad (7.53)$$

When all kets are normalized to 1 and the customary phase conventions are assumed, expression C is the Clebsch-Gordon coefficient for the quantum numbers appearing in the parenthesis.

An operator T_μ^ω that transforms as the ket $|j, m>$, when $j = \omega, m = \mu$, is called an irreducible tensor operator of rank ω. Its behavior is linked to a Clebsch-Gordon coefficient by the equation

$$< j, m| T_\mu^\omega |j', m'> = C(j,m| \omega, \mu; j', m') <j|| T^\omega ||j'>. \quad (7.54)$$

Discussion Questions

7.1   Why is the full rotation group about the origin a key group in discussing atomic structures?

7.2   Why do the standard atomic orbitals involve a homogeneous polynomial in x, y, z as a factor?

7.3   How are the atomic orbitals related to linear combinations of products of the Cartesian coordinates?

7.4   The first particle of a given system is described by state function Ψ_a while the second particle is described by state function Ψ_b. If the ket for the former is $|a>$ and the ket for the latter $|b>$, why is the combination represented by ket $|a>|b>$?

7.5   Why are the symmetry properties of a product generally different from those of its factors?

7.6   How do the products of bases in one set with bases in another form a basis set?

7.7 How is the representation generated by the product set related to the representations generated by the factors?

7.8 How is the character of a product representation related to the characters of the contributing factors?

7.9 When do elements from one group combine with those from another to form a new group? Explain.

7.10 Cite groups that are direct products of simpler groups.

7.11 What eigenvalue equations define quantum numbers j and m? How are the kets for a given j related?

7.12 What series describes how two distinct angular momenta combine?

7.13 Derive an integral expression for a typical coefficient in this series.

7.14 How do the eigenvalue equations enable one to deduce the coupling rule

$$m = m_1 + m_2?$$

7.15 How does the eigenket-shift equation enable one to construct a recursion relation for Clebsch-Gordon coefficients?

7.16 How are the standard tensorial operators defined?

7.17 Why is the matrix element $< j, m|T_\mu^\omega|j', m'>$ proportional to the corresponding Clebsch-Gordon coefficient? Upon what does the constant of proportionality depend?

Problems

7.1 Construct the direct product of the matrices

$$\begin{pmatrix} -\frac{1}{2} & -\frac{\sqrt{3}}{2} \\ \frac{\sqrt{3}}{2} & -\frac{1}{2} \end{pmatrix} \text{ and } \begin{pmatrix} \frac{1}{2} & -\frac{\sqrt{3}}{2} & 0 \\ \frac{\sqrt{3}}{2} & -\frac{1}{2} & 0 \\ 0 & 0 & -1 \end{pmatrix}.$$

What character component does this product yield?

7.2 Form the character components for the direct product of (a) the C representation of group C_3 with itself, and (b) the D representation of group C_3 with itself. To what representations do $C \times C$ and $D \times D$ belong?

7.3 Show that $r\,e^{-i\phi}$ is a basis for the C_1 representation of C_5. Form the square $r^2 e^{-2i\phi}$ and from the character components determine the representation for which it is a basis.

7.4 From character components of C_3 and S_1, determine the character components for the primitive symmetry species of C_{3h}.

7.5 How is the system with $j = 0$, $m = 0$ constructed from two particles with $j_1 = \frac{1}{2}$ and $j_2 = \frac{1}{2}$, respectively?

7.6 Use properties of operators to determine how the angular momentum states $j_1 = 1$, $m_1 = 1$ and $j_2 = 1$, $m_2 = -1$ combine.

7.7 Calculate the first two Clebsch-Gordon coefficients for the coupling

$$j \otimes \tfrac{1}{2} \to j + \tfrac{1}{2}.$$

7.8 Employ the result in Example 7.9 to calculate the matrix element

$$< j, j | J_1 | j, j - 1 >.$$

7.9 Construct two 2×2 matrices with the direct product

$$\begin{pmatrix} -\tfrac{1}{2} & -\tfrac{\sqrt{3}}{2} & 0 & 0 \\ \tfrac{\sqrt{3}}{2} & -\tfrac{1}{2} & 0 & 0 \\ 0 & 0 & \tfrac{1}{2} & \tfrac{\sqrt{3}}{2} \\ 0 & 0 & -\tfrac{\sqrt{3}}{2} & \tfrac{1}{2} \end{pmatrix}$$

7.10 Form the character components for the direct product of (a) the B_1 and B_2 representations of D_2, (b) the B_2 and B_3 representations of D_2, and (c) the B_3 and B_1 representations of D_2 and identify each result.

7.11 Show that z, y, x are bases for the B_1, B_2, B_3 representations of group $\mathbf{D}_2$. Form the products zx, xy, yz and from the character components, determine the representation for which each is a basis.

7.12 From character components of $\mathbf{C}_3$ and $\mathbf{S}_2$, determine the character components for the primitive symmetry species of $\mathbf{S}_6$.

7.13 Use properties of operators to calculate the coefficients in the equation

$$|1, 1 > |\tfrac{1}{2}, -\tfrac{1}{2}> = c_1 | 3/2, \tfrac{1}{2} > + c_2|\tfrac{1}{2}, \tfrac{1}{2}>.$$

7.14 Determine how two particles with angular momentum quantum numbers $j_1 = 3/2$ and $j_2 = 3/2$ combine to form a state with $j = 1$ and $m = 1$.

7.15 Use the result found in Example 7.9 to calculate the Clebsch-Gordon coefficient

$$C(j, j - 1|1, 1; j, j - 2).$$

7.16 Relate $C(j, j - 1|1, 0; j, j - 1)$ to the $C(j, j - 1|1, 1; j, j - 2)$ just found. Then construct an expression for it.

References

Books

Hall, G. G.: 1967, *Applied Group Theory*, American Elsevier, New York, pp. 110–117.
In his simple style, Hall discusses direct product representations and Clebsch-Gordon series.

Articles

Bergia, S., Cannata, F., Ruffo, S., and Savoia, M.: 1979, "Group Theoretical Interpretation of von Neumann's Theorem on Composite Systems," *Am. J. Phys.* **47**, 548–552.

Meunier, J.-L.: 1987, "A Simple Demonstration of the Wigner-Eckart Theorem," *Eur. J. Phys.* **8**, 114–116.

CHAPTER 8 / *Permutation Groups*

8.1
The Role Played by Permutations

We have seen how elements of various groups are represented by reorientations and translations of symmetric physical systems. In each case, the reorientations made up subgroups of the full rotation group. Now, note that each operation for a system permuted equivalent regions, displacements, functions, or bras. So the pertinent group can be considered a permutation group.

In physics, we also have to consider that particles of the same species are indistinguishable. Thus, an atom or ion is not altered essentially by any permutation of its electrons. A nucleus is not altered essentially by any permutation of its protons or by any permutation of its neutrons.

The groups limiting the possible states of the fermions, or the bosons, in a given coherent system are permutation groups. These involve more possibilities than we have considered up to now. So we need to develop a theory for permutation operations and groups.

First, a notation will be established for representing a given permutation, a class of permutations and each possible primitive symmetry species. The nature of spin kets will be considered. Allowable configurations of equivalent particles will be deduced.

8.2
Elements of a Full Permutation Group

The equivalent parts to be considered in a given system may be numbered. Then a possible permutation can be described as a reordering of the numbers. Each operation itself is one element of the covering group. Classes of such operations are identified by the cycles that they effect.

Consider a system containing n interchangeable entities of a given kind. The group of all possible permutations of these entities, the full permutation group, is called the *symmetric group* $\mathscr{S}_n$. Since $n!$ different permutations of n objects exist, the order of $\mathscr{S}_n$ is

$$n! \tag{8.1}$$

The operation that replaces object 1 with object α_1, object 2 with object $\alpha_2, \ldots$, object n with object α_n, is denoted

$$P = \begin{pmatrix} 1 & 2 & \cdots & n \\ \alpha_1 & \alpha_2 & & \alpha_n \end{pmatrix}. \tag{8.2}$$

Because of the closure property of the group, each multiple of this operation is an element of the group.

Now, repeated applications of a given P cause the numbers to move within subsets of various sizes. Each of these subsets makes up a *cycle*. The number of distinct elements in a cycle is called its *degree*. As an example, the operation

$$P = \begin{pmatrix} 1 & 2 & 3 & 4 & 5 & 6 & 7 & 8 & 9 \\ 2 & 5 & 6 & 9 & 8 & 7 & 3 & 1 & 4 \end{pmatrix} \tag{8.3}$$

imposes the changes

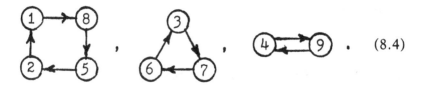 (8.4)

On applying operation (8.3) repeatedly, the numbers 1258 go

through a cycle of degree 4; the numbers 367 go through a cycle of degree 3; the numbers 49 go through a cycle of degree 2.

A step in a cycle is represented by the pertinent numbers placed within parentheses in the order in which they are replaced. Thus, the first change in line (8.4) results from the operation (1258), the second one from operation (367), the third one from operation (49). And we can rewrite operation (8.3) as

$$P = (1258)(367)(49). \tag{8.5}$$

In this reduced form, the cycles have no common numbers. Also, the order in which the cycles are written down does not affect the permutation.

In a given group, a similarity transformation by permutation A changes permutation P to a permutation Q,

$$APA^{-1} = Q. \tag{8.6}$$

From Figure 2.1, this has the same effect as if the objects were renumbered by A, so Q were the same operation as P. Consequently, Q has the same cycle structure as P. In a full permutation group, operations with the same cycle structure, and only these, belong to the same class.

A cycle structure is described by listing each cyclic degree that appears with an exponent that gives the number of times that it appears, within parentheses. When degree one appears a_1 times, degree two, a_2 times, ..., degree n, a_n times, with

$$1a_1 + 2a_2 + \ldots + na_n = n, \tag{8.7}$$

the structure is expressed as

$$(1^{a_1} 2^{a_2} \ldots n^{a_n}). \tag{8.8}$$

In particular, the cycle structure of equation (8.5) is denoted

$$(2^1 3^1 4^1). \tag{8.9}$$

Number (8.1) is obtainable from the fact that n symbols can be ordered in $n!$ different ways. If we allow permutations with various cycle structures, this number gives the number of such permutations. But if one limits oneself to structure (8.8), one has to divide out the number of rearrangements that do not lead to different permutations of this type.

Now, all the a_j cycles of degree j can be rearranged $a_j!$ times without introducing any changes into P. Indeed, (12)(34) and (34)(12) represent the same operation. Furthermore, each of the j cyclic rearrangements within a cycle represents the same operation. Thus, (123), (231), (312) are the same. For a_j cycles of degree j, we have j^{a_j} such rearrangements. Dividing the total number of permuations of n symbols by the product of these numbers yields

$$\frac{n!}{a_1!1^{a_1}a_2!2^{a_2}\ldots a_n!n^{a_n}} \qquad (8.10)$$

for the number of distinct permuation operations with the structure (8.8).

Example 8.1

For what group is the permutation (1234) a generator?
In notation (8.2), the given permutation is

$$A = (1234) = \begin{pmatrix} 1 & 2 & 3 & 4 \\ 2 & 3 & 4 & 1 \end{pmatrix}.$$

Apply this twice to get

$$B = (1234)(1234) = \begin{pmatrix} 1 & 2 & 3 & 4 \\ 2 & 3 & 4 & 1 \end{pmatrix} \begin{pmatrix} 1 & 2 & 3 & 4 \\ 2 & 3 & 4 & 1 \end{pmatrix}$$

$$= \begin{pmatrix} 1 & 2 & 3 & 4 \\ 3 & 4 & 1 & 2 \end{pmatrix} = (13)(24).$$

Apply it thrice to obtain

$$C = (1234)(1234)(1234) = \begin{pmatrix} 1 & 2 & 3 & 4 \\ 2 & 3 & 4 & 1 \end{pmatrix} \begin{pmatrix} 1 & 2 & 3 & 4 \\ 3 & 4 & 1 & 2 \end{pmatrix}$$

$$= \begin{pmatrix} 1 & 2 & 3 & 4 \\ 4 & 1 & 2 & 3 \end{pmatrix} = (4321).$$

Finally, apply it four times to elicit

$$D = (1234)(1234)(1234)(1234) = \begin{pmatrix} 1 & 2 & 3 & 4 \\ 3 & 4 & 1 & 2 \end{pmatrix} \begin{pmatrix} 1 & 2 & 3 & 4 \\ 3 & 4 & 1 & 2 \end{pmatrix}$$

$$= \begin{pmatrix} 1 & 2 & 3 & 4 \\ 1 & 2 & 3 & 4 \end{pmatrix} = (1)(2)(3)(4).$$

Elements A, $A^2 = B$, $A^3 = C$, and $A^4 = I$ constitute the cyclic group of order four. We have labeled it $\mathbf{C}_4$. Although A and C have the same cycle structure (4^1), they here belong to different classes.

Example 8.2

When two objects switch places, they are said to have undergone a *transposition*, or interchange. Show that any given cycle can be represented as a combination of transpositions.

Note that the cyclic permutation

$$P = (123 \ldots n)$$

can be carried out by interchanging objects 1 and 2, then interchanging objects 1 and 3, ..., and finally interchanging objects 1 and n. Indeed if we list the results in order, we have

$$\begin{pmatrix} 1 & 2 & 3 & \ldots & n-1 & n \\ 2 & 1 & 3 & \ldots & n-1 & n \\ 2 & 3 & 1 & \ldots & n-1 & n \\ & & & \vdots & & \\ 2 & 3 & 4 & \ldots & 1 & n \\ 2 & 3 & 4 & \ldots & n & 1 \end{pmatrix}$$

The overall effect is

$$\begin{pmatrix} 1 & 2 & 3 & \ldots & n-1 & n \\ 2 & 3 & 4 & \ldots & n & 1 \end{pmatrix} = (123\ldots n).$$

Thus

$$(12)(13)\ldots(1n) = (123\ldots n).$$

8.3
Symmetry Species for a Symmetric Group

Each of the distinct cycle structures for $\mathscr{S}_n$ corresponds to a different partitioning of number n. And, each partitioning determines a distinct primitive symmetry species.

Descriptive of a species for $\mathscr{S}_n$ are the corresponding eigenkets of a quantum mechanical system containing n identical entities. Now, the kets for the individual entities may be combined to form a completely symmetric function. They may be combined to form a completely antisymmetric function. Since these two combinations are not degenerate, the corresponding species are primitive.

Alternatively, the kets may be multiplied and then superposed on similar products to give a function in which $a_j, a_k \ldots, a_m$ kets are separately bound symmetrically, with

$$a_j \geqslant a_k \geqslant \ldots \geqslant a_m. \tag{8.11}$$

Any interchange of entities between two different symmetric subsets would change the sign of the state function.

Linearly independent functions with the same partitioning make up a degenerate set. This set forms a distinct symmetry species. And since the degeneracy exists as long as the group is applicable, the species is primitive.

The set for distribution (8.11) is represented by a *Young diagram*

$$\tag{8.12}$$

in which there are a_j squares in the first row, a_k squares in the second row, ..., a_m squares in the bottom row. The diagram in which there are b_m rows of m squares, ..., b_2 rows of 2 squares, and b_1 rows of single squares is denoted

$$\{m^{b_m} \cdots 2^{b_2} 1^{b_1}\}. \tag{8.13}$$

An index $b_j = 1$ is generally omitted, however.
Thus, $\{4\}$ represents

$$\tag{8.14}$$

$\{31\}$ represents

$$\tag{8.15}$$

$\{2^2\}$ represents

$$\tag{8.16}$$

$\{21^2\}$ represents

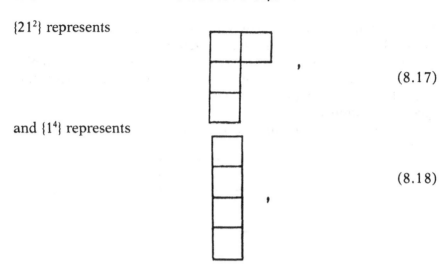

and $\{1^4\}$ represents

(8.17)

(8.18)

Each class of a permutation group is described by its cycle structure (8.8); each primitive symmetry species, by its Young diagram (8.12). The components of the characters follow the laws we have already deduced.

In general, the entities being distributed may be numbered. A particular distribution over a partitioning is then indicated by placing the numbers in pertinent squares. The result is called a *Young table;* however, the entities in a row must be bound symmetrically. If one starts with a single term for the row, one may apply the symmetrizing operator

$$\Pi_S = N \sum_k P_k \qquad (8.19)$$

to construct the appropriate combination. Operator P_k effects the kth permutation. The order here is not important; however, all possible permutations of the entities in the row must be included.

On the other hand, entities in different rows are related antisymmetrically. If one starts with a single term for a column, one may apply the antisymmetrizing operator

$$\Pi_A = N \sum_k (-1)^k P_k \qquad (8.20)$$

to construct the appropriate combination for the column. Now P_k is the kth permutation operator for the entities in the column with k odd when the permutation involves an odd number of interchanges and even when it involves an even number of interchanges.

Example 8.3

How may three identical particles be placed in three different single particle kets?

First, number the particles and the kets. Let $|j\rangle_k$ represent the kth particle in the jth ket. Here $j = 1, 2,$ or 3 and $k = 1, 2,$ or 3. Since the particles are to be identical, the Hamiltonian H is not altered by any operation of the $\mathscr{S}_3$ group. So the eigenkets must belong to primitive symmetry species of this group.

Most of these species occur multiply. Furthermore, the combinations of products generated by equation (6.7)

$$\mathscr{E}^{(r)}|f\rangle = \sum_k y_k^{(r)} \mathscr{C}_{\bar{k}}|f\rangle = \sum_k y_{\bar{k}}^{(r)} \mathscr{C}_k|f\rangle$$

for a given r can be superposed to yield functions corresponding to each independent Young table for the diagram. A particular Young table is associated with a particular energy level and a particular occurrence of the primitive species.

Table 8.1 Components of Characters for the $\mathscr{S}_3$ Group

	(1^3)	$3(12)$	$2(3)$
$\{3\}$	1	1	1
$\{21\}$	2	0	-1
$\{1^3\}$	1	-1	1

Pertinent characters for the $\mathscr{S}_3$ group are described in Table 8.1; results of operations of $\mathscr{S}_3$ acting on the possible single products appear in Table 8.2. Taking the components from the first row of the body of Table 8.1, letting $|f\rangle$ be $|1\rangle_1|2\rangle_2|3\rangle_3$, substituting the appropriate expressions into the formula above, and multiplying by $1/\sqrt{6}$ yields

$$|s\rangle = \frac{1}{\sqrt{6}} (|1\rangle_1|2\rangle_2|3\rangle_3 + |1\rangle_2|2\rangle_1|3\rangle_3 + |1\rangle_1|2\rangle_3|3\rangle_2$$

$$+ |1\rangle_3|2\rangle_2|3\rangle_1 + |1\rangle_2|2\rangle_3|3\rangle_1 + |1\rangle_3|2\rangle_1|3\rangle_2)$$

for symmetry species {3}. Similarly employing the components from the bottom row of Table 8.1 leads to

$$|a\rangle = \frac{1}{\sqrt{6}} (|1\rangle_1|2\rangle_2|3\rangle_3 - |1\rangle_2|2\rangle_1|3\rangle_3 - |1\rangle_1|2\rangle_3|3\rangle_2$$

$$- |1\rangle_3|2\rangle_2|3\rangle_1 + |1\rangle_2|2\rangle_3|3\rangle_1 + |1\rangle_3|2\rangle_1|3\rangle_2)$$

for symmetry species {1³}.

Innumerable combinations can be formed for the doubly degenerate species {21}. Letting $|f\rangle$ in the generating formula be $|1\rangle_1|2\rangle_2|3\rangle_3$ leads to

$$|m_1\rangle = \frac{1}{\sqrt{6}} (2|1\rangle_1|2\rangle_2|3\rangle_3 - |1\rangle_2|2\rangle_3|3\rangle_1 - |1\rangle_3|2\rangle_1|3\rangle_2)$$

after the result is normalized. When $|f\rangle$ is the difference

$$|1\rangle_2|2\rangle_3|3\rangle_1 - |1\rangle_3|2\rangle_1|3\rangle_2,$$

the formula yields

$$2|1\rangle_2|2\rangle_3|3\rangle_1 - |1\rangle_3|2\rangle_1|3\rangle_2 - |1\rangle_1|2\rangle_2|3\rangle_3$$
$$- 2|1\rangle_3|2\rangle_1|3\rangle_2 + |1\rangle_1|2\rangle_2|3\rangle_3 + |1\rangle_2|2\rangle_3|3\rangle_1.$$

Reducing and normalizing this result gives us the orthogonal form

$$|m_2\rangle = \frac{1}{\sqrt{2}} (|1\rangle_2|2\rangle_3|3\rangle_1 - |1\rangle_3|2\rangle_1|3\rangle_2).$$

With respect to $|1\rangle$ and $|2\rangle$, kets $|m_1\rangle$ and $|m_2\rangle$ can be (a) symmetrized or (b) antisymmetrized, resulting in two independent functions of type

1	2
3	

Table 8.2 Effects of the Various Permutation Operations of $\mathcal{S}_3$ on Three-Factor Kets

(1)(2)(3)	(12)(3)	(1)(23)	(2)(31)	(123)	(132)
$\|1\rangle_1\|2\rangle_2\|3\rangle_3$	$\|1\rangle_2\|2\rangle_1\|3\rangle_3$	$\|1\rangle_1\|2\rangle_3\|3\rangle_2$	$\|1\rangle_3\|2\rangle_2\|3\rangle_1$	$\|1\rangle_2\|2\rangle_3\|3\rangle_1$	$\|1\rangle_3\|2\rangle_1\|3\rangle_2$
$\|1\rangle_2\|2\rangle_1\|3\rangle_3$	$\|1\rangle_1\|2\rangle_2\|3\rangle_3$	$\|1\rangle_3\|2\rangle_1\|3\rangle_2$	$\|1\rangle_2\|2\rangle_3\|3\rangle_1$	$\|1\rangle_3\|2\rangle_2\|3\rangle_1$	$\|1\rangle_1\|2\rangle_3\|3\rangle_2$
$\|1\rangle_1\|2\rangle_3\|3\rangle_2$				$\|1\rangle_2\|2\rangle_1\|3\rangle_3$	$\|1\rangle_3\|2\rangle_2\|3\rangle_1$
$\|1\rangle_3\|2\rangle_2\|3\rangle_1$				$\|1\rangle_1\|2\rangle_3\|3\rangle_2$	$\|1\rangle_2\|2\rangle_1\|3\rangle_3$
$\|1\rangle_2\|2\rangle_3\|3\rangle_1$				$\|1\rangle_3\|2\rangle_1\|3\rangle_2$	$\|1\rangle_1\|2\rangle_2\|3\rangle_3$
$\|1\rangle_3\|2\rangle_1\|3\rangle_2$				$\|1\rangle_1\|2\rangle_2\|3\rangle_3$	$\|1\rangle_2\|2\rangle_3\|3\rangle_1$

and two independent functions of type

8.4
Trace for the Reorientation of a Polynomial Ket with Simple Permutational Symmetry

When a multiparticle system possesses symmetry in space, its eigenkets belong to primitive symmetry species of both (a) its permutation group and (b) its geometric group. The kets for each configuration with an allowed permutational symmetry must then be combined to form candidate kets belonging to primitive species of the geometric group. Governing resolution of the configuration is a formula for the trace for each geometric operation.

Suppose that a typical reorientation that changes the system into an equivalent system is R. The complete set of R's make up a geometric group, such as we have already studied. The one-particle kets

$$|1>_j, |2>_k, \ldots, |p>_m \tag{8.21}$$

representing the possible constituent single particle states are bases for representations of this geometric group. As a consequence, an R acting on the array replaces each $|l>_k$ by a linear combination of $|l>_j$'s:

$$R|l>_k = \sum_j |l>_j R_{jk}. \tag{8.22}$$

Number R_{jk} is the coefficient of $|l>_j$ in the sum over j.

Also, suppose that n interchangeable identical particles occupy kets (8.21). A particular arrangement is described by the product

$$|1>_j |2>_k \ldots |p>_m = |f>, \tag{8.23}$$

in which the same ket may be repeated with a different subscript (indicating occupancy by a different particle). The operators effecting the permutations of the particles constitute a permutation group of h elements. Let the uth element in this group be P_u.

Inserting (8.23) and components of the λth character into equation (6.7) yields a ket of the corresponding symmetry species. Since the inverse of a class sum in the permutation group is the same class sum, the normalized result is

$$|\lambda\rangle = \frac{1}{\sqrt{h}} \sum_u y_u^{(\lambda)} P_u (|1\rangle_j |2\rangle_k \ldots |p\rangle_m)$$

$$= \frac{1}{\sqrt{h}} \sum_u y_u^{(\lambda)} (|1\rangle_{P_u j} |2\rangle_{P_u k} \ldots |p\rangle_{P_u m}). \tag{8.24}$$

Substituting $|\lambda\rangle$ into equation (6.7) with the same character components yields $|\lambda\rangle$ in rearranged forms h times; therefore

$$\frac{1}{h} \sum_v y_v^{(\lambda)} P_v \tag{8.25}$$

does not remove anything or add anything to the ket; and

$$|\lambda\rangle = \frac{1}{h} \sum_v y_v^{(\lambda)} P_v |\lambda\rangle. \tag{8.26}$$

Imposing reorientation R on equation (8.26) and introducing equation (8.24) yields

$$R|\lambda\rangle = \frac{R}{h} \sum_v y_v^{(\lambda)} P_v |\lambda\rangle$$

$$= \frac{R}{h} \sum_v y_v^{(\lambda)} P_v \frac{1}{\sqrt{h}} \sum_u y_u^{(\lambda)} P_u (|1\rangle_j |2\rangle_k \ldots |p\rangle_m)$$

$$= \frac{1}{h} \sum_{u,v} y_u^{(\lambda)} \frac{1}{\sqrt{h}} y_v^{(\lambda)} P_v R(|1\rangle_{P_u j} |2\rangle_{P_u k} \ldots |p\rangle_{P_u m}). \tag{8.27}$$

Employing equation (8.22) to implement the action of R on the final expression in parentheses and then rearranging the multiple summation leads to

$R|\lambda\rangle$

$$= \frac{1}{h} \sum_{u,v} y_u^{(\lambda)} \frac{1}{\sqrt{h}} y_v^{(\lambda)} \sum_{q,r,\ldots,t} |1\rangle_q |2\rangle_r \ldots |p\rangle_t P_v R_{qP_uj} R_{rP_uk} \ldots R_{tP_um}$$

$$= \frac{1}{h} \sum_u y_u^{(\lambda)} \sum_{q,r,\ldots,t} \left(\frac{1}{\sqrt{h}} \sum_v y_v^{(\lambda)} P_v^{-1} |1\rangle_q |2\rangle_r \right.$$

$$\left. \ldots |p\rangle_t \right) P_v^{-1} P_v R_{qP_uj} R_{rP_uk} \ldots R_{tP_um}. \qquad (8.28)$$

Since $y_v^{(\lambda)}$ for P_v^{-1} equals that for P_v, the expression in parentheses equals $|\lambda\rangle$ in each term for which

$$q = j, \ r = k, \ \ldots, \ t = m. \qquad (8.29)$$

Consequently, the coefficients contributing to the trace of the matrix for operation R are those in which equations (8.29) hold:

$$\mathrm{tr}\, \mathbf{R}^{(\lambda)} = \frac{1}{h} \sum_u y_u^{(\lambda)} \sum_{j,k,\ldots,m} R_{jP_uj} R_{kP_uk} \ldots R_{mP_um}. \qquad (8.30)$$

Each class of a symmetric group is characterized by its cycle structure

$$(1^a 2^b \ldots n^d). \qquad (8.31)$$

A cycle of degree i in a class contributes

$$\sum_{j,k,\ldots,m} R_{jk} R_{kl} \ldots R_{mj} = \chi(R^i) \qquad (8.32)$$

to the R product for the class. Here $\chi(R^i)$ is the character component in the geometric group for operation R repeated i times.

Substituting result (8.32) into equation (8.30) leads to the formula

$$\mathrm{tr}\mathbf{R}^{(\lambda)} = \frac{1}{h} \sum_u y_u^{(\lambda)} P_u [\chi(R)]^a [\chi(R^2)]^b \ldots [\chi(R^n)]^d. \qquad (8.33)$$

Here h is the order of the permutation group, $y_u^{(\lambda)}$ the component of the λth character for permutation P_u.

8.5
Combining Spin and Orbital Kets

Since a single particle exhibits an intrinsic magnetism, as well as motion through space, its ket contains a spin factor as well as an orbital factor. Because the magnetic moment associated with the spin interacts with the magnetism produced by the orbital motion, the two factors are not completely independent of each other. When the kets for different particles are multiplied as in monomial (8.23), the result factors into a spin part and an orbital part that are also not completely independent of each other.

The spin factors and the orbital factors from the different monomials contributing to a state may be separately combined to form primitive symmetry species of the appropriate permutation group for the given system. A candidate ket is then obtained on multiplying a spin base with an orbital base.

In nature, however, only completely antisymmetric combinations are observed. Other combinations are not. This empirical result is called the *Pauli exclusion principle*.

Each primitive symmetry species of a symmetric group is identified by its Young diagram. The diagram that results from interchanging rows with columns is the *conjugate* of the original one. Following the rules in Section 7.2, we find that only products between conjugate pairs contain $\{1^n\}$ and that each of these products contains $\{1^n\}$ only once. Consequently, a spin function with a certain Young diagram combines only with an orbital function having the conjugate Young diagram. In the character tables, conjugate species are joined by lines.

A spin factor is characterized by its spin quantum number s and its magnetic quantum number m_s. The latter measures the component of spin along the unique axis for the system. A fundamental particle, such as an electron or quark, exhibits an s of $\frac{1}{2}$ and an m_s of either $\pm\frac{1}{2}$.

A ket for m_s equal to $+\frac{1}{2}$ is represented by

$$\boxed{\uparrow} \quad , \tag{8.34}$$

while a ket for m_s equal to $-\tfrac{1}{2}$ is represented by

(8.35)

Since any number of like kets can be combined symmetrically, any number of (8.34) squares and any number of (8.35) squares can occur in a row of a Young table for spin kets. Since only different kets can be combined antisymmetrically, no more than one of (8.34) squares and one of (8.35) squares can occur in a column of a Young table for spin kets. The number of independent different tables that a person can validly construct for a given Young diagram gives the *multiplicity* of the corresponding state.

Example 8.4

Show that the spin basis for {5} is a hextet, the spin basis for {41} is a quartet, and the spin basis for {32} is a doublet.

We assume that the identical particles possess an s of $\tfrac{1}{2}$. Then in

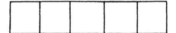

we may have

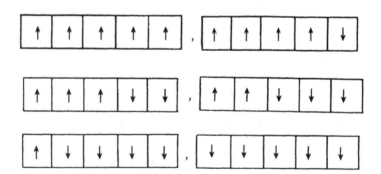

Six independent completely symmetric structures exist.

Similarly in

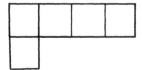

we have

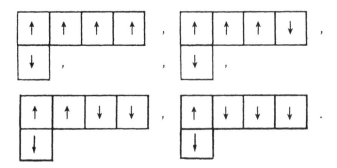

From {41}, we thus obtain four independent spin structures.
In

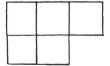

we can have

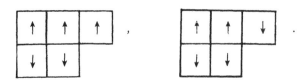

So, two independent {32} spin kets exist.

8.6
States Assumed by Systems of Equivalent Particles

The particles in a given configuration may be described by various sets of simultaneous kets. The kets in each of these sets combine multiplicatively. The products then superpose to form a primitive species of the permutational group for the system. This species need not be a primitive symmetry species of the geometric group for the system. But in any case, equation (8.33) lets us determine the nature of each allowed state without actually carrying out the resolution.

A person can work with the orbital parts alone. A given kind of spin factor is assumed (singlet, doublet, triplet, ...) and its symmetry species ascertained. The orbital factor must then belong to the conjugate species so that the Pauli exclusion principle is satisfied.

Two equivalent spin-$\frac{1}{2}$ particles require a permutational group of order 2; the table for S_2 applies. In the singlet state, the orbital factor must belong to the first row of the character table. Then equation (8.33) yields

$$\chi(R, \text{singlet}) = \frac{1}{2}\left\{\left[\chi(R)\right]^2 + \chi(R^2)\right\} \qquad (8.36)$$

for the R component of the character for this factor. In the triplet state, the orbital ket must belong to the second row of the character table. Equation (8.33) yields

$$\chi(R, \text{triplet}) = \frac{1}{2}\left\{\left[\chi(R)\right]^2 - \chi(R^2)\right\}. \qquad (8.37)$$

Similarly, for three spin-$\frac{1}{2}$ particles, we obtain

$$\chi(R, \text{doublet}) = \frac{1}{6}\left\{2\left[\chi(R)\right]^3 - 2\left[\chi(R^3)\right]\right\}$$

$$= \frac{1}{3}\left\{\left[\chi(R)\right]^3 - \chi(R^3)\right\}. \qquad (8.38)$$

and

$$\chi(R, \text{quartet}) = \frac{1}{6}\left\{\left[\chi(R)\right]^3 - 3\chi(R)\chi(R^2) + 2\chi(R^3)\right\}. \qquad (8.39)$$

For four spin-$\frac{1}{2}$ particles, we obtain

$$\chi(R, \text{singlet}) = \frac{1}{12}\left\{\left[\chi(R)\right]^4 + 3\left[\chi(R^2)\right]^2 - 4\chi(R)\chi(R^3)\right\}, \qquad (8.40)$$

$$\chi(R, \text{triplet}) = \frac{1}{8}\left\{\left[\chi(R)\right]^4 - 2\left[\chi(R)\right]^2\chi(R^2)\right.$$

$$\left. - \left[\chi(R^2)\right]^2 + 2\chi(R^4)\right\}, \tag{8.41}$$

$$\chi(R, \text{quintet}) = \frac{1}{24}\left\{\left[\chi(R)\right]^4 - 6\left[\chi(R)\right]^2\chi(R^2)\right.$$

$$\left. + 3\left[\chi(R^2)\right]^2 + 8\chi(R)\chi(R^3) - 6\chi(R^4)\right\}. \tag{8.42}$$

These results enable a person to calculate each component of the character for the orbital factor corresponding to the given spin multiplicity. Equation (5.24), or equation (5.32), then yields the primitive symmetry species that are present in the composite.

By convention, the symbol for the primitive symmetry species of a constituent particle is written in lower case. Thus, an a_2 electron is an electron in an orbital of A_2 symmetry; an e_2 electron is an electron in an orbital of E_2 symmetry.

Example 8.5

An atom or ion contains three electrons in equivalent f_{2u} orbitals subjected to an octahedrally symmetric field. What different levels can arise when the electronic repulsions and the exclusion principle are taken into account?

We need employ only the subgroup **O**, since the g and u designations can be determined by inspection. For the pertinent binary direct products, we have

$$g \times g = g, \quad g \times u = u \times g = u, \quad u \times u = g.$$

With these, we find that

$$u \times (u \times u) = u \times g = u,$$

any combination of three u orbitals produces a u orbital.

In Table 8.3, components of the characters for each primitive species of **O** are listed. But in equations (8.38) and (8.39), the components for the first, second, and third powers of an operation in each class, on an electron, are needed. These powers are determined and the corresponding components of the characters are identified and tabulated. These components are substituted into equations (8.38) and (8.39), with the listed results.

Equation (5.24) now enables us to determine the primitive symmetry species that are present. A number indicating the multiplicity is added as a leading superscript and the results are written down in the final column of the table.

Adding the u designation, obtained earlier, we find that the combination

$$(f_{2u})^3$$

yields

$$^4A_{2u}, \,^2E_u, \,^2F_{1u}, \text{ and } ^2F_{2u}$$

levels.

Table 8.3 Properties of Characters of the Octahedral Group and for Three f_2 Electrons in an Octahedrally Symmetric Field

O	I	$8C_3$	$6C_2$	$6C_4$	$3C_4^2$	
A_1	1	1	1	1	1	
A_2	1	1	−1	−1	1	
E	2	−1	0	0	2	
F_1	3	0	−1	1	−1	
F_2	3	0	1	−1	−1	
R	I	C_3	C_2	C_4	C_4^2	
R^2	I	C_3	I	C_4^2	I	
R^3	I	I	C_2	C_4	C_4^2	
$\chi(R)$ for F_2	3	0	1	−1	−1	
$\chi(R^2)$ for F_2	3	0	3	−1	3	
$\chi(R^3)$ for F_2	3	3	1	−1	−1	
$\chi(R,\text{ doublet})$	8	−1	0	0	0	$^2E + \,^2F_1 + \,^2F_2$
$\chi(R,\text{ quartet})$	1	1	−1	−1	1	$^4A_2$

8.7 Electronic States

The results obtained for common configurations of equivalent electrons are summarized in Table 8.4. Since a filled shell belongs to the

Table 8.4 States of Equivalent Electrons in Symmetric Molecules

Geometric Group	Configuration	Resulting States
C_{3v}	a_2	$^2A_2$
	a_2^2	$^1A_1$
	e	^2E
	e^2	$^1A_1, {}^1E, {}^3A_2$
	e^3	^2E
	e^4	$^1A_1$
D_{3h}	e'	$^2E'$
$(D_3, D_{3d})^*$	e'^2	$^1A_1', {}^1E', {}^3A_2'$
	e'^3	$^2E'$
	e'^4	$^1A_1'$
	e''	$^2E''$
	e''^2	$^1A_1', {}^1E', {}^3A_2'$
	e''^3	$^2E''$
	e''^4	$^1A_1'$
D_4, C_{4v}, D_{2d}	e	^2E
$(D_{4h})^{**}$	e^2	$^1A_1, {}^1B_1, {}^1B_2, {}^3A_2$
	e^3	^2E
	e^4	$^1A_1$
D_6, C_{6v}	e_1	$^2E_1$
$(D_{6h})^{**}$	e_1^2	$^1A_1, {}^1E_2, {}^3A_2$
	e_1^3	$^2E_1$
	e_1^4	$^1A_1$
	e_2	$^2E_2$
	e_2^2	$^1A_1, {}^1E_2, {}^3A_2$
	e_2^3	$^2E_2$
	e_2^4	$^1A_1$
$C_{\infty v}$	π	$^2\Pi$
$(D_{\infty h})^{**}$	π^2	$^1\Sigma^+, {}^1\Delta, {}^3\Sigma^-$
	π^3	$^2\Pi$
	δ	$^2\Delta$
	δ^2	$^1\Sigma^+, {}^1\Gamma, {}^3\Sigma^-$
	δ^3	$^2\Delta$

Table 8.4 (*continued*).

Geometric Group	Configuration	Resulting States
O, T$_d$	e	^2E
(**O**$_h$)**	e^2	$^1A_1, {}^1E, {}^3A_2$
	e^3	^2E
	e^4	$^1A_1$
	f_1	$^2F_1$
	f_1^2	$^1A_1, {}^1E, {}^1F_2, {}^3F_1$
	f_1^3	$^2F_1, {}^2E, {}^2F_2, {}^4A_1$
	f_1^4	$^1A_1, {}^1E, {}^1F_2, {}^3F_1$
	f_1^5	$^2F_1$
	f_1^6	$^1A_1$
	f_2	$^2F_2$
	f_2^2	$^1A_1, {}^1E, {}^1F_2, {}^3F_1$
	f_2^3	$^2E, {}^2F_1, {}^2F_2, {}^4A_2$
	f_2^4	$^1A_1, {}^1E, {}^1F_2, {}^3F_1$
	f_2^5	$^2F_2$
	f_2^6	$^1A_1$

*For **D**$_3$, omit the primes; for **D**$_{3d}$, omit the primes and add g or u.
For **D$_{4h}$, **D**$_{6h}$, **D**$_{\infty h}$, **O**$_h$, add g or u following the g-u rule.

completely symmetric primitive species, as the corresponding empty shell does, removing one electron from the filled shell produces the same possibilities as placing one electron in the empty shell. And, removing n electrons from the filled shell produces the same symmetry states as adding n electrons to the empty shell.

Lower filled shells are all completely symmetric in the pertinent group; therefore, their presence does not alter the symmetry of the state. However, each additional unfilled shell does. Since each term in the composite ket consists of products of possible constituent kets, the composite species is the direct product of the constituent species. Equation (7.11) applies. Some simple possibilities are listed in Table 8.5.

8.8
Recapitulation

The identical particles in a coherent submicroscopic system may be numbered. Then the operation that replaces particle 1 with particle

Table 8.5 States from Combining Unfilled Shells

Geometric Group	Configuration	Resulting States
C_{2v}	a_1	$^2A_1$
	a_1a_1	$^1A_1, ^3A_1$
	a_1a_2	$^1A_2, ^3A_2$
	b_1b_2	$^1A_2, ^3A_2$
C_{3v}	e	^2E
	a_1e	$^1E, ^3E$
	ee	$^1A_1, ^1A_2, ^1E, ^3A_1, ^3A_2, ^3E$
D_{3h}	$a_2''e'$	$^1E'', ^3E''$
	$e'e''$	$^1A_1'', ^1A_2'', ^1E'', ^3A_1'', ^3A_2'', ^3E''$
	$a_1''e'e''$	$2^2A_1', 2^2A_2', 2^2E', ^4A_1', ^4A_2', ^4E'$
D_{6h}	$a_{1u}e_{1g}e_{2u}e_{2u}$	$2^1B_{1u}, 2^1B_{2u}, 6^1E_{1u}, 3^3B_{1u}, 3^3B_{2u},$ $9^3E_{1u}, ^5B_{1u}, ^5B_{2u}, 3^5E_{1u}$

a_1, particle 2 with particle $a_2, \ldots$, particle n with particle a_n, is written as

$$P = \begin{pmatrix} 1 & 2 & \ldots & n \\ a_1 & a_2 & \ldots & a_n \end{pmatrix}. \quad (8.43)$$

Repeated applications of operation P cause the numbers to circulate in various subsets called cycles. The number of steps in a cycle is called its degree.

The operation containing a_1 cycles of degree 1, a_2 cycles of degree 2, ..., is denoted

$$(1^{a_1} 2^{a_2} \ldots n^{a_n}). \quad (8.44)$$

Now, all operations with the same cycle structure belong to the same class in the full permutation group. Furthermore, each class corresponds to a different partitioning of the number n.

Each distinct partitioning also serves to identify a primitive symmetry species. A particular choice is labeled by

$$\{m^{b_m} \ldots 2^{b_2} 1^{b_1}\} \quad (8.45)$$

and by the corresponding Young diagram

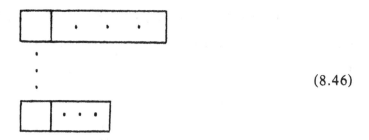

(8.46)

Kets placed in a row of the diagram are bound symmetrically; those placed in a column are bound antisymmetrically.

The submicroscopic system under study may also exhibit symmetry in space. Each operation R of the geometric group then replaces every one-particle ket with a linear combination of these kets. A product of these kets is replaced by a long sum of products with coefficients. The linear combination $|\lambda\rangle$ belonging to the λth primitive symmetry species of the permutation group is replaced by a still longer sum. But this can be rearranged so that the combination $|\lambda\rangle$ factors out of the diagonal terms.

Thus, one can construct an expression for the trace of the matrix representing operation R acting on $|\lambda\rangle$, tr $R^{(\lambda)}$, in terms of the one-particle ket coefficients R_{jk}. Each class of the permutation group is characterized by its structure

$$(1^a 2^b \ldots n^d). \tag{8.47}$$

In this, a cycle of degree i contributes

$$\sum_{j,k,\ldots,m} R_{jk} R_{kl} \ldots R_{mj} = \chi(R^i), \tag{8.47}$$

which is the species character component for operation R repeated i times. The new result is

$$\text{tr} \mathbf{R}^{(\lambda)} = \frac{1}{h} \sum_u y_u^{(\lambda)} P_u \{[\chi(R)]^a [\chi(R^2)]^b \ldots [\chi(R^n)]^d\}, \tag{8.48}$$

where $y_u^{(\lambda)}$ is the component of the λth character for permutation P_u, while h is the order of the permutation group.

Discussion Questions

8.1 Define the symmetric group $\mathscr{S}_n$. What is the order of this group?

8.2 How is a particular permutation represented? How may such a permutation generate cyclic structures within the group?

8.3 What characterizes each class in a symmetric group? How does one calculate the number of elements in such a class?

8.4 What characterizes a primitive symmetry species for a symmetric group? How are such species generated?

8.5 What is (a) a Young diagram, (b) a Young table?

8.6 What conditions does a Young table for a combination ket impose on the combination?

8.7 How does one generate a function meeting these conditions?

8.8 What groups may restrict the eigenkets of a multiparticle system?

8.9 How can kets (8.21) fit into both equations (6.23) and (8.22)?

8.10 When isn't

$$|f\rangle = |1\rangle_j |2\rangle_k \ldots |p\rangle_m$$

a suitable multiparticle ket?

8.11 How may

$$\frac{1}{h} \sum_v y_v^{(\lambda)} P_v$$

act as an identity operator?

8.12 Why can we multiply each term in the expansion of $R|\lambda\rangle$, in equation (8.28), by the inverse permutation operator P_v^{-1}?

8.13 How does equation (8.30) follow from (8.28)?

8.14 How do we obtain equation (8.33) from (8.30)?

8.15 Why does spin interact with orbital motion? Why do the orbital motions of different identical particles interact?

8.16 Why are the orbital kets and the spin kets separately combined to form primitive symmetry species of the pertinent permutation group?

8.17 What are conjugate species? What combinations of these are observed? Why?

8.18 How are Young tables constructed for the independent spin states in a given spin species? How do these yield the mulitiplicity?

8.19 How does one determine the possible configurations of arrays of equivalent particles?

8.20 How are spectroscopic states of molecules labeled by the primitive symmetry species?

Problems

8.1 Find generators for the permutation group $\mathscr{S}_3$. Then construct its Cayley diagram. What geometric groups have the same structure?

8.2 How is the permutation

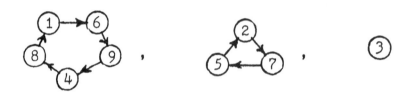

denoted? To what cycle structure does this permutation belong?

8.3 Determine how the (1^4), the three (2^2), and the eight (13) operations affect a product of kets for four identical particles. From the results, generate a superposition for the $\{2^2\}$ species.

8.4 A five-particle orbital system possesses $\{1^5\}$ permutational symmetry. How are components of its character related to the components for a single constituent particle?

8.5 Determine the molecular term symbols arising from the $(g_u)^4$ configuration of an ion with icosahedral symmetry $\mathbf{I}_h$, such as $B_{12}H_{12}^{2-}$.

8.6 Construct a set of generators and defining relations for the permutation group $\mathscr{S}_4$. What geometric groups have the same structure?

8.7 Diagram the permutation

$$(1549)(276)(38)$$

and denote its cycle structure.

8.8 To what species of $\mathscr{S}_4$ does the product of a $\{2^2\}$ orbital factor and a $\{2^2\}$ spin factor belong?

8.9 A five-particle orbital system belongs to the $\{21^3\}$ species. How are the components of its character related to components for a single constituent particle?

8.10 Determine the molecular term symbols arising from the $(\pi_u)^2$ configuration of a molecule with $\mathbf{D}_{\infty h}$ symmetry.

References

Books

Hamermesh, M.: 1962, *Group Theory and its Application to Physical Problems*, Addison-Wesley, Reading, MA, pp. 182-278.
  Hamermesh develops the theory of symmetric groups both intuitively and abstractly, in detail.
Wybourne, B. G.: 1970, *Symmetry Principles and Atomic Spectroscopy*, Wiley-Interscience, New York, pp. 1-27.
  The first part of this book surveys the theory of symmetric groups in a comprehensive useful manner. Procedures and results not often presented are included.

Articles

Ford, D. I.: 1972, "Molecular Term Symbols by Group Theory," *J. Chem. Educ.* 336-340.

CHAPTER 9 / Continuous Groups

9.1
Group-Element Continua

The procedures we have developed so far work with groups whose order is finite or denumerably infinite. The number of elements in a given group, however, may be nondenumerably infinite. Furthermore, the elements may make up one or more discrete continua. Throughout each of these continua, an element is characterized by parameters that vary continuously.

In particular, we will be concerned with sets of elements that form continua in the topological sense. According to this, if element B is in the neighborhood of element A and if element D is in the neighborhood of element C, then element DB is in the neighborhood of element CA. Also, B^{-1} is in the neighborhood of A^{-1}.

A continuum containing the identity element contains one or more infinitesimal generators. By a limiting process analogous to integration, these yield the various elements in the continuum. Each other piece of the given group can be generated by combining one of its elements with the elements in the piece containing the identity. So we will focus our work on this piece, which forms a group by itself.

As noted before, a group is any set of elements with a combinatory law such that the set contains

(a) the identity element,
(b) the inverse of each of its elements,
(c) the binary combination of any two of its elements, in order,

and for which

(d) the associative law holds for larger combinations.

Now, a person may represent group elements by operators

$$A, B, \ldots \tag{9.1}$$

that act on operands. The binary combination BA, of condition (c) above, would be the operator that would effect the same result on a following operand as B acting on the result of A acting on the operand. We call this binary combination *multiplication*. It is the operation with respect to which the elements form the group.

One can also consider the combination $B \pm A$. This would have the same effect as adding or subtracting the effect of A on the operand to or from the effect of B on the operand. The binary combination here is called *addition* or *subtraction*. The principal use of subtraction is in constructing infinitesimal operators and infinitesimal generators.

9.2
Infinitesimal Operators for Single-Parameter Groups

With a single-parameter group, all in one piece, a one-to-one relationship exists between element and parameter. Corresponding to each value of the parameter within the specified bounds is an element. Now, the general properties of groups impose conditions on the parameter. And, when the pertinent operations can be defined, the derivative of an element with respect to the parameter can be constructed.

Let us consider a continuous group with elements denoted by

$$A(\alpha), A(\beta), A(\gamma), \ldots \tag{9.2}$$

where $\alpha, \beta, \gamma, \ldots$ are possible values of the parameter. These form a manifold. Property (c) of a group requires that for any allowed α and β, the manifold also contains a unique γ such that

$$A(\gamma) = A(\beta)A(\alpha). \tag{9.3}$$

How elements $A(\beta)$ and $A(\alpha)$ combine determines γ as a function of β and α:

$$\gamma = f(\beta, \alpha). \tag{9.4}$$

Property (a) of a group requires that one element be the identity element. If the parameter for this element equals ω, then equation (9.3) yields

$$A(\alpha) = A(\omega)A(\alpha) = A(\alpha)A(\omega) \qquad (9.5)$$

and equation (9.4) reduces to

$$\alpha = f(\omega, \alpha) = f(\alpha, \omega). \qquad (9.6)$$

Property (b) of a group requires that each element have an inverse. If we let $A(\bar{\alpha})$ be the inverse of $A(\alpha)$, then equation (9.3) yields

$$A(\omega) = A(\bar{\alpha})A(\alpha) = A(\alpha)A(\bar{\alpha}) \qquad (9.7)$$

and equation (9.4) reduces to

$$\omega = f(\bar{\alpha}, \alpha) = f(\alpha, \bar{\alpha}). \qquad (9.8)$$

From associative law (d), we have

$$A(\gamma)[A(\beta)A(\alpha)] = [A(\gamma)A(\beta)]A(\alpha) \qquad (9.9)$$

and

$$f\left(\gamma, f(\beta, \alpha)\right) = f\left(f(\gamma,\beta),\alpha\right) \qquad (9.10)$$

for all possible values of the parameters.

Since the group elements form a continuum, the parameter does too. If we also require function $f(\alpha,\beta)$ in equation (9.4) to be analytic, then the group is a *Lie* group.

Let us allow linear combinations of elements of a Lie group to be constructed. Then we can form

$$\frac{\Delta A}{\Delta \alpha} = \frac{A(\alpha + \Delta \alpha) - A(\alpha)}{\Delta \alpha}. \qquad (9.11)$$

On letting α decrease to zero, we obtain the derivative

$$\frac{dA}{d\alpha} = \lim_{\Delta \alpha \to 0} \frac{A(\alpha + \Delta \alpha) - A(\alpha)}{\Delta \alpha}. \qquad (9.12)$$

The action of this operator varies with α.

When the derivative is constructed about the identity position, the result is the *infinitesimal operator*

$$B = \left(\frac{dA}{d\alpha}\right)_\omega. \tag{9.13}$$

From the assumed continuity of the elements, an element differing slightly from the identity element converts $A(\alpha)$ to $A(\alpha + \Delta\alpha)$:

$$A(\alpha + \Delta\alpha) = A(\omega + \varepsilon)A(\alpha). \tag{9.14}$$

From equation (9.14), the corresponding relation among the parameters is

$$\alpha + \Delta\alpha = f(\omega + \varepsilon, \alpha). \tag{9.14}$$

In a Lie group, function f is analytic; then we have

$$\alpha + \Delta\alpha = f(\omega, \alpha) + \left(\frac{\partial f(\beta,\alpha)}{\partial \beta}\right)_{\beta=\omega} \varepsilon + 0(\varepsilon^2) \tag{9.15}$$

whence

$$\Delta\alpha = \left(\frac{\partial f(\beta,\alpha)}{\partial \beta}\right)_{\beta=\omega} \varepsilon, \tag{9.16}$$

if terms of higher order in ε are neglected. Now, equation (9.12) can be rewritten in the form

$$\frac{dA}{d\alpha} = \lim_{\varepsilon \to 0} \frac{[A(\omega + \varepsilon) - A(\omega)]A(\alpha)}{\varepsilon \left(\frac{\partial f(\beta,\alpha)}{\partial \beta}\right)_{\beta=\omega}}$$

$$= BA(\alpha) \frac{1}{\left(\frac{\partial f(\beta,\alpha)}{\partial \beta}\right)_{\beta=\omega}}. \tag{9.17}$$

Integration of this equation would allow all elements of the group under consideration to be constructed. Thus, B serves as an *infinitesimal generator* of the group.

9.3
The Canonical Parameter

A person can simplify use of infinitesimal generators by restricting each independent parameter to a canonical form.

Consider a single-parameter group, all in one continuous piece. To it, the discussion in Section 9.2 applies. Now, equation (9.17) tells us how a typical element A varies with the parameter α.

Let us introduce a new parameter τ by the condition

$$\frac{d\tau}{d\alpha} = \frac{1}{\left(\dfrac{\partial f(\beta,\alpha)}{\partial \beta}\right)_{\beta=\omega}}. \qquad (9.18)$$

Furthermore, we will consider

$$\tau = 0 \quad \text{when} \quad \alpha = \omega, \qquad (9.19)$$

that is, when A is the identity element. Also, let us represent this element as unity:

$$A(\omega) = 1. \qquad (9.20)$$

We call τ the *canonical parameter* replacing parameter α.

With transformation (9.18), equation (9.17) reduces to

$$dA = BA\, d\tau. \qquad (9.21)$$

Integrating this for a given operator B leads to the form

$$A = e^{B\tau} = e^{\tau B} \qquad (9.22)$$

in which a general element A is expressed in terms of the infinitesimal operator. A person interprets the exponential as the series

$$e^{\tau B} = 1 + \tau B + \tfrac{1}{2}\tau^2 B^2 + \frac{1}{6}\tau^3 B^3 + \ldots$$

$$= \sum_{n=0}^{\infty} \frac{\tau^n B^n}{n!}, \qquad (9.23)$$

in which the nth term involves operator B iterated n times.

Note that the infinitesimal operator may be determined in the canonical system; then we have

$$B = \left(\frac{dA}{d\tau}\right)_{\tau=0}. \qquad (9.24)$$

Two kinds of groups can be distinguished: those for which τ assumes all values from $-\infty$ to ∞ and those for which τ lies between finite limits. Groups of the second kind are said to be *closed* or *compact*.

Example 9.1

For a one-parameter group, construct an element A infinitesimally close to the identity element 1.

Let A in equation (9.21) be given by equation (9.20). Then we obtain

$$dA = B\, d\tau$$

for an infinitesimal change from the identity element. Letting the infinitesimal change in τ be $\delta\tau$, we consequently have

$$A(\delta a) = 1 + B\, \delta\tau$$

$$= 1 + \left(\frac{dA}{d\tau}\right)_0 \delta\tau.$$

Example 9.2

Show how all elements depending on parameter τ alone can be generated by a limiting process.

Construct the expression

$$\left(1 + B\frac{\tau}{k}\right)^k$$

with a given τ. Note that as k increases, factor τ/k decreases, becoming the infinitesimal $\delta\tau$ as k increases without limit. The expression in parenthesis reduces to that in Example 9.1. But from the definition of the exponential, we have

$$\lim_{k\to\infty}\left(1 + B\frac{\tau}{k}\right)^k = e^{\beta\tau} = A(\tau).$$

Since $1 + B\delta\tau$ is an element of the group, the limit $e^{\beta\tau}$ is. We identify it as $A(\tau)$ in the last equality.

The limiting process carried out here is referred to as *exponentiation* of the infinitesimal generator B.

9.4
Form-Preserving Transformations of Functions

Each element of a Lie group may transform a pertinent function while preserving aspects of its form. The actions may be considered either passive, acting on the reference frame, or active, altering the physical representation of the function. Candidate functions include those describing symmetric gravitational fields, electromagnetic fields, color fields, density fields, velocity fields, actual distributions, and potential distributions. They include symmetric quantum mechanical state functions.

In our discussion, let us consider the independent variables to be Cartesian coordinates in a Euclidean space. Let us consider the coordinates to be three in number, labeled x, y, z. We also suppose the function to be $F(x, y, z)$ before a particular transformation and $F'(x, y, z)$ after it.

Furthermore, an equivalent point in the function is considered to be at (x, y, z) before the transformation and at (x', y', z') after it. Also, the movement from this initial point to its transformation is measured by a canonical parameter τ. Thus

$$x' = X(x, y, z, \tau), \qquad (9.25)$$

$$y' = Y(x, y, z, \tau), \qquad (9.26)$$

Form-Preserving Transformations of Functions

$$z' = Z(x, y, z, \tau). \tag{9.27}$$

As τ approaches zero, x', y', and z' approach x, y, and z. So differentiating equations (9.25), (9.26), (9.27) at $\tau = 0$ yields

$$\frac{\partial x}{\partial \tau} = \xi(x, y, z), \tag{9.28}$$

$$\frac{\partial y}{\partial \tau} = \eta(x, y, z), \tag{9.29}$$

$$\frac{\partial z}{\partial \tau} = \zeta(x, y, z), \tag{9.30}$$

whence

$$\frac{dx}{\xi} = \frac{dy}{\eta} = \frac{dz}{\zeta} = d\tau. \tag{9.31}$$

Equations (9.31) are the *characteristic differential equations* for the transformation.

During a transformation, each infinitesimal change in a function F obeys

$$dF = \frac{\partial F}{\partial x} dx + \frac{\partial F}{\partial y} dy + \frac{\partial F}{\partial z} dz. \tag{9.32}$$

But from equation (9.31), we have

$$dx = \xi \, d\tau, \qquad dy = \eta \, d\tau, \qquad dz = \zeta \, d\tau. \tag{9.33}$$

Substituting these into equation (9.32) yields

$$dF = \left(\xi \frac{\partial}{\partial x} + \eta \frac{\partial}{\partial y} + \zeta \frac{\partial}{\partial z} \right) F \, d\tau. \tag{9.34}$$

The expression in parenthesis is an infinitesimal operator like the B in equation (9.21):

$$B = \xi \frac{\partial}{\partial x} + \eta \frac{\partial}{\partial y} + \zeta \frac{\partial}{\partial z}. \tag{9.35}$$

So equation (9.34) becomes

$$dF = BF\, d\tau. \tag{9.36}$$

A formal integration of equation (9.36) yields

$$F' = e^{B\tau} F. \tag{9.37}$$

For quantum mechanical systems, the product of the state function with its complex conjugate is invariant in a symmetry operation. This condition is met here if

$$F'^*F' = F^*F. \tag{9.38}$$

But equations (9.37) and (9.38) are consistent only if

$$(e^{B\tau})^* e^{B\tau} = e^0 = 1. \tag{9.39}$$

With τ real, a numerical representation of operator B must satisfy

$$B^* + B = 0. \tag{9.40}$$

This holds if the eigenvalues of B are imaginary. We then have

$$B = iC \tag{9.41}$$

with C an operator with real eigenvalues, a *Hermitian* operator. The corresponding B is said to be *anti-Hermitian*.

Example 9.3

Determine the finite transformations of the group whose infinitesimal operator is

$$x\frac{\partial}{\partial x} + y\frac{\partial}{\partial y} + z\frac{\partial}{\partial z}.$$

On comparing this operator with equation (9.35), we see that

$$\xi = x, \quad \eta = y, \quad \zeta = z$$

and equations (9.31) become

$$\frac{dx}{x} = \frac{dy}{y} = \frac{dz}{z} = d\tau.$$

These characteristic equations integrate to yield

$$\ln \frac{x'}{x} = \ln \frac{y'}{y} = \ln \frac{z'}{z} = \tau,$$

whence

$$x' = xe^\tau, \quad y' = ye^\tau, \quad z' = ze^\tau.$$

Thus, the transformation amounts to a uniform scaling of the Cartesian coordinates.

9.5
Canonical Coordinates

With a particular coordinate system, a one-parameter transformation becomes a simple translation.

In section 9.4, we noted that along the path of the transformation

$$x' = X(x, y, z, \tau), \qquad (9.42)$$

$$y' = Y(x, y, z, \tau), \qquad (9.43)$$

$$z' = Z(x, y, z, \tau), \qquad (9.44)$$

we have

$$\frac{dx}{\xi} = \frac{dy}{\eta} = \frac{dz}{\zeta} = d\tau, \qquad (9.45)$$

where ξ, η, ζ are the partial derivatives of X, Y, and Z with respect to τ.

When equations (9.45) can be integrated, they yield independent solutions of the type

$$G(x', y', z') = G(x, y, z) + \tau = A + \tau, \qquad (9.46)$$

$$H(x', y', z') = H(x, y, z) = B, \qquad (9.47)$$

$$I(x', y', z') = I(x, y, z) = C. \qquad (9.48)$$

Now, coordinates u, v, w may be introduced by the definitions

$$u = G(x, y, z), \qquad (9.49)$$

$$v = H(x, y, z), \qquad (9.50)$$

$$w = I(x, y, z). \qquad (9.51)$$

These are *canonical coordinates* for the transformation. In terms of them, the transformation reduces to the form

$$u' = u + \tau, \qquad (9.52)$$

$$v' = v, \qquad (9.53)$$

$$w' = w, \qquad (9.54)$$

which is a simple translation along the u axis.

Of interest in many discussions are the properties or functions that do not change. Here equations (9.47) and (9.48) are such; they are *invariant functions* under the transformations of the group.

Example 9.4

Determine canonical coordinates for the group for which the infinitesimal operator is

$$x \frac{\partial}{\partial x} + y \frac{\partial}{\partial y} + z \frac{\partial}{\partial z}.$$

With equations (9.35) and (9.31), we find that

$$\frac{dx}{x} = \frac{dy}{y} = \frac{dz}{z} = d\tau.$$

Solutions of these equations include

$$\ln x = \tau + \ln a = \tau + A,$$

$$\ln \frac{x}{y} = \ln b = B,$$

$$\ln \frac{x}{z} = \ln c = C.$$

So equations (9.49) through (9.51) become

$$u = \ln x,$$

$$v = \ln \frac{x}{y},$$

$$w = \ln \frac{x}{z}.$$

9.6
Elements of Single-Piece Multiparameter Groups

Independent single-parameter groups can be combined to produce multiparameter groups. Conversely, a given multiparameter group can be broken down into single parameter groups. Each of these has its own infinitesimal operator, behaving as we have seen. For the combination of such groups to be a group, however, the different infinitesimal operators must meet certain conditions. Some of these will be discussed here; others, in section 9.8.

Consider a group, all in one continuous piece, that involves n independent parameters. Let a possible set of these in canonical form be

$$\tau = (\tau_1, \tau_2, \ldots, \tau_n). \tag{9.55}$$

Then a typical element of the group is represented by

$$A(\tau) = A(\tau_1, \tau_2, \ldots, \tau_n) \tag{9.56}$$

Furthermore, the jth infinitesimal operator would be

$$B_j = \left(\frac{\partial A}{\partial \tau_j}\right)_{\tau=0}. \qquad (9.57)$$

Since the τ_j's are independent, the different B_j's are independent infinitesimal operators.

By exponentiation, we find that the element when only parameter τ_j is different from zero is

$$A(\ldots, 0, \tau_j, 0, \ldots) = e^{B_j \tau_j}, \qquad (9.58)$$

while the general element in the continuum containing the identity element is

$$A(\tau_1, \ldots, \tau_n) = e^{B_1 \tau_1} \ldots e^{B_n \tau_n}. \qquad (9.59)$$

Thus, the B_j's serve as infinitesimal generators for the group.

Now, one can keep the ratios between the parameters constant while moving away from the identity element. Thus, we may set

$$\tau_j = c_j \tau \quad \text{where} \quad j = 1, 2, \ldots, n, \qquad (9.60)$$

with each c_j real. Substituting into equation (9.59) then yields

$$A = e^{\Sigma B_j c_j \tau} = e^{(\Sigma c_j B_j)\tau} = 1 + \Sigma c_j B_j \tau + \ldots \qquad (9.61)$$

and

$$\left(\frac{dA}{d\tau}\right)_{\tau=0} = \Sigma c_j B_j. \qquad (9.62)$$

The result is a one-parameter subgroup with $\Sigma c_j B_j$ the infinitesimal operator and τ the parameter. Since there is no limitation on the c_j's, any linear combination of infinitesimal operators is an infinitesimal operator.

Next, consider the parameters of the given group to be plotted as Cartesian coordinates in a Euclidean space. Such a space is called the *manifold* for the parameters. If all elements are represented by points within a finite region of the manifold, the group is said to be *compact*. But if there is no finite boundary in any direction, the group is *noncompact*.

If any closed loop in the manifold can be continuously deformed to a point, the group is said to be *simply connected*. Among the groups with infinitesimal operators having the same algebraic properties, the simply connected one is called the *universal covering group* for the operators.

A subgroup whose classes are classes of the original larger group is called a *normal* or *invariant subgroup*. If S is a simply connected Lie group and N is a discrete normal subgroup, one can construct a factor group G such that

$$S = NG. \qquad (9.63)$$

This group would possess the same infinitesimal operator algebra. But, whenever N contains more than one element, G would be multiply connected. (Examples will be considered later.)

9.7
Important Lie Groups

Lie groups can be classified by their realizations. Thus, they may be represented by closed sets of transformations acting in a vector space. A typical transformation links a new position to an old one by analytic functions.

The simplest relationship is the *homogeneous linear* transformation, which can be expressed by the matrix equation

$$\mathbf{r}' = \mathbf{A}\mathbf{r}. \qquad (9.64)$$

Here $\mathbf{A}$ is a square $n \times n$ matrix while $\mathbf{r}$ and $\mathbf{r}'$ are n-element column matrices containing the coordinates before and after the transformation. The corresponding group element is represented by $\mathbf{A}$.

An *affine* group consists of this transformation combined with a translation. Thus, it involves the nonhomogeneous transformation

$$\mathbf{r}' = \mathbf{A}\mathbf{r} + \mathbf{a} \qquad (9.65)$$

where $\mathbf{a}$ is an additional n-element column matrix.

Continuous Groups

A *fractional linear group* in two variables consists of all transformations of the type

$$x' = \frac{a_{11}x + a_{12}y + a_{13}}{a_{31}x + a_{32}y + a_{33}}, \qquad (9.66)$$

$$y' = \frac{a_{21}x + a_{22}y + a_{23}}{a_{31}x + a_{32}y + a_{33}}. \qquad (9.67)$$

More variables can, of course, be introduced.

The homogeneous linear transformations are classified further, as follows:

A *general linear group* in n dimensions is made up of all nonsingular $n \times n$ matrices with real (r), complex (c), or quaternion (q) components. It is designated **GL**(n, r), **GL**(n, c), or **GL**(n, q), respectively. The number of parameters equals n^2, $2n^2$, and $4n^2$. Each of the parameters would have a specified range, which may be infinite.

Various subgroups arise when conditions are imposed on the transformations. Thus, the transformations might not alter volumes in the vector space or/and they might not alter certain metrics.

The volume-preserving groups are called *special linear groups*. Picking out the matrices of **GL**(n, r) with determinants equal to 1 yields the **SL**(n, r) group. With **GL**(n, c), one may choose those with real determinants equal to 1, getting the **SL**$_1$(n, c) group. Or, one may choose those with complex determinants of modulus unity, obtaining the **SL**$_2$(n, c) group. With **GL**(n, q), only the unimodular subgroup **SL**(n, q) arises. In each case, the number of independent parameters is reduced by 1. The subgroup **SL**(n, c) consists of those matrices in **GL**(n, c) with determinants whose real part equals 1 and whose imaginary part vanishes. Here the number of independent parameters is reduced by 2.

Groups preserving bilinear symmetric metrics are said to be *orthogonal*. When the metric tensor **G** is then diagonalized, positive numbers may occur n_+ times and negative numbers may occur n_- times on the diagonal. Zeros will not occur there as long as the metric is nonsingular. The sum $n_+ + n_- = n$ is the overall dimensionality. The corresponding groups are designated **O**(n, r) = **O**(n_+, n_-, r), **O**(n, c) = **O**(n_+, n_-, c), **O**(n, q) = **O**(n_+, n_-, q), depending on whether they are represented by real (r), complex (c), or quaternionic (q) matrices.

Important Lie Groups

Consider a group represented by the $n \times n$ matrices **A**. Preservation of the metric when matrix **A** acts is then ensured if

$$\mathbf{A}\mathbf{G}\tilde{\mathbf{A}} = \mathbf{G}, \tag{9.68}$$

where $\tilde{\mathbf{A}}$ is the transpose of **A**. In index notation, this equation becomes

$$\sum_{j,k} A_{ij} G_{jk} A_{lk} = G_{il}. \tag{9.69}$$

As long as the metric is symmetric, interchanging i and l in equation (9.69) does not essentially alter the equation. Consequently, only $\frac{1}{2}n(n + 1)$ equations in the set are independent. For the real groups $\mathbf{O}(n, r)$, the number of independent parameters is

$$n^2 - \tfrac{1}{2}n(n + 1) = \tfrac{1}{2}n(n - 1). \tag{9.70}$$

For the complex groups $\mathbf{O}(n, c)$, the number of independent parameters is twice as great, namely

$$n(n - 1). \tag{9.71}$$

From equation (9.68), the determinant of **A** equals either $+1$ or -1. In the former situation, the transformations represented are said to be *proper;* in the latter situation, *improper*. Only the proper operations can be reached continuously from the identity operation.

The *special orthogonal groups* are the subgroups of orthogonal groups that also preserve volumes. They include $\mathbf{SO}(n, r)$ and $\mathbf{SO}(n, c)$, from the real and the complex orthogonal groups. The number of parameters is not altered from equations (9.70) and (9.71), respectively.

Groups preserving sesquilinear symmetric metrics are called *unitary*. They occur in complex and quaternion forms, designated $\mathbf{U}(u, c)$ and $\mathbf{U}(n, q)$, respectively. A unitary group in n dimensions consists of the $n \times n$ matrices **A** satisfying the equation

$$\mathbf{A}\mathbf{G}\mathbf{A}^\dagger = \mathbf{G}, \tag{9.72}$$

where $\mathbf{A}^\dagger$ is the Hermitian adjoint of **A**.

For the complex unitary group, equation (9.72) yields

$$\sum_{j,k} A_{ij} G_{jk} A_{lk}^* = G_{il}. \qquad (9.73)$$

As with equation (9.69), only $\frac{1}{2}n(n + 1)$ of these equations are independent. But for $l = i$, the complex conjugate of each term on the left also appears in the sum. So the imaginary parts cancel regardless of how the elements of **A** are chosen. We are left with only n conditions here. For $l \neq i$, on the other hand, this automatic canceling no longer occurs. The imaginary terms, as well as the real terms, yield conditions on **A**. We obtain

$$2[\tfrac{1}{2}n(n + 1) - n] = n^2 - n \qquad (9.74)$$

conditions. Since each component of the matrix **A** may be complex, we have $2n^2$ parameters in all. Subtracting out the number of conditions,

$$2n^2 - n - (n^2 - n) = n^2, \qquad (9.75)$$

gives us n^2 independent real parameters.

The *special unitary groups* are subgroups of the unitary groups that also preserve volumes. In particular, **SU**(n, c) is the subgroup of **U**(n, c) with unimodular matrices. The number of independent parameters is thus reduced by 1.

Groups preserving bilinear antisymmetric metrics are called *symplectic*. They occur as **Sp**(n, r) and **Sp**(n, c) with n even, and as **Sp**(n, q). A symplectic group in n dimensions consists of the $n \times n$ matrices **A** satisfying the equation

$$\mathbf{A G \tilde{A}} = \mathbf{G} \qquad (9.76)$$

with

$$G_{ij} = -G_{ji}. \qquad (9.77)$$

In index form, we again have equation (9.69). But equation (9.77) ensures that the equations are satisfied when $l = i$, regardless of how the components are chosen. So, we now have only

$$\tfrac{1}{2}n(n + 1) - n = \tfrac{1}{2}n(n - 1) \qquad (9.78)$$

Important Lie Groups

conditions on these. For the real symplectic group, we obtain

$$n^2 - \tfrac{1}{2}n(n-1) = \tfrac{1}{2}n(n+1) \qquad (9.79)$$

independent parameters; for the complex symplectic group, twice this or

$$n(n+1) \qquad (9.80)$$

independent parameters.

The *special symplectic groups* are subgroups of symplectic groups that also preserve volumes. They include **SSp**(n, r) and **SSp**(n, c), from the real and the complex symplectic groups. A possible **G** here has the form

$$\mathbf{G} = \begin{pmatrix} 0 & 1 & 0 & 0 & \\ -1 & 0 & 0 & 0 & \\ 0 & 0 & 0 & 1 & \cdots \\ 0 & 0 & -1 & 0 & \\ & \vdots & & & \cdots \end{pmatrix}. \qquad (9.81)$$

In physics, a person must consider symmetries with respect to Lorentz transformations and space-time translations. The *Lorentz group* consists of all possible homogeneous *Lorentz transformations*, while the *Poincare group* consists of these together with the translations.

The Lorentz group is made up of all real 4×4 matrices **A** satisfying

$$\mathbf{A}\mathbf{G}\tilde{\mathbf{A}} = \mathbf{G} \qquad (9.82)$$

with

$$\mathbf{G} = \begin{pmatrix} 1 & 0 & 0 & 0 \\ 0 & 1 & 0 & 0 \\ 0 & 0 & 1 & 0 \\ 0 & 0 & 0 & -1 \end{pmatrix}, \qquad (9.83)$$

and is labeled **O**(3, 1; r). In index form, we have

$$\sum_{j,k} A_{ij} G_{jk} A_{lk} = G_{il}. \tag{9.84}$$

But since interchanging i and l does not essentially alter the equation (**G** is symmetric), only 10 of these are independent. The number of independent parameter equals

$$16 - 10 = 6. \tag{9.85}$$

Note that the Lorentz group is not simply connected. The universal covering group for it is the **SL**(2, c) group. Since the determinant of **Ã** equals the determinant of **A**, equation (9.82) leads to

$$\det \mathbf{A} = \pm 1. \tag{9.86}$$

The Lorentz group with det **A** = +1 is called **SO**(3, 1; r). This proper group is an invariant subgroup of the Lorentz group.

9.8
Commutators of the Infinitesimal Operators

Infinitesimal operators for multiparameter groups can act on one another. The result, however, is not an infinitesimal operator. Instead the *commutator*, defined as

$$[B_j, B_k] = B_j B_k - B_k B_j \tag{9.87}$$

is.

Consider a group in which all elements vary continuously with n independent canonical parameters, as in section 9.6. Take the two elements

$$A(\ldots, 0, \tau_j, 0, \ldots) = e^{B_j \tau_j} = 1 + B_j \tau_j + \tfrac{1}{2}(B_j \tau_j)^2 + \ldots, \tag{9.88}$$

$$A(\ldots, 0, \tau_k, 0, \ldots) = e^{B_k \tau_k} = 1 + B_k \tau_k + \tfrac{1}{2}(B_k \tau_k)^2 + \ldots, \tag{9.89}$$

together with their inverses

Commutators of the Infinitesimal Operators

$$A^{-1}(\ldots, 0, \tau_j, 0, \ldots) = e^{-B_j\tau_j} = 1 - B_j\tau_j + \tfrac{1}{2}(B_j\tau_j)^2 - \ldots, \quad (9.90)$$

$$A^{-1}(\ldots, 0, \tau_k, 0, \ldots) = e^{-B_k\tau_k} = 1 - B_k\tau_k + \tfrac{1}{2}(B_k\tau_k)^2 - \ldots. \quad (9.91)$$

Respecting the fact that B_j does not generally commute with B_k, construct the product of equations (9.88), (9.89), (9.90), and (9.91):

$$A(\ldots, 0, \tau_j, 0, \ldots) A(\ldots, 0, \tau_k, 0, \ldots) \times$$
$$A^{-1}(\ldots, 0, \tau_j, 0, \ldots) A^{-1}(\ldots, 0, \tau_k, 0, \ldots) =$$
$$1 + (B_j B_k - B_k B_j)\tau_j \tau_k + \ldots. \quad (9.92)$$

Since expression (9.92) is a product of elements of the group, it is an element of the group. But any element may be written in form (9.61).

To transform expression (9.92) to this form, we set

$$\tau_j = b_j \tau^{\tfrac{1}{2}}, \quad \tau_k = b_k \tau^{\tfrac{1}{2}}, \quad (9.93)$$

with b_j and b_k real. Then we have

$$1 + (B_j B_k - B_k B_j) b_j b_k \tau + \ldots = 1 + \sum_l c_l B_l \tau + \ldots. \quad (9.94)$$

Since equation (9.94) is an identity in τ, we find that

$$B_j B_k - B_k B_j = \frac{1}{b_j b_k} \Sigma c_l B_l \quad (9.95)$$

or

$$[B_j, B_k] = \sum_l c^l_{jk} B_l, \quad (9.96)$$

if

$$c^l_{jk} = \frac{c_l}{b_j b_k}. \quad (9.97)$$

Coefficient c^l_{jk} is called the lth *structure constant* for commutator $[B_j, B_k]$. Because c_l, b_j, and b_k are real, c^l_{jk} is real.

From equation (9.87), the structure constants must meet certain conditions. Indeed since

$$[B_j, B_k] = -[B_k, B_j], \qquad (9.98)$$

we have

$$c^l_{jk} = -c^l_{kj}. \qquad (9.99)$$

And since

$$\{[B_j B_k], B_l\} + \{[B_k B_l], B_j\} + \{[B_l, B_j], B_k\} = 0, \qquad (9.100)$$

we have

$$\sum_r (c^r_{jk} c^s_{rl} + c^r_{kl} c^s_{rj} + c^r_{lj} c^s_{rk}) = 0. \qquad (9.101)$$

When the behavior of group elements near the identity element is given, formula (9.57) can be used to construct expressions for the infinitesimal operators. These can be employed in equation (9.96) to get the structure constants. Conversely, when the structure constants are given, matrices of the requisite size (dimensionality) satisfying equation (9.96) may be constructed. These may be employed in equation (9.59) to form the elements. Thus, the structure constants serve to define a single-piece, simply connected Lie group.

9.9
Lie Algebras

A complete set of infinitesimal operators not only generates the piece of a Lie group containing the identity but also forms the basis for an algebra.

Consider the independent infinitesimal operators

$$B_1, B_2, \ldots, B_n. \qquad (9.102)$$

These generate the elements of a single piece Lie group through exponentiation. By equation (9.62), each linear combination

$$\sum_j c_j B_j \qquad (9.103)$$

Lie Algebras

of these is also an infinitesimal generator for the group. However, the combination

$$B_j B_k \tag{9.104}$$

is not an infinitesimal generator; instead, the commutator $[B_j, B_k]$ is. By equation (9.96), we have

$$[B_j, B_k] = B_j B_k - B_k B_j = \sum_l c^l_{jk} B_l. \tag{9.105}$$

The set of elements (9.102) together with composition rules (9.103) and (9.105), analogous to addition and multiplication, form a closed algebra. Because the corresponding group is a Lie group, it is called the *Lie algebra* for the group.

One may factor an i out of each operator (9.102):

$$B_j = iC_j. \tag{9.106}$$

We then have the set of independent operators

$$C_1, C_2, \ldots, C_n \tag{9.107}$$

and the dependent operator

$$\sum_j c_j C_j, \tag{9.108}$$

as in (9.102) and (9.103). But equation (9.105) is replaced with

$$[C_j, C_k] = \sum_l -i c^l_{jk} C_l. \tag{9.109}$$

Since c^l_{jk} is real, the coefficient of C_l on the right is imaginary. Because of the direct relationship (9.106), conditions (9.107), (9.108), and (9.109) define the Lie algebra as well as (9.102), (9.103), and (9.105).

Now, the independent quantum mechanical operators in a given set are characterized by their commutation relations. In general, they also combine linearly to give operators in the set. Furthermore, they are Hermitian so they meet conditions (9.108) and (9.109); they form a Lie algebra.

A given Lie algebra may correspond to various Lie groups with different boundaries on their manifolds. Of these, only one is simply

9.10
The U(1, c) Group

There are two 1-dimensional unitary groups, $\mathbf{U}(1, c)$ and $\mathbf{U}(1, q)$, but only the complex form is common.

For it, matrix $\mathbf{A}$ reduces to a single element and metric $\mathbf{G}$ becomes number 1. With the element of $\mathbf{A}$ a complex number, equation (9.72) then reduces to

$$AA^* = 1. \qquad (9.110)$$

This is satisfied by

$$A = e^{im\alpha} \qquad (9.111)$$

where

$$\alpha = \frac{2\pi}{n} \qquad (9.112)$$

with m and n any integers. Thus, the group is represented by the infinite set of exponentials (9.111).

For a quantum mechanical system, the probability density

$$\rho = \Psi^*\Psi \qquad (9.113)$$

is associated with measurements, not the state function Ψ itself. But in the transformation

$$\Psi' = e^{im\alpha}\Psi, \qquad (9.114)$$

this function ρ is not altered. Consequently, $\mathbf{U}(1, c)$ is a symmetry group for any quantum mechanical system. In discussions where quaternions are not being used, the c may be dropped and the group designated $\mathbf{U}(1)$.

9.11
The U(2, c) Group

The 2-dimensional unitary group is represented by 2×2 matrices symbolized by **A**. Each of these contains four elements. The elements may be complex or quaternion. Only the former will be considered here.

One may consider the matrices to act on vectors in a plane. With rectangular coordinates in a Euclidean plane, the metric would be

$$G_{jk} = \delta_{jk}. \qquad (9.115)$$

Then equation (9.72) reduces to the unitary condition

$$\mathbf{A}\mathbf{A}^\dagger = \mathbf{I}, \qquad (9.116)$$

which imposes four conditions on the four complex numbers. We are left with four independent parameters. So we must have four independent infinitesimal generators.

In the neighborhood of the identity, one may let

$$\mathbf{A} = \mathbf{I} + i\mathbf{M}, \qquad (9.117)$$

with matrix **M** small. Substituting into equation (9.116) yields

$$\mathbf{I} = (\mathbf{I} + i\mathbf{M})(\mathbf{I} - i\mathbf{M}^\dagger) = \mathbf{I} + i\mathbf{M} - i\mathbf{M}^\dagger + \mathbf{M}\mathbf{M}^\dagger. \qquad (9.118)$$

When **M** is infinitesimal, the last term may be neglected and we have

$$\mathbf{M} = \mathbf{M}^\dagger. \qquad (9.119)$$

Thus, **M** is Hermitian.

The general Hermitian 2×2 matrix is expressible as a linear combination of four standard matrices:

$$\begin{pmatrix} w + z & x + iy \\ x - iy & w - z \end{pmatrix} = $$
$$x \begin{pmatrix} 0 & 1 \\ 1 & 0 \end{pmatrix} + y \begin{pmatrix} 0 & i \\ -i & 0 \end{pmatrix} + z \begin{pmatrix} 1 & 0 \\ 0 & -1 \end{pmatrix} + w \begin{pmatrix} 1 & 0 \\ 0 & 1 \end{pmatrix}. \qquad (9.120)$$

For the infinitesimal matrix, quantities x, y, z, and w would be infinitesimal. Correspondingly, the four infinitesimal generators are

$$\sigma_x = \begin{pmatrix} 0 & 1 \\ 1 & 0 \end{pmatrix}, \quad \sigma_y = \begin{pmatrix} 0 & i \\ -i & 0 \end{pmatrix}, \quad \sigma_z = \begin{pmatrix} 1 & 0 \\ 0 & -1 \end{pmatrix}, \quad (9.121)$$

and

$$\mathbf{I} = \begin{pmatrix} 1 & 0 \\ 0 & 1 \end{pmatrix}. \quad (9.122)$$

Expressions (9.121) are called the *Pauli spin matrices*.

The group **SU**(2, c) is subject to the additional constraint that the determinant of **A** is unity:

$$\det \mathbf{A} = 1. \quad (9.123)$$

For operators in the neighborhood of unity, we have to first order

$$\det (\mathbf{I} + i\mathbf{M}) = 1 + i \operatorname{Tr} \mathbf{M}. \quad (9.124)$$

The right side reduces to 1 and constraint (9.123) is satisfied when the trace of **M** vanishes. But, this vanishing eliminates **I** from the generators, leaving σ_x, σ_y, and σ_z.

The commutators for these generators are

$$[\sigma_x, \sigma_y] = -2i\sigma_z, \quad [\sigma_y, \sigma_z] = -2i\sigma_x, \quad [\sigma_z, \sigma_x] = -2i\sigma_y. \quad (9.125)$$

Also note that

$$\sigma_x^2 = \mathbf{I}, \quad \sigma_y^2 = \mathbf{I}, \quad \sigma_z^2 = \mathbf{I},$$

and

$$\sigma_x^2 + \sigma_y^2 + \sigma_z^2 = 3\mathbf{I}. \quad (9.127)$$

Eigenvalue equations can be constructed with the Pauli spin matrices. Indeed, the two element column eigenmatrices of σ_z are

$$\alpha = \begin{pmatrix} 1 \\ 0 \end{pmatrix}, \quad \beta = \begin{pmatrix} 0 \\ 1 \end{pmatrix}, \quad (9.128)$$

because we have

$$\sigma_z \alpha = \alpha, \qquad (9.129)$$

$$\sigma_z \beta = -\beta. \qquad (9.130)$$

Thus, the eigenvalues for σ_z are $+1$ and -1. Furthermore, α and β are eigenmatrices of $\sigma_x^2 + \sigma_y^2 + \sigma_z^2$ with the eigenvalue 3.

In discussions where quaternions are not being used, the 2-dimensional special unitary group is designated **SU**(2). In the next chapter, we will consider the relationship of this group to the special orthogonal group in 3-dimensional Euclidean space, **SO**(3, r), which we will designate **SO**(3). The 2-dimensional subgroup of **SO**(3) will be designated **SO**(2).

Discussion Questions

9.1 Cite examples of continua.
9.2 When is a set of elements (a) 1-dimensional, (b) n-dimensional?
9.3 Under what conditions does a set of elements form a group?
9.4 When does a continuum of elements not containing the identity form a piece of a group?
9.5 When can a set of elements be parameterized?
9.6 If the given elements depend on a single parameter and form a group, how do any two values of the parameter combine?
9.7 What does the associative law tell us about how different elements of a one-parameter group combine?
9.8 How do we interpret the action on an operand of (a) $A(\beta) A(\alpha)$, (b) $A(\beta) - A(\alpha)$?
9.9 How is the derivative of group element A with respect to parameter a defined?
9.10 When is this derivative an infinitesimal generator for the group?
9.11 How is the canonical parameter τ obtained?
9.12 How is an element infinitesimally close to the identity element constructed?
9.13 Explain the exponentiation of an infinitesimal generator.
9.14 Distinguish between passive and active transformations.
9.15 How are the characteristic differential equations for transforming coordinates in physical space obtained?

9.16 How is an infinitesimal operator for a physical function constructed for these characteristic equations?

9.17 Determine the condition that an infinitesimal operator for a quantum mechanical state function must satisfy.

9.18 How are canonical coordinates constructed? How do these behave in a transformation?

9.19 How are multiparameter groups related to single-parameter groups? What one-parameter subgroups exist for a given multiparameter group?

9.20 When are two infinitesimal generators for a group independent?

9.21 Show that each linear combination of independent infinitesimal generators is an infinitesimal generator.

9.22 Define (a) universal covering group, (b) normal or invariant subgroup.

9.23 What kinds of matrices may serve as elements of a Lie group?

9.24 How are these matrices limited for (a) orthogonal groups, (b) unitary groups, (c) symplectic groups?

9.25 What additional limitation is involved for the special groups?

9.26 Using the notion that infinitesimal generators may act as differentiating operators, explain why the product of two infinitesimal generators is not generally an infinitesimal generator.

9.27 Prove that the commutator of two infinitesimal generators is an infinitesimal generator.

9.28 Explain what the Lie algebra for a group is.

9.29 Show that a numerical representation of the **U**(1) group is

$$1, e^{im_1 a}, e^{im_2 a}, \ldots .$$

9.30 Why is **U**(1) a symmetry group for a coherent quantum mechanical system?

9.31 How are the representations of **U**(2) related to the Pauli spin matrices?

9.32 How are these representations altered for group **SU**(2)?

9.33 Show that for

$$\det (\mathbf{I} + i\mathbf{M}) = 1$$

to first order, the trace of **M** must vanish.

Problems

9.1 Construct the general element for a single-parameter group with the infinitesimal generator ik and canonical parameter a.

9.2 A function of angle ϕ has the form

$$\Phi = e^{iM\phi}$$

before a transformation and the form

$$\Phi' = e^{iM(\phi+a)}$$

after. Determine the infinitesimal operator for the group of transformations.

9.3 Find the finite transformations of the one-parameter group whose infinitesimal operator is

$$x\frac{\partial}{\partial y} - y\frac{\partial}{\partial x} + \lambda\frac{\partial}{\partial z}.$$

9.4 A group consists of the transformations

$$x' = ax, \qquad y' = a^2 y.$$

Show how the substitutions

$$u = \ln x, \qquad v = \frac{y}{x^2}$$

reduce them to a canonical form.

9.5 A group consists of the transformations

$$\begin{pmatrix} x' \\ y' \\ z' \end{pmatrix} = \begin{pmatrix} \cos\theta & -\sin\theta & 0 \\ \sin\theta & \cos\theta & 0 \\ 0 & 0 & 1 \end{pmatrix} \begin{pmatrix} x \\ y \\ z \end{pmatrix}.$$

Determine an infinitesimal operator for the group.

Continuous Groups

9.6 The number of independent equations in (9.69) is

$$S = n + (n - 1) + (n - 2) + \ldots + 1.$$

Evaluate this series.

9.7 From the matrix representations of the Pauli spin matrices, show that

$$\sigma_x^2 = \mathbf{I}, \qquad \sigma_y^2 = \mathbf{I}, \qquad \sigma_z^2 = \mathbf{I}.$$

9.8 Show that matrices (9.128) satisfy the eigenvalue equations

$$\sigma_z \alpha = \alpha, \qquad \sigma_z \beta = -\beta.$$

9.9 If the general element of a Lie group has the form

$$\mathbf{U} = e^{iMa},$$

what is its infinitesimal generator?

9.10 A function of distance x has the form

$$\psi = A e^{ikx}$$

before a transformation and the form

$$\psi' = A e^{ik(x+a)}$$

after. Find an infinitesimal operator for the group of transformations.

9.11 If the infinitesimal operator for a group is

$$x \frac{\partial}{\partial y},$$

what are the corresponding finite transformations?

9.12 A group consists of the transformations

$$\begin{pmatrix} x' + iy' \\ x' - iy' \\ z \end{pmatrix} = \begin{pmatrix} e^{i\phi} & 0 & 0 \\ 0 & e^{-i\phi} & 0 \\ 0 & 0 & 1 \end{pmatrix} \begin{pmatrix} x + iy \\ x - iy \\ z \end{pmatrix}.$$

Determine an infinitesimal operator for the group.

9.13 An infinitesimal operator for a one-parameter group is

$$x^2 \frac{\partial}{\partial x} + xy \frac{\partial}{\partial y}.$$

Show that the differential equation

$$x \frac{dy}{dx} - y = F\left(\frac{y}{x}\right)$$

is invariant under operations of the group.

9.14 Verify the equation

$$\exp(B_j \tau_j) \exp(B_k \tau_k) \exp(-B_j \tau_j) \exp(-B_k \tau_k)$$
$$= 1 + [B_j, B_k] \tau_j \tau_k + \text{higher terms}.$$

9.15 Show that

$$[\sigma_x, \sigma_y] = -2i\sigma_z,$$

$$[\sigma_y, \sigma_z] = -2i\sigma_x,$$

$$[\sigma_z, \sigma_x] = -2i\sigma_y.$$

References

Books

Cornwell, J. F.: 1984, *Group Theory in Physics*, vol. 1, Academic Press, London, pp. 44-67.
  In these pages, Cornwell introduces the theory of Lie groups in a thorough, fairly detailed, yet accessible style.
Gilmore, R.: 1974, *Lie Groups, Lie Algebras, and Some of Their Applications*, Wiley, New York, pp. 1-119.
  Gilmore's presentation is more detailed and less accessible. Yet it is quite useful as a reference.
Hall, G. G.: 1967, *Applied Group Theory*, American Elsevier, New York, pp. 76-92.
  Of the texts here, this is the simplest and easiest to follow. Chapter 6 deals with continuous groups, covering the highlights.
Wybourne, B. G.: 1970, *Symmetry Principles and Atomic Spectroscopy*, Wiley-Interscience, New York, pp. 28-48.

Methods developed for the symmetric groups are applied by Wybourne to the general linear group and to some of its subgroups.

Articles

Kim, Y. S., and Noz, M. E.: 1983, "Illustrative Examples of the Symplectic Group," *Am. J. Phys.* **51,** 368-375.

Newman, D. J., and Chan, K. S.: 1986, "Unitary Symmetry and Crystal Tensors," *Am. J. Phys.* **54,** 161-166.

Wilkes, J. M., and Zund, J. D.: 1982, "Group-Theoretic Approach to the Schwarzschild Solution," *Am. J. Phys.* **50,** 25-27.

CHAPTER 10 / Rotation Groups

10.1
Introduction

Some of the most important continuous groups are represented by groups of proper rotations. These may be enlarged by adding various reflections.

The simplest is the group of rotations about a fixed point in a Euclidean plane, the **SO**(2) group. This is a one-piece Abelian group; however, the manifold for its parameter is multiply connected.

A person may consider the Euclidean plane in 3-dimensional space. Then the 2-dimensional rotation group can be enlarged by adding the reflection that leaves the plane unchanged. Instead, or in addition, one may add reflections in perpendicular planes passing through the fixed point.

Next in order of complexity is the group of proper rotations about a fixed point in 3-dimensional Euclidean space, the **SO**(3) group. Since the rotations do not commute, except when they are infinitesimal, the group is not Abelian. But the manifold for its three independent parameters is doubly connected.

Exhibiting the same Lie algebra is the **SU**(2) group. As a consequence, each element thereof corresponds to a physical rotation in physical space. The correspondence, however, is two-to-one; two different elements in the **SU**(2) group correspond to one in the **SO**(3) group. As a result, the manifold is singly connected.

The Lie algebra for **SO**(3) and **SU**(2) exhibits characteristics that can be generalized for any number n of dimensions. In particular, the infinitesimal generators can be combined to yield operators that commute with all the independent infinitesimal generators, Casimir operators. Furthermore, some can be combined to yield step operators.

10.2
Two-Dimensional Rotation Groups

A one-piece continuous group is defined by its independent infinitesimal operators together with the manifold over which they act. Additional pieces can be added by introducing independent discrete operators.

Consider a Euclidean plane with a fixed point as base. On this point erect perpendicular x and y axes. To a general point (x, y) draw radius vector **r** making angle ϕ with the x axis. Let us construct the infinitesimal operator

$$B = x \frac{\partial}{\partial y} - y \frac{\partial}{\partial x} = iJ_z. \qquad (10.1)$$

Comparing this with equation (9.35), we see that

$$\xi = -y, \qquad \eta = y. \qquad (10.2)$$

So the differential equations for the characteristics consist of

$$-\frac{dx}{y} = \frac{dy}{x} = d\alpha, \qquad (10.3)$$

by equation (9.31).

Rearranging the first equation (10.3) yields

$$x\,dx + y\,dy = 0 \qquad (10.4)$$

whence

$$x^2 + y^2 = r^2 = \text{const} \qquad (10.5)$$

and

$$\frac{dy}{\sqrt{r^2 - y^2}} = d\alpha. \qquad (10.6)$$

Projecting r on the x and y axes gives us

$$x = r \cos \phi, \qquad y = r \sin \phi. \qquad (10.7)$$

Two-Dimensional Rotation Groups

Then substituting for y in equation (10.6) leads to

$$d\phi = d\alpha, \tag{10.8}$$

which integrates to

$$\phi = \phi_0 + \alpha. \tag{10.9}$$

Thus, the finite operations generated by equation (10.1) are rotations of the given system or function by angle α about the origin. Whenever rotation by 2π radians restores the initial conditions, a pertinent function would be single valued. The group of possible rotations would make up cyclic group $\mathbf{C}_\infty$, in the notation of Chapter 1. In the notation of section 9.7, the group is the special orthogonal group $\mathbf{SO}(2)$.

With polar coordinates r and ϕ, operator (10.1) simplifies to

$$\frac{\partial}{\partial \phi} = iJ_z. \tag{10.10}$$

But an infinitesimal operator commutes with the elements that it generates. In the one-piece group, these include all elements of the group. As a consequence, they have a common complete set of eigenfunctions.

The eigenvalue equation for (10.10) is

$$iJ_z\psi = \frac{\partial}{\partial \phi} \psi = \lambda \psi. \tag{10.11}$$

The integral of this equation is

$$\psi = Ae^{\lambda\phi}. \tag{10.12}$$

To keep the magnitude of ψ finite as the rotation angle increases without limit, eigenvalue λ must be imaginary:

$$\lambda = im. \tag{10.13}$$

In the simplest case, the eigenfunction is single valued, making

$$m = 0, \pm 1, + 2, \ldots. \tag{10.14}$$

In Section 10.4, we will find that the function can also be double valued, with

$$m = \pm \tfrac{1}{2}, \pm \tfrac{3}{2}, \ldots . \qquad (10.15)$$

The finite operators are obtained on exponentiating operator (10.10), with choice (10.13). For rotation by angle α, we get

$$C(\alpha) = e^{iJ_z\alpha} = e^{\lambda\alpha} = e^{im\alpha}. \qquad (10.16)$$

For choice (10.14), each primitive symmetry species is 1-dimensional. Then for each rotation $C(\alpha)$, the eigenvalue determines the representation and the character. Key forms appear in table 10.1.

To obtain the characters for a $\mathbf{C}_n$ group from these, one lets

$$\alpha = r\frac{2\pi}{n} \text{ where } r = 1, 2, \ldots, n-1 \qquad (10.17)$$

and

$$m = 0, 1, 2, \ldots, n-1. \qquad (10.18)$$

In this way, table 10.1 yields the entries of table 1.4.

Groups $\mathbf{C}_1, \mathbf{C}_2, \ldots, \mathbf{C}_\infty$ may be enlarged by adding the generating element σ_h, reflection in the xy plane, with all space being Euclidean. Each new group then consists of two disconnected pieces. The infinitesimal operator for each part is the same; but the bases split into those that are unaffected by σ_h and those that are changed in sign by it. Correspondingly, the primitive symmetry species split. The results appear in table 10.2 and table 1.6.

When a group contains, in addition to an element A of order n, an element B that does not commute with A, the character table is altered in a more profound manner. Let us enlarge the cyclic groups by adding generating element σ_v, reflection in a plane containing the x and z axes. Again, the group consists of two disconnected pieces. The rotation operations continuous with the identity cause the changes

$$r' = r, \quad \phi' = \phi + \alpha. \qquad (10.19)$$

Those continuous with the reflection operation σ_v involve

$$r' = r, \quad \phi' = -\phi + \alpha = -(\phi - \alpha). \tag{10.20}$$

Table 10.1 Characters for the $\mathbf{C}_\infty$ Group

	$I \ldots C(\alpha) \ldots$
Γ_0	$1 \ldots 1 \ldots$
Γ_1	$1 \ldots e^{i\alpha} \ldots$
$\ldots$	$\ldots \ldots \ldots$
Γ_m	$1 \ldots e^{im\alpha} \ldots$
$\ldots$	$\ldots \ldots \ldots$

Table 10.2 Characters for the $\mathbf{C}_{\infty h}$ Group

	I		$C(\alpha)$		σ_h		$\sigma_h C(\alpha)$	
Γ_0'	1	$\ldots$	1	$\ldots$	1	$\ldots$	1	$\ldots$
$\ldots$	$\ldots$	$\ldots$	$\ldots$	$\ldots$	$\ldots$	$\ldots$	$\ldots$	$\ldots$
Γ_m'	1	$\ldots$	$e^{im\alpha}$	$\ldots$	1	$\ldots$	$e^{im\alpha}$	$\ldots$
$\ldots$								
Γ_0''	1	$\ldots$	1	$\ldots$	-1	$\ldots$	-1	$\ldots$
$\ldots$	$\ldots$	$\ldots$	$\ldots$	$\ldots$	$\ldots$	$\ldots$	$\ldots$	$\ldots$
Γ_m''	1	$\ldots$	$e^{im\alpha}$	$\ldots$	-1	$\ldots$	$-e^{im\alpha}$	$\ldots$
$\ldots$	$\ldots$	$\ldots$	$\ldots$	$\ldots$	$\ldots$	$\ldots$	$\ldots$	$\ldots$

The infinitesimal operator for both these parts is

$$\frac{\partial}{\partial \phi}, \tag{10.21}$$

but the initial conditions differ. The effect of the reflection is to reverse the direction of rotation, to interchange $e^{im\alpha}$ with $e^{-im\alpha}$. Using the same procedure as in section 1.11 yields the results in table 10.3. The corresponding characters appear in table 10.4.

Table 10.3 Irreducible Representations of the $\mathbf{C}_{\infty v}$ Group

	I		$C(\alpha)$		σ_v
Γ_{0+}	1	...	1	...	1
Γ_{0-}	1	...	1	...	−1
...	...	...	...	...	...
Γ_m	$\begin{pmatrix} 1 & 0 \\ 0 & 1 \end{pmatrix}$	...	$\begin{pmatrix} e^{im\alpha} & 0 \\ 0 & e^{-im\alpha} \end{pmatrix}$	...	$\begin{pmatrix} 0 & 1 \\ 1 & 0 \end{pmatrix}$
...	...	...	...	...	...

Table 10.4 Characters for the $\mathbf{C}_{\infty v}$ Group

	I		$2C(\alpha)$		$\infty \sigma_v$
Γ_{0+}	1	...	1	...	1
Γ_{0-}	1	...	1	...	−1
...	...	...	...	...	...
Γ_m	2	...	$2 \cos m\alpha$	...	0
...	...	...	...	...	...

10.3
The SO(3) Spherical Group

On a given point in 3-dimensional Euclidean space, a person can erect three mutually perpendicular directed straight lines and label them the x, y, and z axes. Rotations about these axes are mutually independent.

In section 10.2, we have seen that the infinitesimal operator for rotation around the z axis is

$$x \frac{\partial}{\partial y} - y \frac{\partial}{\partial x} = iJ_z. \qquad (10.22)$$

Replacing x by y, y, by z, and z by x transforms formula (10.22) to

$$y \frac{\partial}{\partial z} - z \frac{\partial}{\partial y} = iJ_x \qquad (10.23)$$

Repeating the substitutions yields

$$z\frac{\partial}{\partial x} - x\frac{\partial}{\partial z} = iJ_y \qquad (10.24)$$

Now, iJ_z, iJ_x, and iJ_y are the infinitesimal operators for rotations about the z axis, the x axis, and the y axis, respectively. They are independent; and the general infinitesimal operator for rotation in 3-dimensional space is a linear combination of these three operators.

From results (10.12) and (10.13), we see that the eigenvalues of these operators are imaginary. The forms yielding real eigenvalues are

$$J_x = -i\left(y\frac{\partial}{\partial z} - z\frac{\partial}{\partial y}\right), \qquad (10.25)$$

$$J_y = -i\left(z\frac{\partial}{\partial x} - x\frac{\partial}{\partial z}\right), \qquad (10.26)$$

$$J_z = -i\left(x\frac{\partial}{\partial y} - y\frac{\partial}{\partial x}\right), \qquad (10.27)$$

By straightforward computation, the commutators of these are

$$[J_x, J_y] = iJ_z, \qquad (10.28)$$

$$[J_y, J_z] = iJ_x, \qquad (10.29)$$

$$[J_z, J_x] = iJ_y, \qquad (10.30)$$

Furthermore, the combination

$$c_1 J_x + c_2 J_y + c_3 J_z \qquad (10.31)$$

is an infinitesimal operator for the group. With natural units, in which

$$\hbar = 1, \qquad (10.32)$$

J_x, J_y, and J_z are the general quantum mechanical angular momentum operators. Note that these form the basis for a Lie algebra. Addition is represented by sum (10.31), multiplication by the commutation rules (10.28), (10.29), (10.30).

For various Lie groups, composite operators that commute with all the independent infinitesimal operators can be constructed. Such an operator is called a *Casimir operator*.

From the angular momentum operators, let us construct the scalar product

$$J^2 = J_x^2 + J_y^2 + J_z^2. \tag{10.33}$$

By straightforward computation, we find that

$$[J^2, J_x] = [J^2, J_y] = [J^2, J_z] = 0. \tag{10.34}$$

Thus, J^2 is a Casimir operator for the **SO**(3) group. There are no other independent operators for this group.

In constructing a manifold for the group, we place the origin at the fixed point. Rotation by ϕ radians about a particular axis is then represented by a point on the axis distance ϕ units from the fixed point. But there is a periodicity of 2π. For symmetry reasons, we consider the rotations from $\phi = -\pi$ to $\phi = \pi$ as distinct. But the rotation by $-\pi$ produces the same effect as rotation by π about the same axis; and rotation by $\pi + \delta$ produces the same effect as rotation by $-\pi + \delta$.

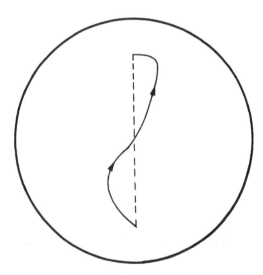

Figure 10.1 Path of elements in the manifold for **SO**(3) involving one jump.

The SO(3) Spherical Group

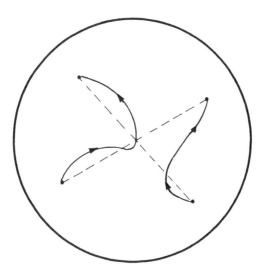

Figure 10.2 Continuous path of elements in the manifold for **SO**(3) involving two jumps.

So, we consider the manifold to be a solid sphere of radius π centered on the fixed point. We recognize that a point on the sphere represents the same transformation as its antipode on the opposite side. Successively larger rotations, starting from the identity position at the center of the sphere, with a varying axis, but returning to the initial configuration, trace out a path such as the one in figure 10.1. At a rotation angle of π radians, the path jumps to the antipodal position.

The transformations from the identity may involve more than one jump before the system returns to its initial configuration. A possibility is illustrated in figure 10.2. Now, a path from the center that does not reach the surface can be contracted to a point by continuous deformation. The corresponding transformations may be considered related. However, a path that involves one jump between antipodes cannot be contracted continuously to a point. Then there must be a related second series of elements in the group. Now, the closed path in figure 10.2 can be deformed so that the two jumps lie beside each other in opposite directions, as figure 10.3 indicates. A further continuous deformation, to figure 10.4, eliminates both jumps and gives a path that can be contracted continuously to a point.

In general, by continuous deformations a person can eliminate an even number of jumps, thus, ending with either a single jump or with

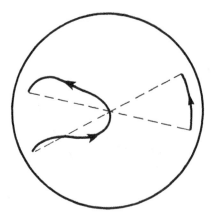

Figure 10.3 Distorton of the paths in figure 10.2.

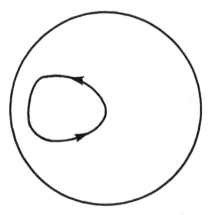

Figure 10.4 A further distortion that eliminates both jumps.

no jump. Accordingly, there are double-valued and single-valued bases for irreducible representations but none of higher dimensionalities.

10.4
Key Eigenvalue Equations for the Spherical Group

Since J_x, J_y, and J_z do not commute, they do not have a common complete set of eigenfunctions. But since J^2 commutes with J_z, these

Key Eigenvalue Equations for the Spherical Group   281

operators do have such a set. Then from J_x and J_y, a person can construct step operators that connect the different eigenfunctions for a given eigenvalue of J^2.

Let the eigenvalue of J^2 be a function of j—with j to be further identified later. From equations (10.11) and (10.13), the eigenvalue for J_z is m. We consider an eigenfunction ψ_{jm}, with a definite j and m, to be a representation of the ket $|jm\rangle$, with the complex conjugate $\langle jm|$. The normalization to 1 is described by the integral

$$\int_R \psi_{jm}{}^* \psi_{jm} d^3\mathbf{r} \equiv \langle jm|jm\rangle = 1. \qquad (10.35)$$

The eigenvalue equation for J_z is now written

$$J_z|jm\rangle = m|jm\rangle. \qquad (10.36)$$

From the independent operators J_x and J_y, let us construct

$$J_\pm = J_x \pm iJ_y. \qquad (10.37)$$

With equations (10.30) and (10.29), we find that

$$[J_z, J_\pm] = [J_z, J_x] \pm i[J_z, J_y] = iJ_y \pm i(-iJ_x) = \pm(J_x \pm iJ_y) = \pm J_\pm. \qquad (10.38)$$

But since

$$J_z J_\pm = J_\pm J_z + [J_z, J_\pm], \qquad (10.39)$$

we have

$$J_z J_\pm |jm\rangle = J_\pm J_z |jm\rangle \pm J_\pm |jm\rangle = (m \pm 1) J_\pm |jm\rangle. \qquad (10.40)$$

Thus, $J_\pm |jm\rangle$ is an eigenket of J_z with the eigenvalue $m \pm 1$ *or* it vanishes identically. When it is not zero, we write

$$J_\pm |jm\rangle = N_\pm(j, m) |j\ m \pm 1\rangle \qquad (10.41)$$

with $N_\pm(j, m)$ the normalization factor.

The eigenvalue equation for J^2 is

$$J^2|jm\rangle = \eta(j)|jm\rangle \tag{10.42}$$

with $\eta(j)$ to be determined. From equations (10.37) and (10.28), we have

$$J_\mp J_\pm = (J_x \mp iJ_y)(J_x \pm iJ_y) = J_x^2 + J_y^2 \mp iJ_y J_x \pm iJ_x J_y$$
$$= J^2 - J_z^2 \mp J_z \tag{10.43}$$

whence

$$J^2 = J_z(J_z \pm 1) + J_\mp J_\pm. \tag{10.44}$$

Let equation (10.44) act on the eigenket in equation (10.36),

$$J^2|jm\rangle = [J_z(J_z \pm 1) + J_\mp J_\pm]|jm\rangle, \tag{10.45}$$

then introduce equations (10.42) and (10.36):

$$\eta(j)|jm\rangle = m(m \pm 1)|jm\rangle + J_\mp J_\pm|jm\rangle. \tag{10.46}$$

Multiply this result by bra $\langle jm|$ and reduce with normalization conditions (10.35) and (10.41) to get

$$\eta(j) = m(m \pm 1) + \langle jm||J_\mp J_\pm|jm\rangle.$$
$$= m(m \pm 1) + \langle J_\pm|jm||J_\pm|jm\rangle$$
$$= m(m \pm 1) + N_\pm^*(jm)N_\pm(jm), \tag{10.47}$$

whence

$$N_\pm^* N_\pm = \eta(j) - m(m \pm 1). \tag{10.48}$$

Now in equation (10.47) the final term must equal zero when m is at its maximum and the positive sign is chosen; also, when m is at its minimum and the negative sign is chosen. Let the most positive m be j:

$$m_{max} = j. \tag{10.49}$$

Then equation (10.47) yields

Key Eigenvalue Equations for the Spherical Group

$$\eta(j) = m_{max}(m_{max} + 1) = j(j + 1). \qquad (10.50)$$

At the most negative m for the given j, we similarly have

$$\eta(j) = m_{min}(m_{min} - 1). \qquad (10.51)$$

On comparing with the final form in equation (10.50), we see that

$$m_{min} = -j. \qquad (10.52)$$

Substituting for $\eta(j)$ in equation (10.47) now leads to

$$N^*_\pm(j, m)N_\pm(j, m) = j(j + 1) - m(m \pm 1). \qquad (10.53)$$

For $N_\pm(j, m)$ itself, one may choose the phase angle to be zero.

In summary, the infinitesimal operators of the **SO**(3) group have eigenkets with the properties

$$J_z|jm\rangle = m|jm\rangle, \qquad (10.54)$$

$$J_+|jm\rangle = [j(j + 1) - m(m + 1)]^{\frac{1}{2}}|jm + 1\rangle, \qquad (10.55)$$

$$J_-|jm\rangle = [j(j + 1) - m(m - 1)]^{\frac{1}{2}}|jm - 1\rangle, \qquad (10.56)$$

$$J^2|jm\rangle = j(j + 1)|jm\rangle. \qquad (10.57)$$

Note that m varies by integers from $-j$ to $+j$. Thus, we may have either

$$j = 0, 1, 2, \ldots \text{ for single-valued eigenfunctions} \qquad (10.58)$$

or

$$j = \tfrac{1}{2}, \tfrac{3}{2}, \tfrac{5}{2}, \ldots \text{ for double-valued eigenfunctions.} \qquad (10.59)$$

No other possibilities exist—in agreement with our conclusion in Section 10.3.

A complete set of normalized orthogonal eigenkets for a given j is said to form a multiplet. Since m varies by integers from $-j$ to $+j$, there are $2j + 1$ members in a multiplet.

10.5
Behavior of Irreducible Tensor Components

Operator multiplets can be defined as the normalized eigenket multiplets are. A given multiplet is said to make up an irreducible tensor. Under operations of the **SO**(3) group, the members transform into linear combinations of one another.

The standard operator that transforms as $|\omega, \mu\rangle$ is labeled T_μ^ω. But since $J_\pm T_\mu^\omega$ is still an operator, the equation analogous to equations (10.55) and (10.56) has the form

$$[J_\pm, T_\mu^\omega] = [\omega(\omega + 1) - \mu(\mu \pm 1)]^{\frac{1}{2}} T_{\mu \pm 1}^\omega, \quad (10.60)$$

while that analogous to equation (10.54) is

$$[J_z, T_\mu^\omega] = \mu T_\mu^\omega. \quad (10.61)$$

10.6
The Relationship of SU(2) to SO(3)

The special unitary group in two dimensions is represented by the 2×2 unitary matrices whose determinants equal 1. Suitable infinitesimal operators were constructed in section 9.11; they consist of the Pauli spin matrices. These satisfy commutation relations (9.125) and yield the Casimir operator (9.127).

But if in formulas (10.28), (10.29), (10.30), and (10.33), we let

$$J_x = -\tfrac{1}{2}\sigma_x, \quad J_y = -\tfrac{1}{2}\sigma_y, \quad J_z = -\tfrac{1}{2}\sigma_z, \quad (10.62)$$

conditions (9.125) and (9.127) are obtained. As a consequence, the Lie algebra for **SU**(2) is not really different from that for **SO**(3). And eigenvalue equations (10.54) through (10.57) apply also to the **SU**(2) group.

As before, we choose the point about which rotations are to take place as the origin of a Cartesian coordinate system. The coordinates of a point in the physical system are then (x, y, z). But these may be employed in equation (9.120) with $w = 0$:

$$\begin{pmatrix} z & x+iy \\ x-iy & -z \end{pmatrix} = x\begin{pmatrix} 0 & 1 \\ 1 & 0 \end{pmatrix} + y\begin{pmatrix} 0 & i \\ -i & 0 \end{pmatrix} + z\begin{pmatrix} 1 & 0 \\ 0 & -1 \end{pmatrix}. \quad (10.63)$$

This equation can be rewritten in the form

$$R = x\sigma_x + y\sigma_y + z\sigma_z. \quad (10.64)$$

Now in a coordinate transformation, an operator such as R is subjected to a similarity transformation. Since the inverse of a unitary matrix $\mathbf{A}$ is $\mathbf{A}^\dagger$, this has the form

$$\mathbf{R}' = \mathbf{ARA}^\dagger. \quad (10.65)$$

10.7
Finite Operators of the SU(2) Group

A matrix representation of each element of the **SU**(2) group may be constructed by iteration of the pertinent infinitesimal generator. A similarity transformation of matrix $\mathbf{R}$ by the resulting operator then yields the corresponding transformation of the Cartesian coordinates of a physical point.

Let the canonical parameter for generator σ_z of set (9.121) be $\tfrac{1}{2}\lambda$. Exponentiation with condition (9.41) then leads to the matrix

$$\mathbf{A}_z = \exp\!\left(\frac{i\lambda}{2}\,\sigma_z\right) = \begin{pmatrix} e^{i\lambda/2} & 0 \\ 0 & e^{-i\lambda/2} \end{pmatrix}. \quad (10.66)$$

Employing this in equation (10.65) produces

$$\begin{pmatrix} z' & x'+iy' \\ x'-iy' & -z' \end{pmatrix} = \begin{pmatrix} ze^{i\lambda/2} & (x+iy)e^{i\lambda/2} \\ (x-iy)e^{-i\lambda/2} & -ze^{-i\lambda/2} \end{pmatrix}\begin{pmatrix} e^{-i\lambda/2} & 0 \\ 0 & e^{i\lambda/2} \end{pmatrix}$$

$$= \begin{pmatrix} z & (x+iy)e^{i\lambda} \\ (x-iy)e^{-i\lambda} & -z \end{pmatrix}. \quad (10.67)$$

whence

$$x' = x\cos\lambda - y\sin\lambda, \quad (10.68)$$

$$y' = x \sin \lambda + y \cos \lambda, \tag{10.69}$$

$$z' = z. \tag{10.70}$$

Thus, group element $\mathbf{A}_z$ rotates the physical system by angle λ about the z axis.

Similarly, let the canonical parameter for generator σ_y of set (9.121) be $\tfrac{1}{2}\mu$. By exponentiation with condition (9.41), we have

$$\mathbf{A}_y = \exp\left(\frac{i\mu}{2}\sigma_y\right) = \begin{pmatrix} \cos\dfrac{\mu}{2} & -\sin\dfrac{\mu}{2} \\ \sin\dfrac{\mu}{2} & \cos\dfrac{\mu}{2} \end{pmatrix}. \tag{10.71}$$

Subsitution into expression (10.65) now leads to

$$x' = x \cos \mu + z \sin \mu, \tag{10.72}$$

$$y' = y, \tag{10.73}$$

$$z' = -x \sin \mu + z \cos \mu. \tag{10.74}$$

Group element $\mathbf{A}_y$ rotates the physical system by angle μ about the y axis.

Also, let the canonical parameter for generator σ_x be $\tfrac{1}{2}\nu$. By exponentiation as before, we find that

$$\mathbf{A}_x = \exp\left(\frac{i\nu}{2}\sigma_x\right) = \begin{pmatrix} \cos\dfrac{\nu}{2} & i\sin\dfrac{\nu}{2} \\ i\sin\dfrac{\nu}{2} & \cos\dfrac{\nu}{2} \end{pmatrix}. \tag{10.75}$$

Employing this in formula (10.65) yields

$$x' = x, \tag{10.76}$$

$$y' = y \cos \nu - z \sin \nu, \tag{10.77}$$

$$z' = y \sin \nu + z \cos \nu. \tag{10.78}$$

We see that group element A_x rotates the physical system by angle v about the x axis.

Distinct elements of the **SU**(2) group arise for

$$-2\pi > \lambda > 2\pi, \quad -2\pi > \mu > 2\pi, \quad -2\pi > v > 2\pi. \quad (10.79)$$

Each of these has twice the range it has for the **SO**(3) group. The corresponding manifold, constructed as for **SO**(3), now has a radius of 2π and the paths corresponding to those in figures 10.1, 10.2, 10.3 would not involve jumps. As a consequence, the **SU**(2) group is the universal covering group for the **SO**(3) group.

10.8
Commutation Relations for a Step Operator

An operator that takes a physical system from one eigenstate to a neighboring one is called a step operator. When the step is constant for the range of states, the step operator satisfies a particular commutation relation.

Consider a quantum mechanical system with its accompanying Hilbert space. Also, consider an observable for which A is a Hermitian operator. If $|i\rangle$ and $|j\rangle$ are two neighboring normalized eigenkets for the observable,

$$A|i\rangle = a_i|i\rangle, \quad (10.80)$$

$$A|j\rangle = a_j|j\rangle, \quad (10.81)$$

and if these are linked by the operator B,

$$B|i\rangle = |j\rangle, \quad (10.82)$$

then B is a normalized step operator.

Let A act on equation (10.82), reduce the result with equation (10.81), and reintroduce equation (10.82):

$$AB|i\rangle = A|j\rangle = a_j|j\rangle = a_j B|i\rangle. \quad (10.83)$$

Also, let B act on equation (10.80) and rearrange the result:

$$BA|i\rangle = Ba_i|i\rangle = a_i B|i\rangle. \quad (10.84)$$

Then subtract equation (10.84) from equation (10.83) to get

$$(AB - BA)|i\rangle = (a_j - a_i)B|i\rangle = kB|i\rangle. \qquad (10.85)$$

If this equation holds for an arbitrary eigenket in the range considered, then we must have

$$[A, B] = AB - MA = kB \qquad (10.86)$$

with

$$k = a_j - a_i. \qquad (10.87)$$

Conversely, assumption of relationship (10.86), with A the operator for the eigenkets, implies that B is a step operator with k the spacing between the pertinent eigenstates. Here B need not be a normalized operator.

The angular momentum operators provide a most important example. Indeed from formula (10.38), we have

$$[J_z, J_+] = J_+ \qquad (10.88)$$

and

$$[J_z, J_-] = -J_-. \qquad (10.89)$$

In addition, from relation (10.43) one has

$$J_+J_- = J^2 - J_z^2 + J_z \qquad (10.90)$$

and

$$J_-J_+ = J^2 - J_z^2 - J_z \qquad (10.91)$$

whence

$$[J_+, J_-] = 2J_z. \qquad (10.92)$$

Since J^2 is the corresponding Casimir operator, we also have

$$[J^2, J_\pm] = 0 \qquad (10.93)$$

and

$$[J^2, J_z] = 0. \tag{10.94}$$

Conditions (10.88) and (10.89), together with conditions (10.92), (10.93), and (10.94), characterize the Lie algebra of the angular momentum step operators.

10.9
Conditions Defining Related Groups

The infinitesimal generators of rotation groups satisfy certain conditions which we will now summarize in a general form.

Consider independent operators

$$A_1, A_2, \ldots, A_n \tag{10.95}$$

such that any linear combination

$$\sum c_j A_j \tag{10.96}$$

is an operator. Also suppose that these obey the commutation rules

$$[A_j, A_k] = \sum_l c_{jk}^l A_l. \tag{10.97}$$

When coefficients c_{jk}^l are real, the operators may be related to group parameters in the canonical manner.

A continuous group of transformations may be constructed from these operators by considering them to effect rotations about a point in a hypothetical n-dimensional space. Let ψ be a function in this space. Consider A_j to be anti-Hermitian. Also let it act as the infinitesimal operator for rotating the function in the jth coordinate plane. For rotation by the infinitesimal angle ε, we then have

$$\psi' = (1 + \varepsilon A_j)\psi. \tag{10.98}$$

By integration, all finite rotations in the coordinate plane can be generated. Combining these with similar rotations in the other coordinate planes yields the Lie group.

In each of our applications, the composition rules (10.96) and (10.97) will represent physical conditions. Thus, physical relationships will determine the Lie algebra in a natural manner. But the Lie group will not have any simple physical interpretation.

Let us construct composite operators like J^2 which commute with all operators in the set. A complete set of independent ones is identified. Its members are called Casimir operators C_j; for these

$$[C_j, A_k] = 0, \quad \begin{matrix} j = 1, 2, 2 \ldots, m, \\ k = 1, 2, \ldots, n. \end{matrix} \quad (10.99)$$

A person may search the original operators for the maximum number that commute with one another. Let us label these H_i. These also commute with the Casimir operators C_j. A complete set of states that are simultaneous eigenstates for C_j and H_i can be constructed, with eigenvalues C_j and H_i,

$$|C_j, H_i\rangle, \quad (10.100)$$

analogous to the states $|J, M\rangle$ for angular momentum.

The remaining operators of set (10.95) may now be combined into independent step operators E_h. These would satisfy the relations

$$[C_j, E_h] = 0, \quad (10.101)$$

$$[H_i, E_h] = h_i E_h, \quad (10.102)$$

analogous to the rules satisfied by $J_x \pm iJ_y$.

For a given physical system, a person may be confronted with operators that take the system from one state to another but do not satisfy the conditions for a Lie group. But if one can combine these into operators that do satisfy these conditions, a Lie algebra is thereby obtained. In the next chapter, we will consider examples.

Discussion Questions

10.1 How may infinitesimal operators define a group?

10.2 What one-piece group is generated by the operator

$$x \frac{\partial}{\partial y} - y \frac{\partial}{\partial x} = iJ_z?$$

Discussion Questions

Explain.

10.3 Why is a complete set of eigenfunctions of the operators of the **SO(2)** group obtained on constructing the suitable eigenfunctions of

$$\frac{\partial}{\partial \phi} ?$$

10.4 How are characters for the **SO(2)** group generated?

10.5 How are the cyclic groups enlarged by adding (a) reflection σ_h, (b) reflection σ_v?

10.6 Use the result from question 10.2 to determine the one-piece group generated by the operators

$$y\frac{\partial}{\partial z} - z\frac{\partial}{\partial y} = iJ_x,$$

$$z\frac{\partial}{\partial x} - x\frac{\partial}{\partial z} = iJ_y,$$

$$x\frac{\partial}{\partial y} - y\frac{\partial}{\partial x} = iJ_z.$$

10.7 Why can the eigenvalues of J_x, J_y, J_z represent physical quantities?

10.8 How is J^2 a Casimir operator?

10.9 Describe the manifold for the **SO(3)** group.

10.10 Consider closed paths of elements of the **SO(3)** group in the manifold. What kinds of paths arise?

10.11 Which combinations of J_x, J_y, J_z, and J^2 have a common set of eigenkets? Explain.

10.12 Show that

$$J_x \pm iJ_y = J_\pm$$

is a step operator for the common eigenkets of J_z and J^2.

10.13 Show that

$$J^2 = J_z(J_z \pm 1) + J_\mp J_\pm.$$

10.14 Relate the normalization constant for $J_\pm |jm\rangle$ to the eigenvalues for J_z and J^2.

10.15 How is the eigenvalue of J^2 obtained from the preceding results?

10.16 Explain why there are two kinds of eigenkets for J^2 and J_z.

10.17 Explain how the irreducible tensor components transform.

10.18 What commutation relation does operator T_μ^ω satisfy? Why?

10.19 Explain the relation of **SU**(2) to **SO**(3).

10.20 How are the finite operators of the **SU**(2) group constructed?

10.21 Explain how these transform physical coordinates.

10.22 Why is **SU**(2) the universal covering group for **SO**(3)?

10.23 Explain what commutation relation is satisfied by a step operator.

10.24 How may a continuous group of transformations be generated by operators satisfying a given Lie algebra?

10.25 What commutation relations characterize the operators for an n-dimensional rotation group?

Problems

10.1 Show how the infinitesimal operator

$$x \frac{\partial}{\partial y} - y \frac{\partial}{\partial x}$$

reduces to

$$\frac{\partial}{\partial \phi}.$$

10.2 Show by differentiation that

$$[J_x, J_y] = iJ_z.$$

10.3 Express iJ_x in spherical coordinates.

10.4 Using the pertinent series expansion, establish the second equality in line (10.66).

10.5 Carry out the matrix multiplications and reductions that lead from condition (10.66) to equations (10.68) through (10.70).

10.6 In equation (10.98), the angle about the origin from a reference axis may be ϕ while A_j is $\partial/\partial\phi$. Function ψ may be composed of terms depending on ϕ as exp $(im\phi)$. Introduce a rotation by angle ε and show that equation (10.98) is satisfied when this angle is small.

10.7 With commutation relations (10.28) through (10.30), show that

$$[J^2, J_x] = 0.$$

10.8 Express iJ_y in spherical coordinates.

10.9 With the appropriate series expansions, establish the second equality in line (10.71).

10.10 Carry out the matrix multiplications and reductions that lead from conditions (10.71) to equations (10.72) through (10.74).

10.11 Justify the second equality in line (10.75).

10.12 From conditions (10.65) and (10.75), derive equations (10.76) through (10.78).

References

Books

Edmonds, A. R.: 1957, *Angular Momentum in Quantum Mechanics*, Princeton University Press, Princeton, NJ, pp. 3-20, 53-55. This is a comprehensive monograph on the 3-dimensional angular momentum operators.

Hall, G. G.: 1967, *Applied Group Theory*, American Elsevier, New York, pp. 93-109. In Chapter 7, Hall introduces the reader to 2-dimensional and 3-dimensional rotation groups.

Lipkin, H. J.: 1966, *Lie Groups for Pedestrians*, North-Holland Publishing Co., Amsterdam, pp. 1-18. Lipkin considers how the familiar angular momentum results can be generalized.

Rose, M. E.: 1957, *Elementary Theory of Angular Momentum*, Wiley, New York, pp. 3-31, 48-106. This monograph is similar to Edmond's, employing operator methods to treat physical angular momenta.

CHAPTER 11 / *Physical Lie Algebras*

11.1
General Considerations

According to quantum mechanics, physical systems exist in states represented by normalized eigenkets. The complete set of eigenkets for a given particle forms the basis for a Hilbert space. Going from one possible state to another is represented as a transformation in the Hilbert space.

Now, the Hilbert space may be enlarged to include the state of physical space in which the particle is absent. Then an operator would exist for introducing the particle in a given state. Furthermore, another operator would exist for eliminating the particle, starting with it in a given state.

To the vacuum state, other particles in a family may be added. Accordingly, the Hilbert space would be enlarged to include the possible states of these particles. But if a person is solely interested in relationships among the particles in the family, only one state (a corresponding state) for each need be considered, with the Hilbert space thus truncated.

The properties that must be given to the introduction and elimination operators (we call them creation and annihilation operators) depend on whether the particles involved are bosons or fermions. In either case, the operators are characterized by an algebra; however, the algebra generated by them is not a Lie algebra. On the other hand, a complete set of bilinear products of creation and annihilation operators do make up a Lie algebra. In this chapter, we will see how such algebras arise and consider certain key applications.

11.2
Characterizing Creation and Annihilation Operators

A submicroscopic particle with definite properties can be represented as an exited state of the continuum. Introducing it is a step-up operation proceeding from a lower state. This would be effected by a *creation operator* a_j^+. Removing it is a step-down operation acting on the excited state. This would be effected by an *annihilation operator* a_j^-. In each case, the subscript identifies the particle state involved.

Any number of bosons of a given kind can occupy a given particle state. Initially in the jth state, there may be n_j bosons; finally there may be $n_j + 1$ or $n_j - 1$ bosons. Furthermore, a model for the jth state is a harmonic oscillator of angular frequency ω_j. From the way ladder operators for such an oscillator act, we write

$$a_j^+ |\ldots n_j \ldots\rangle = \sqrt{n_j + 1}\,|\ldots n_j + 1 \ldots\rangle \qquad (11.1)$$

and

$$a_j^- |\ldots n_j \ldots\rangle = \sqrt{n_j}\,|\ldots n_j - 1 \ldots\rangle. \qquad (11.2)$$

With equations (11.1) and (11.2), a person obtains the commutation relations for bosons

$$[a_j^+, a_k^+] \equiv a_j^+ a_k^+ - a_k^+ a_j^+ = 0, \qquad (11.3)$$

$$[a_j^-, a_k^-] \equiv a_j^- a_k^- - a_k^- a_j^- = 0, \qquad (11.4)$$

and

$$[a_j^-, a_k^+] \equiv a_j^- a_k^+ - a_k^+ a_j^- = \delta_{jk}. \qquad (11.5)$$

Since relation (11.5) violates form (10.97), these operators do not generate a Lie algebra.

On the other hand, only one fermion of a given kind can occupy a given particle state. So step operations (11.1) and (11.2) are replaced by

$$a_j^+ |\ldots n_j \ldots\rangle = (-1)^{N_j}\sqrt{1 - n_j}\,|\ldots n_j + 1 \ldots\rangle \qquad (11.6)$$

and

$$a_j^- |\ldots n_j \ldots \rangle = (-1)^{N_j-1}\sqrt{n_j}|\ldots n_j - 1 \ldots\rangle, \quad (11.7)$$

where

$$N_j = \sum_{i=1}^{j} n_i. \quad (11.8)$$

The first factor on the right of equations (11.6) and (11.7) has been chosen to maintain the required exchange antisymmetry. The radical has been chosen so the jth particle state contains at most one of the fermions.

With equations (11.6) and (11.7), one obtains the anticommutation relations for fermions

$$\{a_j^+, a_k^+\} \equiv a_j^+ a_k^+ + a_k^+ a_j^+ = 0, \quad (11.9)$$

$$\{a_j^-, a_k^-\} \equiv a_j^- a_k^- + a_k^- a_j^- = 0, \quad (11.10)$$

and

$$\{a_j^-, a_k^+\} \equiv a_j^- a_k^+ + a_k^+ a_j^- = \delta_{jk}. \quad (11.11)$$

Clearly, these operators do not generate a Lie algebra.

Example 11.1

Consider the j, k, m, n states of a boson. Let a_j^+ and a_m^+ be the creation operators for the particle in the j and m states, while a_k^- and a_n^- are the annihilation operators for the particle in the k and n states. Show that the commutator of the bilinear products $a_j^+ a_k^-$ and $a_m^+ a_n^-$ satisfies condition (10.97).

Construct the commutator; then rearrange factors employing relations (11.5), (11.3), and (11.4):

$$[a_j^+ a_k^-, a_m^+ a_n^-] = a_j^+ a_k^- a_m^+ a_n^- - a_m^+ a_n^- a_j^+ a_k^-$$

$$= a_j^+ (\delta_{km} + a_m^+ a_k^-) a_n^- - a_m^+ (\delta_{nj} + a_j^+ a_n^-) a_k^-$$

$$= \delta_{km} a_j^+ a_n^- + a_m^+ a_j^+ a_k^- a_n^- - \delta_{nj} a_m^+ a_k^- - a_m^+ a_j^+ a_k^- a_n^-$$

$$= \delta_{km} a_j^+ a_n^- - \delta_{nj} a_m^+ a_k^-.$$

The result is a linear combination of the bilinear products $a_j^+ a_n^-$ and $a_m^+ a_k^-$.

Example 11.2

Consider the j, k, m, n states of a fermion. Let a_j^+ and a_m^+ be the creation operators for the particle in the j and m states, while a_k^- and a_n^- are the annihilation operators for the particle in the k and n states. Show that the commutator of the bilinear products $a_j^+ a_k^-$ and $a_m^+ a_n^-$ satisfies condition (10.97).

Construct the commutator; then rearrange factors employing relations (11.11), (11.9), and (11.10):

$$[a_j^+ a_k^-, a_m^+ a_n^-] = a_j^+ a_k^- a_m^+ a_n^- - a_m^+ a_n^- a_j^+ a_k^-$$

$$= a_j^+ (\delta_{km} - a_m^+ a_k^-) a_n^- - a_m^+ (\delta_{nj} - a_j^+ a_n^-) a_k^-$$

$$= \delta_{km} a_j^+ a_n^- + a_m^+ a_j^+ a_k^- a_n^- - \delta_{nj} a_m^+ a_k^- - a_m^+ a_j^+ a_k^- a_n^-$$

$$= \delta_{km} a_j^+ a_n^- - \delta_{nj} a_m^+ a_k^-.$$

Again, the result is a linear combination of the bilinear products $a_j^+ a_n^-$ and $a_m^+ a_k^-$.

11.3
Commutators of Bilinear Products of the Operators

Bilinear products of the creation and annihilation operators in a set do combine as the elements of a Lie algebra.

First, consider a system of *boson* states. Label four of these with the indices j, k, m, n. The corresponding creation and annihilation operators, $a_j^\pm$, $a_k^\pm$, $a_m^\pm$, $a_n^\pm$, combine as conditions (11.3), (11.4), and (11.5) dictate.

Typical bilinear products include $a_j^+ a_k^+$, $a_j^- a_k^-$, $a_j^+ a_k^-$, and $a_j^- a_k^+$. Using the commutation rules (11.3), (11.4), and (11.5) as in example 11.1, we find that

$$[a_j^+ a_k^+, a_m^+ a_n^+] = 0, \tag{11.12}$$

$$[a_j^+ a_k^-, a_m^+ a_n^+] = \delta_{km} a_j^+ a_n^+ + \delta_{kn} a_j^+ a_m^+ \tag{11.13}$$

$$[a_j^+ a_k^-, a_m^+ a_n^-] = \delta_{km} a_j^+ a_n^- - \delta_{jn} a_m^+ a_k^-, \tag{11.14}$$

$$[a_j^- a_k^-, a_m^+ a_n^+] = \delta_{km} a_j^- a_n^+ + \delta_{jm} a_n^+ a_k^- + \delta_{kn} a_j^- a_m^+ + \delta_{jn} a_m^+ a_k^-. \tag{11.15}$$

Other combinations behave similarly; the commutator of any two bilinear products of the boson operators is either zero or a linear combination of bilinear products. Thus, such products may generate a Lie algebra.

In like manner, consider a system of *fermion* states. Let four of these be labeled by the indices j, k, m, n. The corresponding creation and annihilation operators, $a_j^\pm$, $a_k^\pm$, $a_m^\pm$, $a_n^\pm$, combine following conditions (11.9), (11.10), and (11.11).

The commutators of typical bilinear products are now constructed. Employing the anticommutation rules as in example 11.2, we find that

$$[a_j^+ a_k^+, a_m^+ a_n^+] = 0, \tag{11.16}$$

$$[a_j^+ a_k^-, a_m^+ a_n^+] = \delta_{km} a_j^+ a_n^+ - \delta_{kn} a_j^+ a_m^+ \tag{11.17}$$

$$[a_j^+ a_k^-, a_m^+ a_n^-] = \delta_{km} a_j^+ a_n^- - \delta_{jn} a_m^+ a_k^-, \tag{11.18}$$

$$[a_j^- a_k^-, a_m^+ a_n^+] = \delta_{km} a_j^- a_n^+ + \delta_{jm} a_n^+ a_k^- - \delta_{kn} a_j^- a_m^+ - \delta_{jn} a_m^+ a_k^-. \tag{11.19}$$

Other combinations behave similarly; the commutator of any two bilinear products of the fermion operators is either zero or a linear combination of bilinear products. Consequently, the bilinear products may generate a Lie algebra.

11.4
Proton-Neutron Systems

Protons and neutrons are held together in nuclei by strong interactions, the so-called nuclear forces. These overwhelm the electromagnetic interactions, which are also present, but the electric charge on a proton allows it to occupy the same state as a neutron. Remember, both are fermions.

With respect to the strong interaction, a proton behaves exactly like a neutron. In the approximation that other interactions can be neglected, they appear merely as two different states of the same particle. Thus, creation and annihilation operators can be defined for them, bilinear products constructed, and the corresponding Lie algebra and group identified.

Let a_p^+ and a_p^- be the creation and annihilation operators for the proton in a certain state, while a_n^+ and a_n^- are the creation and annihilation operators for the neutron in the same state. For a system, the bilinear products that do not change the number of particles include

$$a_p^+ a_p^-, \ a_n^+ a_n^-, \ a_p^+ a_n^-, \ a_n^+ a_p^- \qquad (11.20)$$

Other such bilinear products can be constructed; but they are dependent on these.

The first operator annihilates a proton and then creates it again; the second operator does the same thing for a neutron. So these are number operators, counting the number of protons and neutrons in the given state when they act on the pertinent ket. We write

$$a_p^+ a_p^- + a_n^+ a_n^- = B, \qquad (11.21)$$

with B the baryon number for the given state. The third operator annihilates a neutron and creates a proton in the same state; with respect to charge it acts as a step-up operator. We write

$$a_p^+ a_n^- = \tau_+. \qquad (11.22)$$

In contrast, the fourth operator annihilates a proton and creates a neutron in the same state; it acts as a step-down operator. We write

$$a_n^+ a_p^- = \tau_-. \qquad (11.23)$$

To complete the set, we construct

$$\tfrac{1}{2}(a_p^+ a_p^- - a_n^+ a_n^-) = \tau_0 = Q - \tfrac{1}{2}B. \qquad (11.24)$$

Since a proton carries one unit of charge and a neutron no charge, this operator acting on the ket for the state yields the total charge Q minus half the baryon number B for the given state.

Commutators of these operators can be reduced with formula (11.18). We find that

$$[B, \tau_+] = 0, \tag{11.25}$$

$$[B, \tau_-] = 0, \tag{11.26}$$

$$[B, \tau_0] = 0. \tag{11.27}$$

Thus B is a Casimir operator for the set. In addition, we find that

$$[\tau_0, \tau_+] = \tau_+, \tag{11.28}$$

$$[\tau_0, \tau_-] = -\tau_-, \tag{11.29}$$

$$[\tau_+, \tau_-] = 2\tau_0. \tag{11.30}$$

On comparing these relations with conditions (10.88), (10.89), and (10.92), we see that the tau operators behave like the angular momentum operators.

Whenever more than one particle state is involved, formulas (11.21) through (11.24) need to be generalized. So we add index k to each operator to designate the spatial-spin state of the corresponding particle. Condition (11.21) is replaced by

$$\sum_k (a_{pk}^+ a_{pk}^- + a_{nk}^+ a_{nk}^-) = B, \tag{11.31}$$

with B the number of nucleons in the system. Similarly, conditions (11.22), (11.23) and (11.24) are replaced by

$$\sum_k a_{pk}^+ a_{nk}^- = \tau_+, \tag{11.32}$$

$$\sum_k a_{nk}^+ a_{pk}^- = \tau_-, \tag{11.33}$$

$$\tfrac{1}{2}\sum_k (a_{pk}^+ a_{pk}^- - a_{nk}^+ a_{nk}^-) = \tau_0 = Q - \tfrac{1}{2}B. \tag{11.34}$$

In relation (11.34), Q is the total electric charge in the system.

Since bilinear products corresponding to orthogonal quantum states commute, each quantum state k acts independently. Thus, commutation rules (11.25) through (11.30) still apply. For simplicity, we may drop the summation over k in our further discussions.

But the commutation rules are those for rotation groups **SO**(3) and **SU**(2). Furthermore, the transformations between the neutron and the proton states are unitary. So the corresponding Lie group is a unimodular unitary group. For the proton-neutron systems, it is the **SU**(2) group.

Note that the eigenvalues for τ_0 are given by $Q - \frac{1}{2}B$. Since Q equals $0, 1, \ldots, B$, a given baryon number B yields the values $-\frac{1}{2}B, -\frac{1}{2}B + 1, \ldots, \frac{1}{2}B$. One can compare these with the eigenvalues for J_z, namely $-J, -J + 1, \ldots, J$. The operator analogous to J^2 is labeled T^2; it has the eigenvalue

$$T(T + 1) = \tfrac{1}{2}B(\tfrac{1}{2}B + 1). \tag{11.35}$$

Two different nuclei with the same T are called isobars. A complete isobaric multiplet contains $2T + 1 = B + 1$ different nuclei. Because of the similarity to the results for spin, the property considered here is called *isospin*, the multiplet an isospin multiplet.

11.5
The Quark Lattice

On bombarding nucleons with high-velocity particles, various short-lived excited states can be produced. Existence of these states suggests the presence of at least one additional quantization. Besides existing as fermions with spin, the submicroscopic building blocks of nature can carry electric charge Q and hypercharge Y. The latter is related to the baryon number B and the strangeness S.

We now have to deal with shifts in both charge and hypercharge or strangeness. At the least, three different fermion building blocks are needed together with three different mirror images of these. We call the former quarks, the latter antiquarks.

The two nonstrange quarks separated by one unit of electric charge are labeled u (up) and d (down). The quark with one unit of strangeness is labeled s (strange). By convention, its strangeness is given a negative sign.

Just as nucleons combine to form composite systems (the various nuclei), the quarks combine to form composite systems. The source of the bonding may be considered a three aspect charge labeled red, green, and blue. For the moment, we neglect the effect of this charge and the antiparticle effect, except insofar as they allow composites to form.

For the quarks (and for the antiquarks), we have the creation operators a_u^+, a_d^+, a_s^+, and the annihilation operators a_u^-, a_d^-, a_s^-. The bilinear products that do not change the number of particles include

$$a_u^+ a_u^-,\ a_d^+ a_d^-,\ a_s^+ a_s^-,\ a_u^+ a_d^-,\ a_d^+ a_u^-,\ a_d^+ a_s^-,\ a_s^+ a_d^-,\ a_s^+ a_u^-,\ a_u^+ a_s^- \quad (11.36)$$

Other such bilinear products can be constructed; however, they are either dependent on these or they alter the number of particles.

Generalizing how the bilinear products were combined in conditions (11.21) through (11.24), we write

$$\tfrac{1}{3}(a_u^+ a_u^- + a_d^+ a_d^- + a_s^+ a_s^-) = B, \quad (11.37)$$

$$a_u^+ a_d^- = I_+, \qquad a_d^+ a_u^- = I_-, \quad (11.38)$$

$$\tfrac{1}{2}(a_u^+ a_u^- - a_d^+ a_d^-) = I_0, \quad (11.39)$$

$$a_d^+ a_s^- = U_+, \qquad a_s^+ a_d^- = U_-, \quad (11.40)$$

$$\tfrac{1}{2}(a_d^+ a_d^- - a_s^+ a_s^-) = U_0, \quad (11.41)$$

$$a_s^+ a_u^- = V_+, \qquad a_u^+ a_s^- = V_-, \quad (11.42)$$

$$\tfrac{1}{2}(a_s^+ a_s^- - a_u^+ a_u^-) = V_0. \quad (11.43)$$

Formula (11.37) is an expansion of condition (11.21) with a different normalization. The eigenvalue B is called the *baryon number*. Operators (11.38) and (11.39) are isotopic spin operators; I_+ acts to increase the electric charge by one unit, while I_- decreases the electric charge by one unit. Operator U_+ relieves the strangeness by one unit, while U_- has the opposite effect. Operator V_- similarly relieves the strangeness by one unit, while V_+ has the opposite effect.

We have constructed ten operators. But since they were formed from nine independent bilinear operators, only nine of them can be independent. Indeed, we find that

$$I_0 + U_0 + V_0 = 0. \tag{11.44}$$

The independent combination of U_0 and V_0 may be taken as

$$\tfrac{2}{3}(U_0 - V_0) = Y, \tag{11.45}$$

whence

$$\tfrac{1}{3}(a_u^+ a_u^- + a_d^+ a_d^- - 2a_s^+ a_s^-) = Y = B + S. \tag{11.46}$$

The eigenvalue obtained with operator (11.46) is called the *hypercharge Y*.

Since these operators do not alter the number of particles, they do not alter the baryon number B. On the other hand, U_+ and V_- acting on the same state produce states one charge unit apart, one strangeness unit higher. If at the original Y there was only one state, its I_3 would be 0. At the transformed Y there would then be two states, with $I_3 = +\tfrac{1}{2}$ and $I_3 = -\tfrac{1}{2}$.

This argument can be generalized for any number of states. If we take the unit along the Y axis to be $\sqrt{3}/2$ times as long as a unit on the I_3 axis, the states form a regular hexagonal lattice. A representation of the shift operators then appears as in figure 11.1.

Each multiplet of composite particles can be plotted on the hexagonal lattice. But in such a plot, reflection through the Y axis would interchange u and d quarks in corresponding particles. Since these quarks are equivalent in the strong-interaction approximation, we consider this reflection to be a symmetry operation.

Similarly, reflection in the 30 degree diagonal of figure 11.2 interchanges s and d quarks in corresponding particles. Also, reflection in the 150 degree diagonal interchanges u and s quarks in corresponding particles. These reflections are symmetry operations in the strong-interaction approximation.

As a consequence, the boundary of the hexagonal lattice of systems must have the form shown in figure 11.2. It is characterized by the parameters λ and μ. Indeed, note that

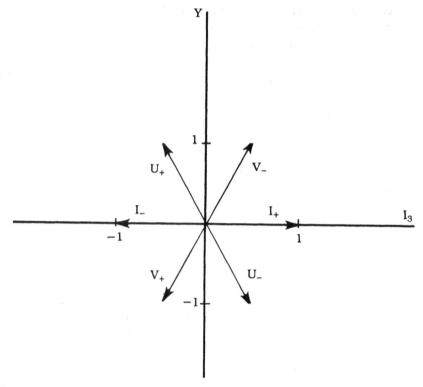

Figure 11.1 Basic step-up and step-down operations of the group, arranged symmetrically.

$$\lambda = 2I \quad \text{at} \quad Y = Y_{max} \quad (11.47)$$

and

$$\mu = 2I \quad \text{at} \quad Y = Y_{min}. \quad (11.48)$$

Here I is the isospin quantum number for the states at the given Y.

A multiplet exists for each integral λ and integral μ. The number of states at each lattice point is governed by the rules:

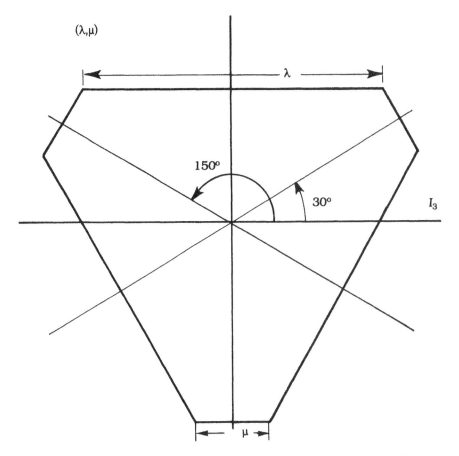

Figure 11.2 Boundary of the hexagonal lattice for a (λ, μ) multiplet.

(a) The outer ring of lattice points has one state at each point.
(b) Going inward, each consecutive ring of lattice points has one more state at each point. This increase continues until one arrives at a ring that is a triangle or a point.
(c) The number of states at a lattice point within a triangle is the same as on its perimeter.

(For examples, see figures 11.3 through 11.8.)

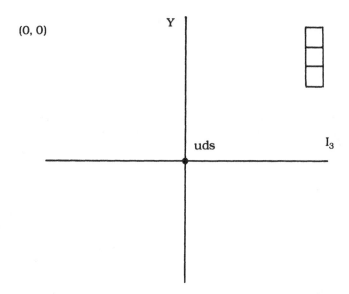

Figure 11.3 A singlet. This may be formed from a completely antisymmetric arrangement of the quarks.

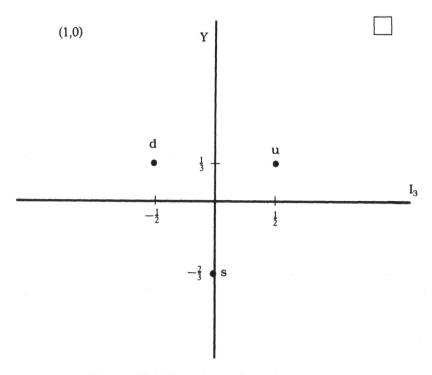

Figure 11.4 The triplet of single quarks.

306

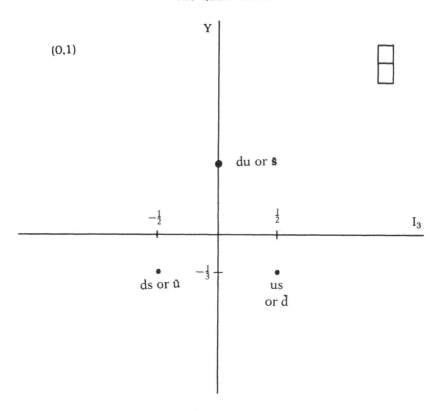

Figure 11.5 A hypothetical triplet of quark pairs. Or, the triplet of single antiquarks.

Example 11.3

In the common realization of figure 11.7, composite particle *uud* is identified as the proton while composite particle *udd* is identified as the neutron. What is the electric charge on (a) quark *u*, (b) quark *d*?

Let x be the charge on *u* while y is the charge on *d*. Setting the charge on *uud* equal to 1 unit gives us

$$2x + y = 1,$$

while setting the charge on *udd* equal to 0 yields

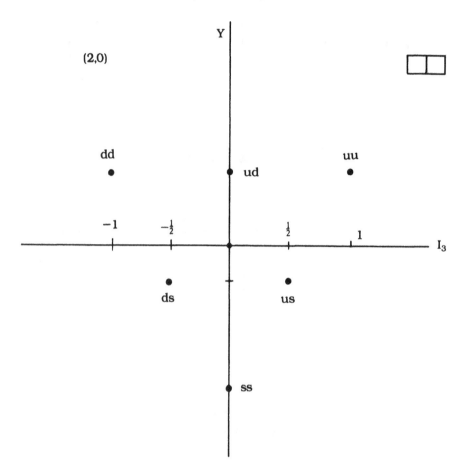

Figure 11.6 A hypothetical sextet of quark pairs.

$$x + 2y = 0.$$

Solving these equations simultaneously yields

$$x = \tfrac{2}{3}, \quad y = -\tfrac{1}{3}.$$

Example 11.4

For the composite particle *uds* in figure 11.3 to be vacuum-like, its net charge must vanish. What is the electric charge on quark *s*?

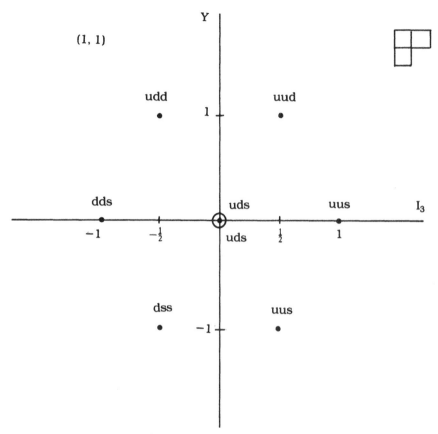

Figure 11.7 The octet from quark triplets.

Letting the charge on u be x, that on d be y, and that on s be z, we have

$$z = -x - y.$$

Substituting in the results from Example 11.3 gives

$$z = -\tfrac{2}{3} + \tfrac{1}{3} = -\tfrac{1}{3}.$$

Example 11.5

How many independent *uds* permutations exist?

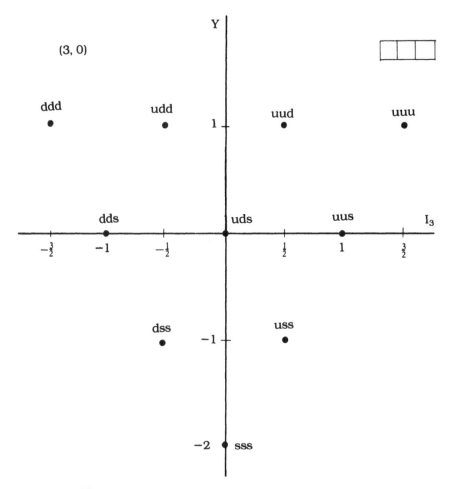

Figure 11.8 The decuplet from quark triplets.

The place of the first quark in the composite can be chosen in three different ways: the place of the second quark, in two different ways; the place of the last one, in one way. We have

$$n = 3 \cdot 2 \cdot 1 \text{ ways} = 6 \text{ ways}.$$

One of these structures appears in the (3, 0) multiplet and one in (0, 0). That leaves four to appear in the two, (1, 1) multiplets. So in each (1, 1) multiplet there are two *uds* states.

11.6
Classifying Fundamental Particles

In the laboratory, the triplets and the sextet have not been observed. (We will consider why this happens later.) However, all members of common octets and decuplets have been identified.

From properties of such members, properties of the u, d, and s quarks can be induced. (See examples 11.3, 11.4, and 11.5.) The corresponding quantum numbers are listed in table 11.1. Analogous to the s quark, we have the excited quarks c, b, and t.

The composite particles making up the common doubly degenerate ($\alpha-\beta$) octet of spin $\frac{1}{2}$ and the common decuplet of spin 3/2 are listed in table 11.2.

From figure 11.5, we see that the antiquarks behave in the groups as antisymmetric quark pairs. So a family of quark-antiquark pairs in mesons forms a doubly degenerate ($\alpha-\beta$) octet and a singlet. See table 11.3 for the common multiplets with spin 0 and 1.

Table 11.1 Quantum Numbers for Quarks and Antiquarks

Particle	B	Q	I_3	S	C	B	C
u	$\frac{1}{3}$	$\frac{2}{3}$	$\frac{1}{2}$	0	0	0	0
d	$\frac{1}{3}$	$-\frac{1}{3}$	$-\frac{1}{2}$	0	0	0	0
s	$\frac{1}{3}$	$-\frac{1}{3}$	0	-1	0	0	0
c	$\frac{1}{3}$	$\frac{2}{3}$	0	0	1	0	0
b	$\frac{1}{3}$	$-\frac{1}{3}$	0	0	0	-1	0
t	$\frac{1}{3}$	$\frac{2}{3}$	0	0	0	0	1
$\tilde{u}$	$-\frac{1}{3}$	$-\frac{2}{3}$	$-\frac{1}{2}$	0	0	0	0
$\tilde{d}$	$-\frac{1}{3}$	$\frac{1}{3}$	$\frac{1}{2}$	0	0	0	0
$\tilde{s}$	$-\frac{1}{3}$	$\frac{1}{3}$	0	1	0	0	0
$\tilde{c}$	$-\frac{1}{3}$	$-\frac{2}{3}$	0	0	-1	0	0
$\tilde{b}$	$-\frac{1}{3}$	$\frac{1}{3}$	0	0	0	1	0
$\tilde{t}$	$-\frac{1}{3}$	$-\frac{2}{3}$	0	0	0	0	-1

Note: Symbols B, Q, I_3, S, C, B, T are the baryon number, electric charge, zth component of isospin, strangeness, charm, bottom, and top numbers

Table 11.2 Properties of Some Composite Systems of Quarks, the Common Baryons

Quark Composition	Q	I_3	S	Spin-$\frac{1}{2}$ Octet		Spin-$\frac{3}{2}$ Decuplet	
				Particle	Mass (MeV)	Particle	Mass (MeV)
uuu	2	$\frac{3}{2}$	0			N^{*++}	1230
uud	1	$\frac{1}{2}$	0	p	938.28	N^{*+}	1231
udd	0	$-\frac{1}{2}$	0	n	939.57	N^{*0}	1232
ddd	-1	$-\frac{3}{2}$	0			N^{*-}	1234
uds	0	0	-1	Λ^0	1115.6		
uus	1	1	-1	Σ^+	1189.4	Σ^{*+}	1382
uds	0	0	-1	Σ^0	1192.5	Σ^{*0}	1382
dds	-1	-1	-1	Σ^-	1197.3	Σ^{*-}	1387
uss	0	$\frac{1}{2}$	-2	Ξ^0	1314.9	Ξ^{*0}	1532
dss	-1	$-\frac{1}{2}$	-2	Ξ^-	1321.3	Ξ^{*-}	1535
sss	-1	0	-3			Ω^-	1672

Table 11.3 Properties of Some Composite Quark-Antiquark Systems, the Common Mesons

Quark Composition	Q	I_3	Y	Spin 0		Spin 1	
				Particle	Mass (MeV)	Particle	Mass (MeV)
$u\tilde{s}$	1	$\frac{1}{2}$	1	K^+	493.67	K^{*+}	892
$d\tilde{s}$	0	$\frac{1}{2}$	1	K_a^0	497.67	K_a^{*0}	899
$u\tilde{d}$	1	1	0	π^+	139.57	ρ^+	769
$\frac{1}{2}u\tilde{u} + \frac{1}{2}d\tilde{d}$	0	0	0	π^0	134.96	ρ^0	769
$d\tilde{u}$	-1	-1	0	π^-	139.57	ρ^-	769
$s\tilde{d}$	0	$\frac{1}{2}$	-1	K_b^0	497.67	K_b^{*0}	899
$s\tilde{u}$	-1	$-\frac{1}{2}$	-1	K^-	493.67	K^{*-}	892
$\frac{1}{6}u\tilde{u} + \frac{1}{6}d\tilde{d} + \frac{2}{3}s\tilde{s}$	0	0	0	η	548.8	ω	782.6
$\frac{1}{3}u\tilde{u} + \frac{1}{3}d\tilde{d} + \frac{1}{3}s\tilde{s}$	0	0	0	η'	957.6	ϕ	1019.5

11.7
Commutators of the SU(3) Operators

Creation and annihilation operators for the three principal quarks were defined in section 11.5. A complete set of bilinear products was constructed. Certain of these were identified as step operators. Others were combined to form the baryon number operator and the null operators for the step ones.

The commutators of the baryon, step, and null operators are operators in the set. As a consequence, these operators generate a Lie algebra. But in figure 11.1, we have one more dimension, one more degree of freedom, than we had for the proton-neutron systems where the group was **SU**(2). Still, the transformations among the quark states are unitary. So the group here is the **SU**(3) group.

Proceeding as with isospin, we obtain the relations

$$[B, I_\pm] = 0, \quad [B, I_0] = 0, \qquad (11.49)$$

$$[B, U_\pm] = 0, \quad [B, U_0] = 0, \qquad (11.50)$$

$$[B, V_\pm] = 0, \quad [B, V_0] = 0, \qquad (11.51)$$

$$[I_0, I_\pm] = \pm I_\pm, \quad [I_+, I_-] = 2I_0, \qquad (11.52)$$

$$[U_0, U_\pm] = \pm U_\pm, \quad [U_+, U_-] = 2U_0, \qquad (11.53)$$

$$[V_0, V_\pm] = \pm V_\pm \quad [V_+, V_-] = 2V_0. \qquad (11.54)$$

Also with formula (11.18), we find that

$$[I_0, U_0] = 0, \quad [I_0, V_0] = 0, \quad [U_0, V_0] = 0, \qquad (11.55)$$

$$[I_0, U_\pm] = \mp\tfrac{1}{2}U_\pm, \quad [I_0, V_\pm] = \mp\tfrac{1}{2}V_\pm, \qquad (11.56)$$

$$[U_0, I_\pm] = \mp\tfrac{1}{2}I_\pm, \quad [U_0, V_\pm] = \mp\tfrac{1}{2}V_\pm, \qquad (11.57)$$

$$[V_0, I_\pm] = \mp\tfrac{1}{2}I_\pm, \quad [V_0, U_\pm] = \mp\tfrac{1}{2}U_\pm, \qquad (11.58)$$

$$[I_\pm, U_\mp] = 0, \quad [I_\pm, V_\mp] = 0, \quad [U_\pm, V_\mp] = 0, \qquad (11.59)$$

$$[I_\pm, U_\pm] = \pm V_\mp, \quad [I_\pm, V_\pm] = \pm U_\mp, \quad [U_\pm, V_\pm] = \pm I_\mp. \qquad (11.60)$$

11.8
SU(3) Young Diagrams

According to the Pauli exclusion principle, the overall state function for a composite system must be antisymmetric with respect to an exchange of any two equivalent fermions. The permutations occur among the particle spatial-spin states. In the approximation that the three quarks are equivalent, all permutations among the quark states must be considered. These states may be treated as the spin states were in chapter 8, with Young diagrams.

A single quark system is represented by a single square:

$$\square \qquad (11.61)$$

There are three such states, as figure 11.4 shows.

Two quarks would yield $3 \times 3 = 9$ states. The antisymmetric set is represented by two vertically arranged squares:

$$(11.62)$$

The three states of this type are identified in figure 11.5. The symmetric set is represented by two squares arranged horizontally:

$$\square\square \qquad (11.63)$$

The six states of this type are identified in figure 11.6.

To diagram (11.62) one can add a square below to get the antisymmetric set

$$(11.64)$$

plotted in figure 11.3. Alternatively, one can add a square to the right, getting the hybrid states

 (11.65)

plotted in figure 11.7. There are a total of eight states here.

To diagram (11.63), one can similarly add a square below to get an independent set of hybrid states

 (11.66)

Again there are eight states, as in figure 11.7. Alternatively, one can add a square to the right, getting the set of 10 symmetric states

 (11.67)

plotted in figure 11.8.

These results are summarized in table 11.4. The general Young diagram for systems containing three different quarks has the form

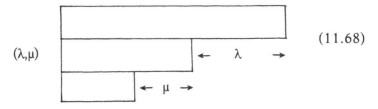 (11.68)

The three-quark columns are vacuum-like. The number of two-quark columns equals μ; the number of one-quark columns equals λ.

In constructing the corresponding Young tables, a person fills in the diagrams with u, d, s in order starting with the earliest one in the upper left corner. Any number of each may occur in a row. But all entries in a column must differ.

Table 11.4 How Quarks May Combine Together with the Number of States Involved

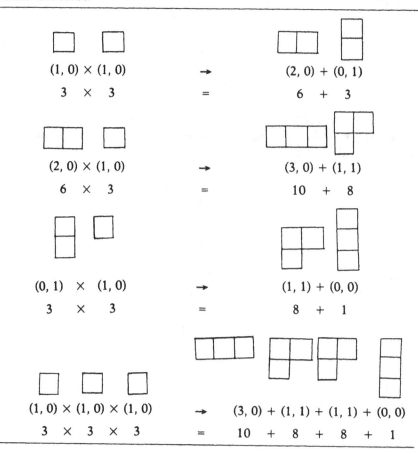

Example 11.6

Construct the Young tables for the multiplet in figure 11.7.

Fill in three-square $(1, 1)$ diagrams with u, d, s following the rules just enunciated to get

The Octet Degeneracy 317

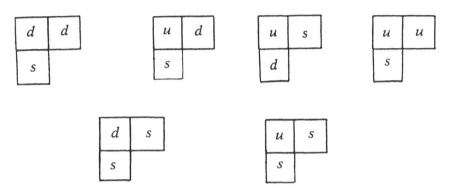

Note how the rules yield the number of states at each lattice point.

11.9
The Octet Degeneracy

For three-quark systems, we have constructed Young diagrams (11.64), (11.65), (11.66), and (11.67). These give rise to the hexagonal lattices of figures 11.3, 11.7, and 11.8. The octet in figure 11.7 is doubly degenerate, as we will now determine.

When all the quarks are different, possible combinations in the $Y - I_3$ plane are governed by the argument in example 8.3. Thus, we employ the completely antisymmetric configuration $\{1^3\}$ for figure 11.3, the completely symmetric configuration $\{3\}$ for the center point in figure 11.8, and the two $\{21\}$ forms, each doubly degenerate, for the center point in figure 11.7.

When two of the quarks are the same and one is different, the completely antisymmetric configuration $\{1^3\}$ and the $\{21\}$ forms antisymmetric in the two identical quarks drop out. There are two independent remaining $\{21\}$ forms.

As an example, consider the proton structure. One may consider its three quarks to occupy the corners of an equilateral triangle in a permutation plane. We then employ components of the C and D character vectors for the $\mathbf{C}_3$ group in the projection operator to construct the unitary spin forms

$$|p\rangle_a = |uud\rangle_a = \frac{1}{\sqrt{3}} (|uud\rangle + \omega|udu\rangle + \omega^2|duu\rangle)$$
(11.69)

and

$$|p>_b = |uud>_b = \frac{1}{\sqrt{3}}(|uud> + \omega^2|udu> + \omega|duu>). \quad (11.70)$$

On adding form (11.70) to (11.69) and renormalizing, we get

$$|p>_a = |uud>_a = \frac{1}{\sqrt{6}}(2|uud> - |udu> - |duu>). \quad (11.71)$$

On subtracting form (11.70) from (11.69) and renormalizing, we obtain

$$|p>_\beta = |uud>_\beta = \frac{1}{\sqrt{2}}(|udu> - |duu>). \quad (11.72)$$

These correspond to the $|m_1>$ and $|m_2>$ forms of example 8.3. The other off-center points in the octet lattice can be considered similarly.

It is found that the kets for the $Y - I_3$ lattice (the unitary-spin kets) combine with the spatial kets and the spin kets to yield a form symmetric in the constituents. But since the quarks are fermions, the complete ket must be antisymmetric. To meet this condition, one introduces an additional quantum number that takes on three aspects. By convention, these are labeled red, green, and blue.

11.10
A Representation of the Color Field

Each quark presumably carries one unit of color charge. We consider this to be the source of the strong interaction that the quark exerts on neighboring quarks. But because there are three kinds of the charge, this interaction is more complicated than that between two equal masses or that between an electron and a positron.

Nevertheless, we may follow Faraday in representing these fields with interaction lines ("lines of force"). From a property with a single sign, mass, extend nondirectional lines. Their density over a cross sectional area would represent the strength of the field at the point under consideration.

The Octet Degeneracy

Two kinds of electric charges exist. Faraday pictured lines running from positive charges to negative charges, with their density over cross sectional area representing the field intensity at a given point, their direction the direction of the field.

For the color field, we may consider the lines to run from a red unit to a green unit, from the green unit to a blue unit, and from the blue unit to the red, in a three-member cycle. (See figure 11.9.)

For each unobserved particle (figures 11.4, 11.5 and 11.6) the cycle would be incomplete. For each observed particle (figures 11.7 and 11.8) the cycle is complete.

Presumably, particles and antiparticles possess mass of the same type, producing similar gravitational fields. But with electric charge, transforming a particle into the corresponding antiparticle reverses the sign and so the direction of lines emanating. With color charge, going from particle to antiparticle similarly reverses the directions of the lines. (See figure 11.10).

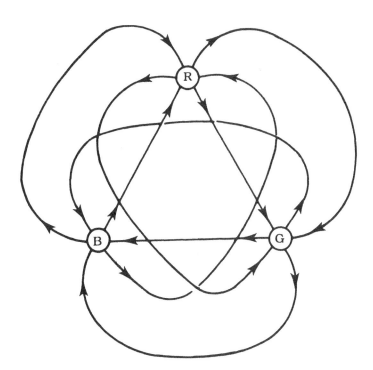

Figure 11.9 A representation of the color field in a baryon.

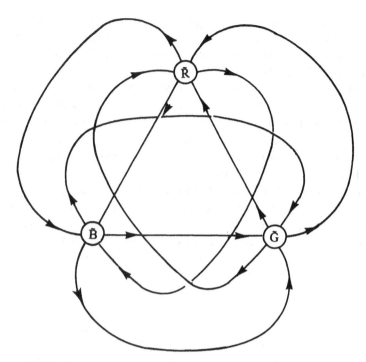

Figure 11.10 A representation of the color field in an antibaryon.

Furthermore, pairs of quarks behave as the corresponding antiquark. Thus, an anti-red quark behaves as a green-blue combination. And quark-antiquark pairs exist as mesons. These form an octet and a decuplet. (An example appears in figure 11.11.)

When one tries to knock a quark out of a baryon, the field energy builds up until it equals or exceeds the energy of a quark- antiquark pair. This appears and the antiquark goes with the leaving quark; the new quark remains behind to keep the three-member cycle unbroken.

Discussion Questions

11.1 Justify the construction of creation and annihilation operators.

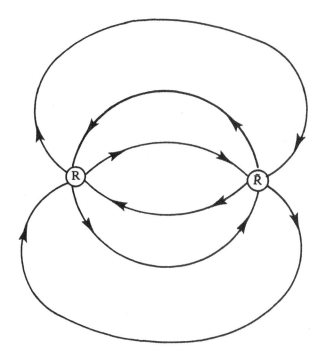

Figure 11.11 A representation of the color field in a meson.

11.2 How do the creation and annihilation operators for a given set of bosons fail to form a Lie algebra?

11.3 How do the creation and annihilation operators for a given set of fermions fail to form a Lie algebra?

11.4 How do the bilinear products for bosons form a Lie algebra?

11.5 How do the bilinear products for fermions form a Lie algebra?

11.6 Why do we need only bilinear products of the form $a_j^+ a_k^-$ in many applications?

11.7 In what approximation does a proton behave as a neutron?

11.8 How do $a_p^+ a_n^-$, $a_n^+ a_p^-$, and $\tfrac{1}{2}(a_p^+ a_p^- - a_n^+ a_n^-)$ act as angular momentum operators?

11.9 Explain how the eigenvalues for τ_0 arise. What ones occur in a multiplet with a given B?

11.10 What is isospin?

11.11 How do the isospin transformations belong to the **SU**(2) group?

11.12 Why is hypercharge Y introduced in addition to the isospin quantum number I_3?

11.13 What are the quarks?

11.14 What roles are played by I_+, I_-, U_+, U_-, V_+, V_-?

11.15 Why do we take the unit along the Y axis to be $\sqrt{3/2}$ times as long as a unit along the I_3 axis?

11.16 What form does the boundary of the hexagonal lattice for a given multiplet assume? Why?

11.17 How many states are there at each lattice point with a given multiplet?

11.18 How do we obtain the charge on each quark?

11.19 How do the antiquarks behave in the $Y-I_3$ plane?

11.20 Why are the common meson families octets?

11.21 How do the operators in the $Y-I_3$ plane generate a Lie algebra?

11.22 Why is the associated group the **SU**(3) group?

11.23 How do Young diagrams describe a given multiplet in the $Y-I_3$ plane?

11.24 How are Young tables constructed for the independent quark states in a multiplet?

11.25 For a system of three different quarks, what is the composition of the corresponding reducible representation of group $\mathscr{S}_3$?

11.26 How is this breakdown affected when two of the quarks are the same?

11.27 What is the degeneracy of the structures represented in example 11.6? Explain.

11.28 What permutations may be represented by the elements of (a) group $\mathbf{C}_{3v}$, (b) group $\mathbf{C}_3$? Explain.

11.29 Why is the color quantum number introduced?

11.30 How may one represent the color field?

Problems

11.1 Show how commutation relations (11.3), (11.4), and (11.5) with $j = k$ follow from conditions (11.1) and (11.2).

11.2 Show how commutation relations (11.3), (11.4), and (11.5) with $j \neq k$ follow from conditions (11.1) and (11.2).

11.3 With the pertinent commutation relations, derive equation (11.12).

11.4 With the pertinent anticommutation relations, derive equation (11.19).

11.5 Establish equations (11.25) and (11.26).
11.6 Establish equation (11.30).
11.7 Deduce the **SU**(3) commutator relations (11.49).
11.8 Deduce the **SU**(3) commutator relations (11.52).
11.9 Derive the **SU**(3) commutator relations

$$[I_\pm\ U_\mp] = 0 \quad \text{and} \quad [I_\pm, U_\pm] = \pm V_\mp.$$

11.10 Derive the **SU**(3) commutator relations (11.56).
11.11 Show how anticommutation relations (11.9), (11.10), and (11.11) with $j = k$ follow from conditions (11.6) and (11.7).
11.12 Show how anticommutation relations (11.9), (11.10), and (11.11) with $j \neq k$ follow from conditions (11.6) and (11.7).
11.13 With the pertinent anticommutation relations, derive equation (11.16).
11.14 With the pertinent commutation relations, derive equation (11.15).
11.15 Establish equation (11.27).
11.16 Establish equations (11.28) and (11.29).
11.17 Deduce the **SU**(3) commutator relations (11.50).
11.18 Deduce the **SU**(3) commutator relations (11.53).
11.19 Derive the **SU**(3) commutator relations

$$[I_\pm, V_\pm] = \mp U_\mp \quad \text{and} \quad [I_\pm, V_\mp] = 0.$$

11.20 Derive the **SU**(3) commutator relations (11.57).

References

Books

Kim, Y. S., and Noz, M. E.: 1986, *Theory and Applications of the Poincare Group*, Reidel, Dordrecht, pp. 255-285. Quantum mechanics acts to explain the discrete spectra occurring (a) in atoms and molecules, (b) in nuclei, and (c) in hadrons. Here the mass spectra of hadrons are considered in some detail using the quark model.

Lipkin, H. J.: 1966, *Lie Groups for Pedestrians*, North-Holland Publishing Co., Amsterdam, pp. 19-176. Lipkin shows how bilinear products of creation and annihilation operators lead to Lie algebras. He applies the resulting theory to elementary particles and to the three-dimensional harmonic oscillator. He considers operators that alter the number of particles. He also introduces the reader to groups of higher rank.

Matthew, P. T.: 1967, "Unitary Symmetry," in Burhop, E. H. S (editor), *High Energy Physics, Volume I*, Academic Press, New York, pp. 391481. This is a review article of considerable scope and depth. The mathematical ideas are developed in a physical manner.

Article

Duffey, G.H.:1982, "Interaction-Line Descriptions of Fields," *Found. Phys.* **12**, 499–508

APPENDIX: *Characters and Bases for the Primitive Symmetry Species of Important Groups*

A.1
Bases for Representations of Geometric Symmetry Groups

A set composed of expressions that are transformed into linear combinations of the expressions by each and every operation of a given group makes up a basis for a matrix representation of the group. The expressions may measure some attribute or property associated with equivalent parts of the given physical system. Furthermore, they may be represented by standard displacements, rotations, coordinates, functions of these, kets, vectors, dyads, polyads, tensors.

In constructing appropriate bases, one needs to choose an origin and standard directions from this origin. These referents are generally chosen to make the bases as simple as possible. Whenever operations of the group leave a point unchanged, this point may well be chosen as the origin. Whenever some operations of the group leave an axis unchanged, this axis may be chosen as a coordinate axis. A coordinate plane may well be identified with a reflection plane. (And so on.)

For the full rotation group and its subgroups, we place the origin at the center, the point that is left unchanged by the operations. With the full group, there are no preferred directions from this point. For particular subgroups, however, there are preferred axes to be chosen.

In the tables, vectors **i**, **j**, and **k** are the conventional Cartesian base vectors drawn from the center along the pertinent x, y, and z axes. In some tables, we employ a unit vector drawn perpendicular to the z axis labeled $\mathbf{e}_1$ and rotations of it by C_n^m.

In table A.1, correspondence between the more instructive vector and polyadic bases and the conventional functional bases are presented.

Table A.1 Corresponding Bases

Polyads	Functions
i	x
j	y
k	z
i × **j**	R_z
j × **k**	R_x
k × **i**	R_y
ii	x^2
jj	y^2
kk	z^2
$\frac{1}{\sqrt{2}}(\mathbf{ij} + \mathbf{ji})$	xy
$\frac{1}{\sqrt{2}}(\mathbf{jk} + \mathbf{kj})$	yz
$\frac{1}{\sqrt{2}}(\mathbf{ki} + \mathbf{ik})$	zx
$\frac{1}{\sqrt{2}}(\mathbf{i} + i\mathbf{j})$	$re^{i\phi}$
$\frac{1}{\sqrt{2}}(\mathbf{i} - i\mathbf{j})$	$re^{-i\phi}$

A.2
Properties of the Primitive Symmetry Species of Geometric Symmetry Groups

The groups whose elements are represented by reorientations about a point in space are subgroups of the group of operations that transform into itself a sphere centered on the invariant point. Key characters and bases for the most important of these appear in the following tables.

The boldface symbol in the upper left corner of a table identifies the symmetry group. Each letter under this symbol labels a different symmetry species. The numbers in a row are the components of the character vector for the species; the polyads are possible bases for the species. The heading for each column of numbers labels the class, by a typical operation, and gives the number of covering operations in the class.

Note that the following trigonometric relationships are useful for groups containing C_5,

$$2 \cos 144° = -\tfrac{1}{2}(1 + \sqrt{5}),$$

$$2 \cos 72° = -\tfrac{1}{2}(1 - \sqrt{5}).$$

C_1	I			
A	1		i, j, k	j×k, k×i, i×j

$C_s \equiv S_1$	I	σ_h		
A'	1	1	i, j	ii, jj, kk, ij, ji jk, kj, ki, ik
A''	1	−1	k	i×j j×k, k×i

$C_i \equiv S_2$	I	i		
A_g	1	1		ii, jj, kk, jk, kj, ki, ik, ij, ji
A_u	1	−1	i, j, k	j×k, k×i, i×j

S_4	I	S_4	C_2	S_4^3		
A	1	1	1	1	i×j	$\frac{1}{\sqrt{2}}(\mathrm{ii}+\mathrm{jj}), \mathrm{kk}, \frac{1}{\sqrt{2}}(\mathrm{ij}-\mathrm{ji})$
B	1	−1	1	−1	k	$\frac{1}{\sqrt{2}}(\mathrm{ii}-\mathrm{jj}), \frac{1}{\sqrt{2}}(\mathrm{ij}+\mathrm{ji})$
C	1	i	−1	$-i$	$\frac{1}{\sqrt{2}}(\mathrm{i}-i\mathrm{j})$	$\frac{1}{\sqrt{2}}(\mathrm{j}\times\mathrm{k}+i\mathrm{k}\times\mathrm{i})$	$\frac{1}{\sqrt{2}}\mathrm{k}(1+i\mathrm{j}), \frac{1}{\sqrt{2}}(1+i\mathrm{j})\mathrm{k}$
D	1	$-i$	−1	i	$\frac{1}{\sqrt{2}}(\mathrm{i}+i\mathrm{j})$	$\frac{1}{\sqrt{2}}(\mathrm{j}\times\mathrm{k}-i\mathrm{k}\times\mathrm{i})$	$\frac{1}{\sqrt{2}}\mathrm{k}(1-i\mathrm{j}), \frac{1}{\sqrt{2}}(1-i\mathrm{j})\mathrm{k}$

S_6	I	C_3	C_3^2	i	S_6^5	S_6		
A_g	1	1	1	1	1	1		$\mathbf{i}\times\mathbf{j}$
C_g	1	ω	ω^2	1	ω	ω^2	$\frac{1}{\sqrt{3}}(\mathbf{e}_1' + \omega^2\mathbf{e}_2' + \omega\mathbf{e}_3')$	$\frac{1}{\sqrt{3}}(\mathbf{e}_1\mathbf{e}_1 + \omega^2\mathbf{e}_2\mathbf{e}_2 + \omega\mathbf{e}_3\mathbf{e}_3)$
								$\frac{1}{\sqrt{3}}\mathbf{k}(\mathbf{e}_1 + \omega^2\mathbf{e}_2 + \omega\mathbf{e}_3)$
D_g	1	ω^2	ω	1	ω^2	ω	$\frac{1}{\sqrt{3}}(\mathbf{e}_1' + \omega\mathbf{e}_2' + \omega^2\mathbf{e}_3')$	$\frac{1}{\sqrt{3}}(\mathbf{e}_1\mathbf{e}_1 + \omega\mathbf{e}_2\mathbf{e}_2 + \omega^2\mathbf{e}_3\mathbf{e}_3)$
								$\frac{1}{\sqrt{3}}\mathbf{k}(\mathbf{e}_1 + \omega\mathbf{e}_2 + \omega^2\mathbf{e}_3)$
A_u	1	1	1	−1	−1	−1	$\mathbf{k}$	$\frac{1}{\sqrt{2}}(\mathbf{i}\mathbf{i}+\mathbf{j}\mathbf{j}),\ \mathbf{k}\mathbf{k},\ \frac{1}{\sqrt{2}}(\mathbf{i}\mathbf{j}-\mathbf{j}\mathbf{i})$
C_u	1	ω	ω^2	−1	$-\omega$	$-\omega^2$	$\frac{1}{\sqrt{3}}(\mathbf{e}_1 + \omega^2\mathbf{e}_2 + \omega\mathbf{e}_3)$	
D_u	1	ω^2	ω	−1	$-\omega^2$	$-\omega$	$\frac{1}{\sqrt{3}}(\mathbf{e}_1 + \omega\mathbf{e}_2 + \omega^2\mathbf{e}_3)$	

$\omega = \exp(2\pi i/3)$ $\mathbf{e}_1 = \mathbf{i},\ \mathbf{e}_2 = C_3\mathbf{i},\ \mathbf{e}_1' = \mathbf{j}\times\mathbf{k},\ \mathbf{e}_2' = C_3\mathbf{e}_1',\ \mathbf{e}_3' = C_3^2\mathbf{e}_1'$

C_2	I	C_2		
A	1	1	$\mathbf{k}$	$\mathbf{i}\times\mathbf{j}$	$\mathbf{i}\mathbf{i},\ \mathbf{j}\mathbf{j},\ \mathbf{k}\mathbf{k},\ \mathbf{i}\mathbf{j},\ \mathbf{j}\mathbf{i}$
B	1	−1	$\mathbf{i},\mathbf{j}$	$\mathbf{j}\times\mathbf{k},\ \mathbf{k}\times\mathbf{i}$	$\mathbf{k}\mathbf{i},\ \mathbf{i}\mathbf{k},\ \mathbf{j}\mathbf{k},\ \mathbf{k}\mathbf{j}$

C_3	I	C_3	C_3^2			
A	1	1	1	$\mathbf{k}$	$\mathbf{i}\times\mathbf{j}$	$\frac{1}{\sqrt{2}}(\mathbf{ii}+\mathbf{jj}), \mathbf{kk}, \frac{1}{\sqrt{2}}(\mathbf{ij}-\mathbf{ji})$
C	1	ω	ω^2	$\frac{1}{\sqrt{3}}(\mathbf{e}_1+\omega^2\mathbf{e}_2+\omega\mathbf{e}_3)$	$\frac{1}{\sqrt{3}}(\mathbf{e}_1'+\omega^2\mathbf{e}_2'+\omega\mathbf{e}_3')$	$\frac{1}{\sqrt{3}}(\mathbf{e}_1\mathbf{e}_1+\omega^2\mathbf{e}_2\mathbf{e}_2+\omega\mathbf{e}_3\mathbf{e}_3)$
						$\frac{1}{\sqrt{3}}\mathbf{k}(\mathbf{e}_1+\omega^2\mathbf{e}_2+\omega\mathbf{e}_3)$
						$\frac{1}{\sqrt{3}}(\mathbf{e}_1+\omega^2\mathbf{e}_2+\omega\mathbf{e}_3)\mathbf{k}$
D	1	ω^2	ω	$\frac{1}{\sqrt{3}}(\mathbf{e}_1+\omega\mathbf{e}_2+\omega^2\mathbf{e}_3)$	$\frac{1}{\sqrt{3}}(\mathbf{e}_1+\omega\mathbf{e}_2'+\omega^2\mathbf{e}_3')$	$\frac{1}{\sqrt{3}}(\mathbf{e}_1\mathbf{e}_1+\omega\mathbf{e}_2\mathbf{e}_2+\omega^2\mathbf{e}_3\mathbf{e}_3)$
						$\frac{1}{\sqrt{3}}\mathbf{k}(\mathbf{e}_1+\omega\mathbf{e}_2+\omega^2\mathbf{e}_3)$
						$\frac{1}{\sqrt{3}}(\mathbf{e}_1+\omega\mathbf{e}_2+\omega^2\mathbf{e}_3)\mathbf{k}$

$\omega = \exp(2\pi i/3)$ $\mathbf{e}_1 = \mathbf{i}, \mathbf{e}_2 = C_3\mathbf{i}, \mathbf{e}_3 = C_3^2\mathbf{i}, \mathbf{e}_1' = \mathbf{j}\times\mathbf{k}, \mathbf{e}_2' = C_3\mathbf{e}_1', \mathbf{e}_3' = C_3^2\mathbf{e}_1'$

C_4	I	C_4	C_2	C_4^3			
A	1	1	1	1	$\mathbf{k}$	$\mathbf{i}\times\mathbf{j}$	$\frac{1}{\sqrt{2}}(\mathbf{ii}+\mathbf{jj}), \mathbf{kk}, \frac{1}{\sqrt{2}}(\mathbf{ij}-\mathbf{ji})$
B	1	-1	1	-1			$\frac{1}{\sqrt{2}}(\mathbf{ii}-\mathbf{jj}), \frac{1}{\sqrt{2}}(\mathbf{ij}+\mathbf{ji})$
C	1	i	-1	$-i$	$\frac{1}{\sqrt{2}}(\mathbf{i}-i\mathbf{j})$	$\frac{1}{\sqrt{2}}(\mathbf{j}\times\mathbf{k}-i\mathbf{k}\times\mathbf{i})$	$\frac{1}{\sqrt{2}}\mathbf{k}(\mathbf{i}-i\mathbf{j}), \frac{1}{\sqrt{2}}(\mathbf{i}-i\mathbf{j})\mathbf{k}$
D	1	$-i$	-1	i	$\frac{1}{\sqrt{2}}(\mathbf{i}+i\mathbf{j})$	$\frac{1}{\sqrt{2}}(\mathbf{j}\times\mathbf{k}+i\mathbf{k}\times\mathbf{i})$	$\frac{1}{\sqrt{2}}\mathbf{k}(\mathbf{i}+i\mathbf{j}), \frac{1}{\sqrt{2}}(\mathbf{i}+i\mathbf{j})\mathbf{k}$

C_5	I	C_5	C_5^2	C_5^3	C_5^4		
A	1	1	1	1	1	$\mathbf{k}$	$\mathbf{i} \times \mathbf{j}$
C_1	1	ω	ω^2	ω^3	ω^4	$\dfrac{1}{\sqrt{5}}(\mathbf{e}_1 + \omega^4 \mathbf{e}_2 + \omega^3 \mathbf{e}_3 + \omega^2 \mathbf{e}_4 + \omega \mathbf{e}_5) = \mathbf{f}_1$	
D_1	1	ω^4	ω^3	ω^2	ω	$\dfrac{1}{\sqrt{5}}(\mathbf{e}_1 + \omega \mathbf{e}_2 + \omega^2 \mathbf{e}_3 + \omega^3 \mathbf{e}_4 + \omega^4 \mathbf{e}_5) = \mathbf{f}_2$	
C_2	1	ω^2	ω^4	ω	ω^3		$\dfrac{1}{\sqrt{5}}(\mathbf{e}_1\mathbf{e}_1 + \omega^3\mathbf{e}_2\mathbf{e}_2 + \omega\mathbf{e}_3\mathbf{e}_3 + \omega^4\mathbf{e}_4\mathbf{e}_4 + \omega^2\mathbf{e}_5\mathbf{e}_5)$
D_2	1	ω^3	ω	ω^4	ω^2		$\dfrac{1}{\sqrt{5}}(\mathbf{e}_1\mathbf{e}_1 + \omega^2\mathbf{e}_2\mathbf{e}_2 + \omega^4\mathbf{e}_3\mathbf{e}_3 + \omega\mathbf{e}_4\mathbf{e}_4 + \omega^3\mathbf{e}_5\mathbf{e}_5)$

$\omega = \exp(2\pi i/5)$ $\mathbf{e}_1 = \mathbf{i}$ or $\mathbf{e}_1 = \mathbf{j} \times \mathbf{k}$, $\mathbf{e}_2 = C_5\mathbf{e}_1$, $\mathbf{e}_3 = C_5^2\mathbf{e}_1$, $\mathbf{e}_4 = C_5^3\mathbf{e}_1$, $\mathbf{e}_5 = C_5^4\mathbf{e}_1$

C_6	I	C_6	C_3	C_2	C_3^2	C_6^5		
A	1	1	1	1	1	1	$\mathbf{k}$	$\mathbf{i} \times \mathbf{j}$
B	1	-1	1	-1	1	-1		
C_1	1	$-\omega^2$	ω^2	-1	ω	$-\omega$	$\dfrac{1}{\sqrt{3}}(\mathbf{e}_1 + \omega^2\mathbf{e}_2 + \omega\mathbf{e}_3)$	$\dfrac{1}{\sqrt{3}}(\mathbf{e}_1' + \omega^2\mathbf{e}_2' + \omega\mathbf{e}_3')$
D_1	1	$-\omega$	ω^2	-1	ω	$-\omega^2$	$\dfrac{1}{\sqrt{3}}(\mathbf{e}_1 + \omega\mathbf{e}_2 + \omega^2\mathbf{e}_3)$	$\dfrac{1}{\sqrt{3}}(\mathbf{e}_1' + \omega\mathbf{e}_2' + \omega^2\mathbf{e}_3')$
C_2	1	ω	ω^2	1	ω	ω^2	$\dfrac{1}{\sqrt{3}}(\mathbf{e}_1\mathbf{e}_1 + \omega\mathbf{e}_2\mathbf{e}_2 + \omega^2\mathbf{e}_3\mathbf{e}_3)$	
D_2	1	ω^2	ω	1	ω^2	ω	$\dfrac{1}{\sqrt{3}}(\mathbf{e}_1\mathbf{e}_1 + \omega^2\mathbf{e}_2\mathbf{e}_2 + \omega^2\mathbf{e}_3\mathbf{e}_3)$	

$\omega = \exp(2\pi i/3)$ $\mathbf{e}_1 = \mathbf{i}, \mathbf{e}_2 = C_3\mathbf{i}, \mathbf{e}_3 = C_3^2\mathbf{i}, \mathbf{e}_1' = \mathbf{j} \times \mathbf{k}, \mathbf{e}_2' = C_3\mathbf{e}_3', \mathbf{e}_3' = C_3^2\mathbf{e}_1'$

Note: The last column of the C_5 table shows basis functions: $\mathbf{kf}_1, \mathbf{f}_1\mathbf{k}$ and $\mathbf{kf}_2, \mathbf{f}_2\mathbf{k}$, with top row $\dfrac{1}{\sqrt{2}}(\mathbf{ii}+\mathbf{jj}), \mathbf{kk}, \dfrac{1}{\sqrt{2}}(\mathbf{ij}-\mathbf{ji})$.

Last column of C_6 table (top row): $\dfrac{1}{\sqrt{2}}(\mathbf{ii}+\mathbf{jj}), \mathbf{kk}, \dfrac{1}{\sqrt{2}}(\mathbf{ij}-\mathbf{ji})$; rows for C_1, D_1: $\dfrac{1}{\sqrt{3}}\mathbf{k}(\mathbf{e}_1+\omega^2\mathbf{e}_2+\omega\mathbf{e}_3)$ and $\dfrac{1}{\sqrt{3}}\mathbf{k}(\mathbf{e}_1+\omega\mathbf{e}_2+\omega^2\mathbf{e}_3)$.

C_{2v}	I	C_2	$\sigma_v(zx)$	$\sigma_v'(yz)$		
A_1	1	1	1	1	**k**	**ii, jj, kk**
A_2	1	1	−1	−1		**ij, ji**
B_1	1	−1	1	−1	**i**	**i×j** **ki, ik**
B_2	1	−1	−1	1	**j**	**k×i** **j×k** **jk, kj**

C_{3v}	I	$2C_3$	$3\sigma_v$		
A_1	1	1	1	**k**	$\frac{1}{\sqrt{2}}(\mathbf{ii+jj})$, **kk**
A_2	1	1	−1		$\frac{1}{\sqrt{2}}(\mathbf{ij-ji})$
E	2	−1	0	(**i,j**)	$\left[\frac{1}{\sqrt{2}}(\mathbf{ii-jj}), \frac{1}{\sqrt{2}}(\mathbf{ij+ji})\right]$, (**ik, jk**), (**ki, kj**)
				(**j×k, k×i**)	**i×j**

C_{4v}	I	$2C_4$	C_2	$2\sigma_v$	$2\sigma_d$		
A_1	1	1	1	1	1	**k**	$\frac{1}{\sqrt{2}}(\mathbf{ii+jj})$, **kk**
A_2	1	1	1	−1	−1		$\frac{1}{\sqrt{2}}(\mathbf{ij-ji})$
B_1	1	−1	1	1	−1		$\frac{1}{\sqrt{2}}(\mathbf{ii-jj})$
B_2	1	−1	1	−1	1	**i×j**	$\frac{1}{\sqrt{2}}(\mathbf{ij+ji})$
E	2	0	−2	0	0	(**i,j**)	(**j×k, k×i**)	(**ik, jk**), (**ki, kj**)

$\mathbf{C}_{5v}$	I	$2C_5$	$2C_5^2$	$5\sigma_v$		
A_1	1	1	1	1	$\mathbf{k}$	$\frac{1}{\sqrt{2}}(\mathbf{ii}+\mathbf{jj}),\ \mathbf{kk}$
A_2	1	1	1	−1		$\frac{1}{\sqrt{2}}(\mathbf{ij}-\mathbf{ji})$
E_1	2	$2\cos 72°$	$2\cos 144°$	0	$(\mathbf{i},\mathbf{j})$	$(\mathbf{ik},\mathbf{jk}),\ (\mathbf{ki},\mathbf{kj})$
E_2	2	$2\cos 144°$	$2\cos 72°$	0		$\left[\frac{1}{\sqrt{2}}(\mathbf{ii}-\mathbf{jj}),\ \frac{1}{\sqrt{2}}(\mathbf{ij}+\mathbf{ji})\right]$

$\mathbf{C}_{6v}$	I	$2C_6$	C_3	C_2	$3\sigma_v$	$3\sigma_d$		
A_1	1	1	1	1	1	1	$\mathbf{k}$	$\frac{1}{\sqrt{2}}(\mathbf{ii}+\mathbf{jj}),\ \mathbf{kk}$
A_2	1	1	1	1	−1	−1		$\frac{1}{\sqrt{2}}(\mathbf{ij}-\mathbf{ji})$
B_1	1	−1	1	−1	1	−1		
B_2	1	−1	1	−1	−1	1		
E_1	2	1	−1	−2	0	0	$(\mathbf{i},\mathbf{j})$	$(\mathbf{ik},\mathbf{jk}),\ \mathbf{ki},\ \mathbf{kj})$
E_2	2	−1	−1	2	0	0		$\left[\frac{1}{\sqrt{2}}(\mathbf{ii}-\mathbf{jj}),\ \frac{1}{\sqrt{2}}(\mathbf{ij}+\mathbf{ji})\right]$

C_{2h}	I	C_2	i	σ_h		
A_g	1	1	1	1		$\mathbf{ii, jj, kk, ij, ji}$
B_g	1	−1	1	−1		$\mathbf{ki, ik, jk, kj}$
A_u	1	1	−1	−1	$\mathbf{k}$	
B_u	1	−1	−1	1	$\mathbf{i, j}$	

C_{3h}	I	C_3	C_3^2	σ_h	S_3	S_3^5		
A'	1	1	1	1	1	1		$\mathbf{i \times j}$
C'	1	ω	ω^2	1	ω	ω^2	$\frac{1}{\sqrt{3}}(\mathbf{e}_1 + \omega^2\mathbf{e}_2 + \omega\mathbf{e}_3)$	$\frac{1}{\sqrt{3}}(\mathbf{e}_1'+ 2\omega^2\mathbf{e}_2' + \omega\mathbf{e}_3')$
D'	1	ω^2	ω	1	ω^2	ω	$\frac{1}{\sqrt{3}}(\mathbf{e}_1 + \omega\mathbf{e}_2 + \omega^2\mathbf{e}_3)$	$\frac{1}{\sqrt{3}}(\mathbf{e}_1' + \omega\mathbf{e}_2' + \omega^2\mathbf{e}_3')$
A''	1	1	1	−1	−1	−1	$\mathbf{k}$	
C''	1	ω	ω^2	−1	$-\omega$	$-\omega^2$	$\frac{1}{\sqrt{3}}\mathbf{k}(\mathbf{e}_1 + \omega^2\mathbf{e}_2 + \omega\mathbf{e}_3)$	$\frac{1}{\sqrt{3}}(\mathbf{e}_1 + \omega^2\mathbf{e}_2 + \omega\mathbf{e}_3)\mathbf{k}$
D''	1	ω^2	ω	−1	$-\omega^2$	$-\omega$	$\frac{1}{\sqrt{3}}\mathbf{k}(\mathbf{e}_1 + \omega^2\mathbf{e}_2 + \omega^2\mathbf{e}_3)$	$\frac{1}{\sqrt{3}}(\mathbf{e}_1 + \omega\mathbf{e}_2 + \omega^2\mathbf{e}_3)\mathbf{k}$

Additional assignments for A' row: $\frac{1}{\sqrt{2}}(\mathbf{ii + jj}), \mathbf{kk}$; $\mathbf{j \times k, k \times i}$

$\omega = \exp(2\pi i/3)$ $\mathbf{e}_1 = \mathbf{i}, \mathbf{e}_2 = C_3\mathbf{i}, \mathbf{e}_3 = C_3^2\mathbf{i}, \mathbf{e}_1' = \mathbf{j} \times \mathbf{k}, \mathbf{e}_2' = C_3\mathbf{e}_1', \mathbf{e}_3' = C_3^2\mathbf{e}_1'$

C_{4h}	I	C_4	C_2	C_4^3	i	S_4^3	σ_h	S_4		
A_g	1	1	1	1	1	1	1	1		$\frac{1}{\sqrt{2}}(\mathbf{ii}+\mathbf{jj}), \mathbf{kk}, \frac{1}{\sqrt{2}}(\mathbf{ij}-\mathbf{ji})$
B_g	1	-1	1	-1	1	-1	1	-1		$\frac{1}{\sqrt{2}}(\mathbf{ii}-\mathbf{jj}), \frac{1}{\sqrt{2}}(\mathbf{ij}+\mathbf{ji})$
C_g	1	i	-1	$-i$	1	i	-1	$-i$	$\frac{1}{\sqrt{2}}(\mathbf{j}\times\mathbf{k}-i\mathbf{k}\times\mathbf{i})$	$\frac{1}{\sqrt{2}}(\mathbf{ik}-i\mathbf{jk}), \frac{1}{\sqrt{2}}(\mathbf{ki}-i\mathbf{kj})$
D_g	1	$-i$	-1	i	1	$-i$	-1	i	$\frac{1}{\sqrt{2}}(\mathbf{j}\times\mathbf{k}+i\mathbf{k}\times\mathbf{i})$	$\frac{1}{\sqrt{2}}(\mathbf{ik}+i\mathbf{jk}), \frac{1}{\sqrt{2}}(\mathbf{ki}+i\mathbf{kj})$
A_u	1	1	1	1	-1	-1	-1	-1		
B_u	1	-1	1	-1	-1	1	-1	1	$\mathbf{k}$	
C_u	1	i	-1	$-i$	-1	$-i$	1	i	$\frac{1}{\sqrt{2}}(\mathbf{i}-i\mathbf{j})$	
D_u	1	$-i$	-1	i	-1	i	1	$-i$	$\frac{1}{\sqrt{2}}(\mathbf{i}+i\mathbf{j})$	

335

C_{5h}	I	C_5	C_5^2	C_5^3	C_5^4	σ_h	S_5	S_5^7	S_5^3	S_5^9		
A'	1	1	1	1	1	1	1	1	1	1	$\mathbf{i} \times \mathbf{j}$	$\frac{1}{\sqrt{2}}(\mathbf{ii}+\mathbf{jj}), \mathbf{kk}, \frac{1}{\sqrt{2}}(\mathbf{ij}-\mathbf{ji})$
C_1'	1	ω	ω^2	ω^3	ω^4	1	ω	ω^2	ω^3	ω^4	$\frac{1}{\sqrt{5}}(\mathbf{e}_1+\omega^4\mathbf{e}_2+\omega^3\mathbf{e}_3+\omega^2\mathbf{e}_4+\omega\mathbf{e}_5) = \mathbf{f}_1$	
D_1'	1	ω^4	ω^3	ω^2	ω	1	ω^4	ω^3	ω^2	ω	$\frac{1}{\sqrt{5}}(\mathbf{e}_1+\omega\mathbf{e}_2+\omega^2\mathbf{e}_3+\omega^3\mathbf{e}_4+\omega^4\mathbf{e}_5) = \mathbf{f}_2$	
C_2'	1	ω^2	ω^4	ω	ω^3	1	ω^2	ω^4	ω	ω^3	$\frac{1}{\sqrt{5}}(\mathbf{e}_1\mathbf{e}_1+\omega^3\mathbf{e}_2\mathbf{e}_2+\omega_3\mathbf{e}_3\mathbf{e}_3+\omega^4\mathbf{e}_4\mathbf{e}_4+\omega^2\mathbf{e}_5\mathbf{e}_5)$	
D_2'	1	ω^3	ω	ω^4	ω^2	1	ω^3	ω	ω^4	ω^2	$\frac{1}{\sqrt{5}}(\mathbf{e}_1\mathbf{e}_1+\omega^2\mathbf{e}_2\mathbf{e}_2+\omega^4\mathbf{e}_3\mathbf{e}_3+\omega\mathbf{e}_4\mathbf{e}_4+\omega^3\mathbf{e}_5\mathbf{e}_5)$	
A''	1	1	1	1	1	-1	-1	-1	-1	-1	$\mathbf{k}$	
C_1''	1	ω	ω^2	ω^3	ω^4	-1	$-\omega$	$-\omega^2$	$-\omega^3$	$-\omega^4$	$\frac{1}{\sqrt{5}}(\mathbf{e}_1'+\omega^4\mathbf{e}_2'+\omega^3\mathbf{e}_3'+\omega^2\mathbf{e}_4'+\omega\mathbf{e}_5')$	$\mathbf{kf}_1, \mathbf{f}_1\mathbf{k}$
D_1''	1	ω^4	ω^3	ω^2	ω	-1	$-\omega^4$	$-\omega^3$	$-\omega^2$	$-\omega$	$\frac{1}{\sqrt{5}}(\mathbf{e}_1'+\omega\mathbf{e}_2'+\omega^2\mathbf{e}_3'+\omega^3\mathbf{e}_4'+\omega^4\mathbf{e}_5')$	$\mathbf{kf}_2, \mathbf{f}_2\mathbf{k}$
C_2''	1	ω^2	ω^4	ω	ω^3	-1	$-\omega^2$	$-\omega^4$	$-\omega$	$-\omega^3$		
D_2''	1	ω^3	ω	ω^4	ω^2	-1	$-\omega^3$	$-\omega$	$-\omega^4$	$-\omega^2$		

$$\omega = \exp(2\pi i/5)$$

$$\mathbf{e}_1 = \mathbf{i}, \; \mathbf{e}_2 = C_5\mathbf{e}_1, \; \mathbf{e}_3 = C_5^2\mathbf{e}_1, \; \mathbf{e}_4 = C_5^3\mathbf{e}_1, \; \mathbf{e}_5 = C_5^4\mathbf{e}_1$$

$$\mathbf{e}_1' = \mathbf{j} \times \mathbf{k}, \; \mathbf{e}_2' = C_5\mathbf{e}_1', \; \mathbf{e}_3' = C_5^2\mathbf{e}_1', \; \mathbf{e}_4' = C_5^3\mathbf{e}_1', \; \mathbf{e}_5' = C_5^4\mathbf{e}_1'$$

													basis
A_g	1	1	1	1	1	1	1	1	1	1	1	1	$\mathbf{i} \times \mathbf{j}$
B_g	1	−1	1	−1	1	−1	1	−1	1	−1	1	−1	$\frac{1}{\sqrt{2}}(\mathbf{ii}+\mathbf{jj}),\ \mathbf{kk},\ \frac{1}{\sqrt{2}}(\mathbf{ij}-\mathbf{ji})$
C_{1g}	1	−ω²	ω	−1	ω²	−ω	1	−ω²	ω	−1	ω²	−ω	$\frac{1}{\sqrt{3}}(\mathbf{e}'_1+\omega^2\mathbf{e}'_2+\omega\mathbf{e}'_3)$, $\frac{1}{\sqrt{3}}\mathbf{k}(\mathbf{e}_1+\omega^2\mathbf{e}_2+\omega\mathbf{e}_3)\mathbf{k}$
D_{1g}	1	−ω	ω²	−1	ω	−ω²	1	−ω	ω²	−1	ω	−ω²	$\frac{1}{\sqrt{3}}(\mathbf{e}'_1+\omega\mathbf{e}'_2+\omega^2\mathbf{e}'_3)$, $\frac{1}{\sqrt{3}}\mathbf{k}(\mathbf{e}_1+\omega\mathbf{e}_2+\omega^2\mathbf{e}_3)\mathbf{k}$
C_{2g}	1	ω	ω²	1	ω	ω²	1	ω	ω²	1	ω	ω²	$\frac{1}{\sqrt{3}}(\mathbf{e}_1\mathbf{e}_1+\omega\mathbf{e}_2\mathbf{e}_2+\omega^2\mathbf{e}_3\mathbf{e}_3)$
D_{2g}	1	ω²	ω	1	ω²	ω	1	ω²	ω	1	ω²	ω	$\frac{1}{\sqrt{3}}(\mathbf{e}_1\mathbf{e}_1+\omega^2\mathbf{e}_2\mathbf{e}_2+\omega\mathbf{e}_3\mathbf{e}_3)$
A_u	1	1	1	1	1	1	−1	−1	−1	−1	−1	−1	$\mathbf{k}$
B_u	1	−1	1	−1	1	−1	−1	1	−1	1	−1	1	
C_{1u}	1	−ω²	ω	−1	ω²	−ω	−1	ω²	−ω	1	−ω²	ω	$\frac{1}{\sqrt{3}}(\mathbf{e}_1+\omega^2\mathbf{e}_2+\omega\mathbf{e}_3)$
D_{1u}	1	−ω	ω²	−1	ω	−ω²	−1	ω	−ω²	1	−ω	ω²	$\frac{1}{\sqrt{3}}(\mathbf{e}_1+\omega\mathbf{e}_2+\omega^2\mathbf{e}_3)$
C_{2u}	1	ω	ω²	1	ω	ω²	−1	−ω	−ω²	−1	−ω	−ω²	
D_{2u}	1	ω²	ω	1	ω²	ω	−1	−ω²	−ω	−1	−ω²	−ω	

$\omega = \exp(2\pi i/3)$ $\mathbf{e}_1 = \mathbf{i},\ \mathbf{e}_2 = C_3\mathbf{i},\ \mathbf{e}_3 = C_3^2\mathbf{i},\ \mathbf{e}'_1 = \mathbf{j} \times \mathbf{k},\ \mathbf{e}'_2 = C_3\mathbf{e}'_1,\ \mathbf{e}'_3 = C_3^2\mathbf{e}'_1$

$\mathbf{D}_2$	I	$C_2(z)$	$C_2(y)$	$C_2(x)$		
A	1	1	1	1		$\mathbf{ii, jj, kk}$
B_1	1	1	-1	-1	$\mathbf{k}$	$\mathbf{ij, ji}$
B_2	1	-1	1	-1	$\mathbf{j}$	$\mathbf{ki, ik}$
B_3	1	-1	-1	1	$\mathbf{i}$	$\mathbf{jk, kj}$

$\mathbf{D}_3$	I	$2C_3$	$3C_2$		
A_1	1	1	1		$\frac{1}{\sqrt{2}}(\mathbf{ii + jj}),\ \mathbf{kk}$
A_2	1	1	-1	$\mathbf{k}$	$\frac{1}{\sqrt{2}}(\mathbf{ij - ji})$
E	2	-1	0	$(\mathbf{i, j})$, $(\mathbf{j \times k, k \times i})$	$\left[\frac{1}{\sqrt{2}}(\mathbf{ii - jj}),\ \frac{1}{\sqrt{2}}(\mathbf{ij + ji})\right]$, $(\mathbf{ik, jk}), (\mathbf{ki, kj})$

$\mathbf{D}_4$	I	$2C_4$	C_2	$2C_2'$	$2C_2''$		
A_1	1	1	1	1	1		$\frac{1}{\sqrt{2}}(\mathbf{ii + jj}),\ \mathbf{kk}$
A_2	1	1	1	-1	-1	$\mathbf{k}$	$\frac{1}{\sqrt{2}}(\mathbf{ij - ji})$
B_1	1	-1	1	1	-1		$\frac{1}{\sqrt{2}}(\mathbf{ii - jj})$
B_2	1	-1	1	-1	1		$\frac{1}{\sqrt{2}}(\mathbf{ij + ji})$
E	2	0	-2	0	0	$(\mathbf{i, j})$, $(\mathbf{j \times k, k \times i})$	$(\mathbf{ik, jk}), (\mathbf{ki, kj})$

D_5	I	$2C_5$	$2C_5^2$	$5C_2$		
A_1	1	1	1	1		$\frac{1}{\sqrt{2}}(\mathbf{ii}+\mathbf{jj}),\ \mathbf{kk}$
A_2	1	1	1	−1	$\mathbf{k}$	$\mathbf{i}\times\mathbf{j}$	$\frac{1}{\sqrt{2}}(\mathbf{ij}-\mathbf{ji})$
E_1	2	$2\cos 72°$	$2\cos 144°$	0	$(\mathbf{i},\mathbf{j})$	$(\mathbf{j}\times\mathbf{k},\mathbf{k}\times\mathbf{i})$	$(\mathbf{ik},\mathbf{jk}),\ (\mathbf{ki},\mathbf{kj})$
E_2	2	$2\cos 144°$	$2\cos 72°$	0			$\left[\frac{1}{\sqrt{2}}(\mathbf{ii}-\mathbf{jj}),\ \frac{1}{\sqrt{2}}(\mathbf{ij}+\mathbf{ji})\right]$

D_6	I	$2C_6$	$2C_3$	C_2	$3C_2'$	$3C_2''$		
A_1	1	1	1	1	1	1		$\frac{1}{\sqrt{2}}(\mathbf{ii}+\mathbf{jj}),\ \mathbf{kk}$
A_2	1	1	1	1	−1	−1	$\mathbf{k}$	$\mathbf{i}\times\mathbf{j}$	$\frac{1}{\sqrt{2}}(\mathbf{ij}-\mathbf{ji})$
B_1	1	−1	1	−1	1	−1		
B_2	1	−1	1	−1	−1	1		
E_1	2	1	−1	−2	0	0	$(\mathbf{i},\mathbf{j})$	$(\mathbf{j}\times\mathbf{k},\mathbf{k}\times\mathbf{i})$	$(\mathbf{ik},\mathbf{jk}),\ (\mathbf{ki},\mathbf{kj})$
E_2	2	−1	−1	2	0	0			$\left[\frac{1}{\sqrt{2}}(\mathbf{ii}-\mathbf{jj}),\ \frac{1}{\sqrt{2}}(\mathbf{ij}+\mathbf{ji})\right]$

$\mathbf{D}_{2d}$	I	$2S_4$	C_2	$2C_2'$	$2\sigma_d$		
A_1	1	1	1	1	1		$\frac{1}{\sqrt{2}}(\mathbf{ii}+\mathbf{jj}),\mathbf{kk}$
A_2	1	1	1	−1	−1		$\frac{1}{\sqrt{2}}(\mathbf{ij}-\mathbf{ji})$
B_1	1	−1	1	1	−1	$\mathbf{i}\times\mathbf{j}$	$\frac{1}{\sqrt{2}}(\mathbf{ii}-\mathbf{jj})$
B_2	1	−1	1	−1	1	$\mathbf{k}$	$\frac{1}{\sqrt{2}}(\mathbf{ij}+\mathbf{ji})$
E	2	0	−2	0	0	$(\mathbf{i},\mathbf{j})$	$(\mathbf{j}\times\mathbf{k},\mathbf{k}\times\mathbf{i})$	$(\mathbf{ik},\mathbf{jk}),(\mathbf{ki},\mathbf{kj})$

$\mathbf{D}_{3d}$	I	$2C_3$	$3C_2$	i	$2S_6$	$3\sigma_d$		
A_{1g}	1	1	1	1	1	1		$\frac{1}{\sqrt{2}}(\mathbf{ii}+\mathbf{jj}),\mathbf{kk}$
A_{2g}	1	1	−1	1	1	−1		$\frac{1}{\sqrt{2}}(\mathbf{ij}-\mathbf{ji})$
E_g	2	−1	0	2	−1	0		$\mathbf{i}\times\mathbf{j}$	$(\mathbf{j}\times\mathbf{k},\mathbf{k}\times\mathbf{i})$	$\left[\frac{1}{\sqrt{2}}(\mathbf{ii}-\mathbf{jj}),\frac{1}{\sqrt{2}}(\mathbf{ij}+\mathbf{ji})\right]$
A_{1u}	1	1	−1	−1	−1	1		
A_{2u}	1	1	−1	−1	−1	1	$\mathbf{k}$	
E_u	2	−1	0	−2	1	0	$(\mathbf{i},\mathbf{j})$	$(\mathbf{ik},\mathbf{jk}),(\mathbf{ki},\mathbf{kj})$

$\mathbf{D}_{4d}$	I	$2S_8$	$2C_4$	$2S_8^3$	C_2	$4C_2'$	$4\sigma_d$			
A_1	1	1	1	1	1	1	1			$\frac{1}{\sqrt{2}}(\mathbf{ii}+\mathbf{jj}),\ \mathbf{kk}$
A_2	1	1	1	1	1	-1	-1	$\mathbf{i}\times\mathbf{j}$		$\frac{1}{\sqrt{2}}(\mathbf{ij}-\mathbf{ji})$
B_1	1	-1	1	-1	1	1	-1			
B_2	1	-1	1	-1	1	-1	1		$\mathbf{k}$	
E_1	2	$\sqrt{2}$	0	$-\sqrt{2}$	-2	0	0		$(\mathbf{i},\mathbf{j})$	
E_2	2	0	-2	0	2	0	0	$(\mathbf{j}\times\mathbf{k},\mathbf{k}\times\mathbf{i})$		$\left[\frac{1}{\sqrt{2}}(\mathbf{ii}-\mathbf{jj}),\ \frac{1}{\sqrt{2}}(\mathbf{ij}+\mathbf{ji})\right]$
E_3	2	$-\sqrt{2}$	0	$\sqrt{2}$	-2	0	0			$(\mathbf{ik},\mathbf{jk}),\ (\mathbf{ki},\mathbf{kj})$

$\mathbf{D}_{5d}$	I	$2C_5$	$2C_5^2$	$5C_2$	i	$2S_{10}^3$	$2S_{10}$	$5\sigma_d$			
A_{1g}	1	1	1	1	1	1	1	1			$\frac{1}{\sqrt{2}}(\mathbf{ii}+\mathbf{jj}),\ \mathbf{kk}$
A_{2g}	1	1	1	-1	1	1	1	-1	$\mathbf{i}\times\mathbf{j}$		$\frac{1}{\sqrt{2}}(\mathbf{ij}-\mathbf{ji})$
E_{1g}	2	$2\cos 72°$	$2\cos 144°$	0	2	$2\cos 144°$	$2\cos 72°$	0	$(\mathbf{j}\times\mathbf{k},\mathbf{k}\times\mathbf{i})$		$(\mathbf{ik},\mathbf{jk}),\ (\mathbf{ki},\mathbf{kj})$
E_{2g}	2	$2\cos 144°$	$2\cos 72°$	0	2	$2\cos 72°$	$2\cos 144°$	0			$\left[\frac{1}{\sqrt{2}}(\mathbf{ii}-\mathbf{jj}),\ \frac{1}{\sqrt{2}}(\mathbf{ij}+\mathbf{ji})\right]$
A_{1u}	1	1	1	1	-1	-1	-1	-1			
A_{2u}	1	1	1	-1	-1	-1	-1	1		$\mathbf{k}$	
E_{1u}	2	$2\cos 72°$	$2\cos 144°$	0	-2	$-2\cos 144°$	$-2\cos 72°$	0		$(\mathbf{i},\mathbf{j})$	
E_{2u}	2	$2\cos 144°$	$2\cos 72°$	0	-2	$-2\cos 72°$	$-2\cos 144°$	0			

$\mathbf{D}_{6d}$	I	$2S_{12}$	$2C_6$	$2S_4$	$2C_3$	$2S_{12}^5$	C_2	$6C_2'$	$6\sigma_d$	
A_1	1	1	1	1	1	1	1	1	1	$\frac{1}{\sqrt{2}}(\mathbf{ii}+\mathbf{jj}),\ \mathbf{kk}$
A_2	1	1	1	1	1	1	1	-1	-1	
B_1	1	-1	1	-1	1	-1	1	1	-1	$\mathbf{i}\times\mathbf{j}$
B_2	1	-1	1	-1	1	-1	1	-1	1	$\mathbf{k}$; $\frac{1}{\sqrt{2}}(\mathbf{ij}-\mathbf{ji})$
E_1	2	$\sqrt{3}$	1	0	-1	$-\sqrt{3}$	-2	0	0	$(\mathbf{i},\mathbf{j})$
E_2	2	1	-1	-2	-1	1	2	0	0	$\left[\frac{1}{\sqrt{2}}(\mathbf{ii}-\mathbf{jj}),\ \frac{1}{\sqrt{2}}(\mathbf{ij}+\mathbf{ji})\right]$
E_3	2	0	-2	0	2	0	-2	0	0	
E_4	2	-1	-1	2	-1	-1	2	0	0	$(\mathbf{j}\times\mathbf{k},\ \mathbf{k}\times\mathbf{i})$
E_5	2	$-\sqrt{3}$	1	0	-1	$\sqrt{3}$	-2	0	0	$(\mathbf{ik},\mathbf{jk}),\ (\mathbf{ki},\mathbf{kj})$

$\mathbf{D}_{2h}$	I	$C_2(z)$	$C_2(y)$	$C_2(x)$	i	$\sigma(xy)$	$\sigma(zx)$	$\sigma(yz)$		
A_g	1	1	1	1	1	1	1	1		$\mathbf{ii},\mathbf{jj},\mathbf{kk}$
B_{1g}	1	1	-1	-1	1	1	-1	-1		$\mathbf{ij},\mathbf{ji}$
B_{2g}	1	-1	1	-1	1	-1	1	-1		$\mathbf{ki},\mathbf{ik}$
B_{3g}	1	-1	-1	1	1	-1	-1	1		$\mathbf{jk},\mathbf{kj}$
A_u	1	1	1	1	-1	-1	-1	-1		
B_{1u}	1	1	-1	-1	-1	-1	1	1	$\mathbf{k}$	$\mathbf{i}\times\mathbf{j}$
B_{2u}	1	-1	1	-1	-1	1	-1	1	$\mathbf{j}$	$\mathbf{k}\times\mathbf{i}$
B_{3u}	1	-1	-1	1	-1	1	1	-1	$\mathbf{i}$	$\mathbf{j}\times\mathbf{k}$

$\mathbf{D}_{3h}$	I	$2C_3$	$3C_2$	σ_h	$2S_3$	$3\sigma_v$		
A_1'	1	1	1	1	1	1		$\tfrac{1}{\sqrt{2}}(\mathbf{ii}+\mathbf{jj}),\ \mathbf{kk}$
A_2'	1	1	-1	1	1	-1	$\mathbf{i}\times\mathbf{j}$	$\tfrac{1}{\sqrt{2}}(\mathbf{ij}-\mathbf{ji})$
E'	2	-1	0	2	-1	0	$(\mathbf{i,j})$	$\left[\tfrac{1}{\sqrt{2}}(\mathbf{ii}-\mathbf{jj}),\ \tfrac{1}{\sqrt{2}}(\mathbf{ij}+\mathbf{ji})\right]$
A_1''	1	1	1	-1	-1	-1		
A_2''	1	1	-1	-1	-1	1	$\mathbf{k}$	
E''	2	-1	0	-2	1	0	$(\mathbf{j}\times\mathbf{k},\ \mathbf{k}\times\mathbf{i})$	$(\mathbf{ik,jk}),\ (\mathbf{ki,kj})$

$\mathbf{D}_{4h}$	I	$2C_4$	C_2	$2C_2'$	$2C_2''$	i	$2S_4$	σ_h	$2\sigma_v$	$2\sigma_d$		
A_{1g}	1	1	1	1	1	1	1	1	1	1		$\tfrac{1}{\sqrt{2}}(\mathbf{ii}+\mathbf{jj}),\ \mathbf{kk}$
A_{2g}	1	1	1	-1	-1	1	1	1	-1	-1	$\mathbf{i}\times\mathbf{j}$	$\tfrac{1}{\sqrt{2}}(\mathbf{ij}-\mathbf{ji})$
B_{1g}	1	-1	1	1	-1	1	-1	1	1	-1		$\tfrac{1}{\sqrt{2}}(\mathbf{ii}-\mathbf{jj})$
B_{2g}	1	-1	1	-1	1	1	-1	1	-1	1		$\tfrac{1}{\sqrt{2}}(\mathbf{ij}+\mathbf{ji})$
E_g	2	0	-2	0	0	2	0	-2	0	0	$(\mathbf{j}\times\mathbf{k},\ \mathbf{k}\times\mathbf{i})$	$(\mathbf{ik,jk}),\ (\mathbf{ki,kj})$
A_{1u}	1	1	1	1	1	-1	-1	-1	-1	-1		
A_{2u}	1	1	1	-1	-1	-1	-1	-1	1	1	$\mathbf{k}$	
B_{1u}	1	-1	1	1	-1	-1	1	-1	-1	1		
B_{2u}	1	-1	1	-1	1	-1	1	-1	1	-1		
E_u	2	0	-2	0	0	-2	0	2	0	0	$(\mathbf{i,j})$	

D_{5h}	I	$2C_5$	$2C_5^2$	$5C_2$	σ_h	$2S_5$	$2S_5^3$	$5\sigma_v$		
A_1'	1	1	1	1	1	1	1	1		$\frac{1}{\sqrt{2}}(\mathbf{ii}+\mathbf{jj}),\ \mathbf{kk}$
A_2'	1	1	1	-1	1	1	1	-1	$\mathbf{i}\times\mathbf{j}$	$\frac{1}{\sqrt{2}}(\mathbf{ij}-\mathbf{ji})$
E_1'	2	$2\cos 72°$	$2\cos 144°$	0	2	$2\cos 72°$	$2\cos 144°$	0	$(\mathbf{i},\mathbf{j})$	
E_2'	2	$2\cos 144°$	$2\cos 72°$	0	2	$2\cos 144°$	$2\cos 72°$	0		$\left[\frac{1}{\sqrt{2}}(\mathbf{ii}-\mathbf{jj}),\ \frac{1}{\sqrt{2}}(\mathbf{ij}+\mathbf{ji})\right]$
A_1''	1	1	1	1	-1	-1	-1	-1		
A_2''	1	1	1	-1	-1	-1	-1	1	$\mathbf{k}$	
E_1''	2	$2\cos 72°$	$2\cos 144°$	0	-2	$-2\cos 72°$	$-2\cos 144°$	0	$(\mathbf{j}\times\mathbf{k},\mathbf{k}\times\mathbf{i})$	$(\mathbf{ik},\mathbf{jk}),\ (\mathbf{ki},\mathbf{kj})$
E_2''	2	$2\cos 144°$	$2\cos 72°$	0	-2	$-2\cos 144°$	$-2\cos 72°$	0		

D_{6h}	I	$2C_6$	$2C_3$	C_2	$3C_2'$	$3C_2''$	i	$2S_3$	$2S_6$	σ_h	$3\sigma_d$	$3\sigma_v$		
A_{1g}	1	1	1	1	1	1	1	1	1	1	1	1		$\frac{1}{\sqrt{2}}(\mathbf{ii}+\mathbf{jj}),\ \mathbf{kk}$
A_{2g}	1	1	1	1	-1	-1	1	1	1	1	-1	-1	$\mathbf{i}\times\mathbf{j}$	$\frac{1}{\sqrt{2}}(\mathbf{ij}-\mathbf{ji})$
B_{1g}	1	-1	1	-1	1	-1	1	-1	1	-1	1	-1		
B_{2g}	1	-1	1	-1	-1	1	1	-1	1	-1	-1	1		
E_{1g}	2	1	-1	-2	0	0	2	1	-1	-2	0	0	$(\mathbf{j}\times\mathbf{k},\mathbf{k}\times\mathbf{i})$	$(\mathbf{ik},\mathbf{jk}),\ (\mathbf{ki},\mathbf{kj})$
E_{2g}	2	-1	-1	2	0	0	2	-1	-1	2	0	0		$\left[\frac{1}{\sqrt{2}}(\mathbf{ii}-\mathbf{jj}),\ \frac{1}{\sqrt{2}}(\mathbf{ij}+\mathbf{ji})\right]$

A_{1u}	1	1	1	1	1	1	−1	−1	−1	−1	−1	−1	
A_{2u}	1	1	1	−1	−1	−1	−1	−1	−1	1	1	1	**k**
B_{1u}	1	−1	−1	1	−1	−1	−1	1	1	−1	1	1	
B_{2u}	1	−1	−1	−1	1	1	−1	1	1	1	−1	−1	
E_{1u}	2	1	−1	−2	0	0	−2	−1	1	2	0	0	(**i**, **j**)
E_{2u}	2	−1	−1	2	0	0	−2	1	1	−2	0	0	

T	I	$4C_3$	$4C_3^2$	$3C_2$			
A	1	1	1	1			$\frac{1}{\sqrt{3}}(\mathbf{ii}+\mathbf{jj}+\mathbf{kk})$
C	1	ω	ω^2	1			$\frac{1}{\sqrt{3}}(\mathbf{ii}+\omega^2\mathbf{jj}+\omega\mathbf{kk})$
D	1	ω^2	ω	1			$\frac{1}{\sqrt{3}}(\mathbf{ii}+\omega\mathbf{jj}+\omega^2\mathbf{kk})$
F	3	0	0	−1	(**i**, **j**, **k**)	(**j**×**k**, **k**×**i**, **i**×**j**)	(**jk**, **ki**, **ij**), (**kj**, **ik**, **ji**)

$\omega = \exp(2\pi i/3)$

$\mathbf{T}_d$	I	$8C_3$	$3C_2$	$6S_4$	$6\sigma_d$		
A_1	1	1	1	1	1		$\frac{1}{\sqrt{3}}(\mathbf{ii}+\mathbf{jj}+\mathbf{kk})$
A_2	1	1	1	-1	-1		
E	2	-1	2	0	0		$\left[\frac{1}{\sqrt{6}}(2\mathbf{kk}-\mathbf{ii}-\mathbf{jj}),\ \frac{1}{\sqrt{2}}(\mathbf{ii}-\mathbf{jj})\right]$
F_1	3	0	-1	1	-1	$(\mathbf{j}\times\mathbf{k},\mathbf{k}\times\mathbf{i},\mathbf{i}\times\mathbf{j})$	$\left[\frac{1}{\sqrt{2}}(\mathbf{jk}-\mathbf{kj}),\ \frac{1}{\sqrt{2}}(\mathbf{ki}-\mathbf{ik}),\ \frac{1}{\sqrt{2}}(\mathbf{ij}-\mathbf{ji})\right]$
F_2	3	0	-1	-1	1	$(\mathbf{i},\mathbf{j},\mathbf{k})$	$\left[\frac{1}{\sqrt{2}}(\mathbf{jk}+\mathbf{kj}),\ \frac{1}{\sqrt{2}}(\mathbf{ki}+\mathbf{ik}),\ \frac{1}{\sqrt{2}}(\mathbf{ij}+\mathbf{ji})\right]$

$\mathbf{T}_h$	I	$4C_3$	$4C_3^2$	$3C_2$	i	$4S_6$	$4S_6^2$	$3\sigma_d$		
A_g	1	1	1	1	1	1	1	1		$\frac{1}{\sqrt{3}}(\mathbf{ii}+\mathbf{jj}+\mathbf{kk})$
C_g	1	ω	ω^2	1	1	ω	ω^2	1		$\frac{1}{\sqrt{3}}(\mathbf{ii}+\omega^2\mathbf{jj}+\omega\mathbf{kk})$
D_g	1	ω^2	ω	1	1	ω^2	ω	1		$\frac{1}{\sqrt{3}}(\mathbf{ii}+\omega\mathbf{jj}+\omega^2\mathbf{kk})$
F_g	3	0	0	-1	3	0	0	-1	$(\mathbf{j}\times\mathbf{k},\mathbf{k}\times\mathbf{i},\mathbf{i}\times\mathbf{j})$	$(\mathbf{jk},\mathbf{ki},\mathbf{ij}),(\mathbf{kj},\mathbf{ik},\mathbf{ji})$
A_u	1	1	1	1	-1	-1	-1	-1		
C_u	1	ω	ω^2	1	-1	$-\omega$	$-\omega^2$	-1		
D_u	1	ω^2	ω	1	-1	$-\omega^2$	$-\omega$	-1		
F_u	3	0	0	-1	-3	0	0	1	$(\mathbf{i},\mathbf{j},\mathbf{k})$	

$\omega = \exp(2\pi i/3)$

O	$8C_3$	$6C_2$	$6C_4$	$3C_4^2$			
A_1	1	1	1	1	1		$\frac{1}{\sqrt{3}}(\mathbf{ii}+\mathbf{jj}+\mathbf{kk})$
A_2	1	1	-1	-1	1		
E	2	-1	0	0	2		$\left[\frac{1}{\sqrt{6}}(2\mathbf{kk}-\mathbf{ii}-\mathbf{jj}),\ \frac{1}{\sqrt{2}}(\mathbf{ii}-\mathbf{jj})\right]$
F_1	3	0	-1	1	-1	$(\mathbf{i},\mathbf{j},\mathbf{k})\quad(\mathbf{j}\times\mathbf{k},\mathbf{k}\times\mathbf{i},\mathbf{i}\times\mathbf{j})$	$\left[\frac{1}{\sqrt{2}}(\mathbf{jk}-\mathbf{kj}),\ \frac{1}{\sqrt{2}}(\mathbf{ki}-\mathbf{ik}),\ \frac{1}{\sqrt{2}}(\mathbf{ij}-\mathbf{ji})\right]$
F_2	3	0	1	-1	-1		$\left[\frac{1}{\sqrt{2}}(\mathbf{jk}+\mathbf{kj}),\ \frac{1}{\sqrt{2}}(\mathbf{ki}+\mathbf{ik}),\ \frac{1}{\sqrt{2}}(\mathbf{ij}+\mathbf{ji})\right]$

$\mathbf{O}_h$	I	$8C_3$	$6C_2$	$6C_4$	$3C_4^2$	i	$6S_4$	$8S_6$	$3\sigma_h$	$6\sigma_d$		
A_{1g}	1	1	1	1	1	1	1	1	1	1		$\frac{1}{\sqrt{3}}(\mathbf{ii}+\mathbf{jj}+\mathbf{kk})$
A_{2g}	1	1	-1	-1	1	1	-1	1	1	-1		
E_g	2	-1	0	0	2	2	0	-1	2	0		$\left[\frac{1}{\sqrt{6}}(2\mathbf{kk}-\mathbf{ii}-\mathbf{jj}),\ \frac{1}{\sqrt{2}}(\mathbf{ii}-\mathbf{jj})\right]$
F_{1g}	3	0	-1	1	-1	3	1	0	-1	-1	$(\mathbf{j}\times\mathbf{k},\mathbf{k}\times\mathbf{i},\mathbf{i}\times\mathbf{j})$	$\left[\frac{1}{\sqrt{2}}(\mathbf{jk}-\mathbf{kj}),\ \frac{1}{\sqrt{2}}(\mathbf{ki}-\mathbf{ik}),\ \frac{1}{\sqrt{2}}(\mathbf{ij}-\mathbf{ji})\right]$
F_{2g}	3	0	1	-1	-1	3	-1	0	-1	1		$\left[\frac{1}{\sqrt{2}}(\mathbf{jk}+\mathbf{kj}),\ \frac{1}{\sqrt{2}}(\mathbf{ki}+\mathbf{ik}),\ \frac{1}{\sqrt{2}}(\mathbf{ij}+\mathbf{ji})\right]$
A_{1u}	1	1	1	1	1	-1	-1	-1	-1	-1		
A_{2u}	1	1	-1	-1	1	-1	1	-1	-1	1		
E_u	2	-1	0	0	2	-2	0	1	-2	0		
F_{1u}	3	0	-1	1	-1	-3	-1	0	1	1	$(\mathbf{i},\mathbf{j},\mathbf{k})$	
F_{2u}	3	0	1	-1	-1	-3	1	0	1	-1		

I group

I	I	$12C_5$	$12C_5^2$	$20C_3$	$15C_2$	
A	1	1	1	1	1	$\frac{1}{\sqrt{3}}(\mathbf{ii}+\mathbf{jj}+\mathbf{kk})$
F_1	3	$-2\cos 144°$	$-2\cos 72°$	0	-1	$(\mathbf{i},\mathbf{j},\mathbf{k})\ \ (\mathbf{j}\times\mathbf{k},\mathbf{k}\times\mathbf{i},\mathbf{i}\times\mathbf{j})$ $\left[\frac{1}{\sqrt{2}}(\mathbf{jk}-\mathbf{kj}),\ \frac{1}{\sqrt{2}}(\mathbf{ki}-\mathbf{ik}),\ \frac{1}{\sqrt{2}}(\mathbf{ij}-\mathbf{ji})\right]$
F_2	3	$-2\cos 72°$	$-2\cos 144°$	0	-1	
G	4	-1	-1	1	0	
H	5	0	0	-1	1	$\left[\frac{1}{\sqrt{6}}(2\mathbf{kk}-\mathbf{ii}-\mathbf{jj}),\ \frac{1}{\sqrt{2}}(\mathbf{ii}-\mathbf{jj}),\ \frac{1}{\sqrt{2}}(\mathbf{jk}+\mathbf{kj}),\ \frac{1}{\sqrt{2}}(\mathbf{ki}+\mathbf{ik}),\ \frac{1}{\sqrt{2}}(\mathbf{ij}+\mathbf{ji})\right]$

I$_h$ group

I_h	I	$12C_5$	$12C_5^2$	$20C_3$	$15C_2$	i	$12S_{10}$	$12S_{10}^3$	$20S_6$	15σ	
A_g	1	1	1	1	1	1	1	1	1	1	$\frac{1}{\sqrt{3}}(\mathbf{ii}+\mathbf{jj}+\mathbf{kk})$
F_{1g}	3	$-2\cos 144°$	$-2\cos 72°$	0	-1	3	$-2\cos 72°$	$-2\cos 144°$	0	-1	$(\mathbf{j}\times\mathbf{k},\mathbf{k}\times\mathbf{i},\mathbf{i}\times\mathbf{j})$
F_{2g}	3	$-2\cos 72°$	$-2\cos 144°$	0	-1	3	$-2\cos 144°$	$-2\cos 72°$	0	-1	$\left[\frac{1}{\sqrt{2}}(\mathbf{jk}-\mathbf{kj}),\ \frac{1}{\sqrt{2}}(\mathbf{ki}-\mathbf{ik}),\ \frac{1}{\sqrt{2}}(\mathbf{ij}-\mathbf{ji})\right]$
G_g	4	-1	-1	1	0	4	-1	-1	1	0	
H_g	5	0	0	-1	1	5	0	0	-1	1	$\left[\frac{1}{\sqrt{6}}(2\mathbf{kk}-\mathbf{ii}-\mathbf{jj}),\ \frac{1}{\sqrt{2}}(\mathbf{ii}-\mathbf{jj}),\ \frac{1}{\sqrt{2}}(\mathbf{jk}+\mathbf{kj}),\ \frac{1}{\sqrt{2}}(\mathbf{ki}+\mathbf{ik}),\ \frac{1}{\sqrt{2}}(\mathbf{ij}+\mathbf{ji})\right]$
A_u	1	1	1	1	1	-1	-1	-1	-1	-1	
F_{1u}	3	$-2\cos 144°$	$-2\cos 72°$	0	-1	-3	$2\cos 72°$	$2\cos 144°$	0	1	$(\mathbf{i},\mathbf{j},\mathbf{k})$
F_{2u}	3	$-2\cos 72°$	$-2\cos 144°$	0	-1	-3	$2\cos 144°$	$2\cos 72°$	0	1	
G_u	4	-1	-1	1	0	-4	1	1	-1	0	
H_u	5	0	0	-1	1	-5	0	0	1	-1	

$\mathbf{C}_\infty$	I	...	$C_{2\pi/\phi}$	...			
A	1	...	1	...	$\mathbf{k}$	$\mathbf{i} \times \mathbf{j}$	$\frac{1}{\sqrt{2}}(\mathbf{ii} + \mathbf{jj})$, $\mathbf{kk}$, $\frac{1}{\sqrt{2}}(\mathbf{ij} - \mathbf{ji})$
C_1	1	...	$e^{i\phi}$	...	$(\mathbf{i}, \mathbf{j})$	$(\mathbf{i}\times\mathbf{k}, \mathbf{j}\times\mathbf{k})$, $(\mathbf{k}\mathbf{i}, \mathbf{k}\mathbf{j})$	
D_1	1	...	$e^{-i\phi}$	...			
C_2	1	...	$e^{2i\phi}$	...			$\left[\frac{1}{\sqrt{2}}(\mathbf{ii} - \mathbf{jj}), \frac{1}{\sqrt{2}}(\mathbf{ij} + \mathbf{ji})\right]$
D_2	1	...	$e^{-2i\phi}$	...			
C_3	1	...	$e^{3i\phi}$	...			
D_3	1	...	$e^{-3i\phi}$	...			
...			...				

$\mathbf{C}_{\infty v}$	I	...	$2C_{2\pi/\phi}$	...	$\infty \sigma_v$		
$A_1 \equiv \Sigma^+$	1	...	1	...	1	$\mathbf{k}$	$\frac{1}{\sqrt{2}}(\mathbf{ii} + \mathbf{jj})$, $\mathbf{kk}$
$A_2 \equiv \Sigma^-$	1	...	1	...	-1		$\frac{1}{\sqrt{2}}(\mathbf{ij} - \mathbf{ji})$
$E_1 \equiv \Pi$	2	...	$2\cos\phi$	...	0	$(\mathbf{i}, \mathbf{j})$	$(\mathbf{i}\times\mathbf{k}, \mathbf{j}\times\mathbf{k})$, $(\mathbf{k}\mathbf{i}, \mathbf{k}\mathbf{j})$
$E_2 \equiv \Delta$	2	...	$2\cos 2\phi$	...	0		$\left[\frac{1}{\sqrt{2}}(\mathbf{ii} - \mathbf{jj}), \frac{1}{\sqrt{2}}(\mathbf{ij} + \mathbf{ji})\right]$
$E_3 \equiv \Phi$	2	...	$2\cos 3\phi$	...	0		
...			...				

$D_{\infty h}$	I	...	$2C_{2\pi/\phi}$	...	C_2	i	...	$2S_{2\pi/\phi}$	...	σ_h	$\infty\sigma_v$		
$A_{1g} \equiv \Sigma_g^+$	1	...	1	...	1	1	...	1	...	1	1		$\frac{1}{\sqrt{2}}(\mathbf{ii}+\mathbf{jj}), \mathbf{kk}$
$A_{2g} \equiv \Sigma_g^-$	1	...	1	...	-1	1	...	1	...	1	-1		$\frac{1}{\sqrt{2}}(\mathbf{ij}-\mathbf{ji})$
$E_{1g} \equiv \Pi_g$	2	...	$2\cos\phi$	...	0	2	...	$-2\cos\phi$	...	-2	0		$\mathbf{i}\times\mathbf{j}$
$E_{2g} \equiv \Delta_g$	2	...	$2\cos 2\phi$	...	0	2	...	$2\cos 2\phi$	...	2	0		$(\mathbf{j}\times\mathbf{k}, \mathbf{k}\times\mathbf{i})$ $(\mathbf{ik}, \mathbf{jk}), (\mathbf{ki}, \mathbf{kj})$
$E_{3g} \equiv \Phi_g$	2	...	$2\cos 3\phi$	...	0	2	...	$-2\cos 3\phi$	...	-2	0		$\left[\frac{1}{\sqrt{2}}(\mathbf{ii}-\mathbf{jj}), \frac{1}{\sqrt{2}}(\mathbf{ij}+\mathbf{ji})\right]$
...													
$A_{1u} \equiv \Sigma_u^+$	1	...	1	...	1	-1	...	-1	...	-1	1	$\mathbf{k}$	
$A_{2u} \equiv \Sigma_u^-$	1	...	1	...	-1	-1	...	-1	...	-1	-1		
$E_{1u} \equiv \Pi_u$	2	...	$2\cos\phi$	...	0	-2	...	$2\cos\phi$	...	2	0	$(\mathbf{i}, \mathbf{j})$	
$E_{2u} \equiv \Delta_u$	2	...	$2\cos 2\phi$	...	0	-2	...	$-2\cos 2\phi$	...	-2	0		
$E_{3u} \equiv \Phi_u$	2	...	$2\cos 3\phi$	...	0	-2	...	$2\cos 3\phi$	...	2	0		
...													

SU(2)	I	$\infty C_{2\pi/\phi}$		
$A \equiv \Gamma_0$	1	... 1	... $\frac{1}{\sqrt{3}}(\mathbf{ii} + \mathbf{jj} + \mathbf{kk})$	
$E \equiv \Gamma_{1/2}$	2	... $2\cos\frac{1}{2}\phi$	...	
$F \equiv \Gamma_1$	3	... $\dfrac{\sin\left(\dfrac{3}{2}\right)\phi}{\sin\frac{1}{2}\phi}$	... $(\mathbf{i}, \mathbf{j}, \mathbf{k})$ $(\mathbf{j} \times \mathbf{k}, \mathbf{k} \times \mathbf{i}, \mathbf{i} \times \mathbf{j})$	$\left[\dfrac{1}{\sqrt{2}}(\mathbf{jk} - \mathbf{kj}), \dfrac{1}{\sqrt{2}}(\mathbf{ki} - \mathbf{ik}), \dfrac{1}{\sqrt{2}}(\mathbf{ij} - \mathbf{ji})\right]$
$G \equiv \Gamma_{3/2}$	4	... $4\cos\frac{1}{2}\phi\cos\phi$	...	
$H \equiv \Gamma_2$	5	... $\dfrac{\sin\left(\dfrac{5}{2}\right)\phi}{\sin\frac{1}{2}\phi}$	... $\left[\dfrac{1}{\sqrt{6}}(2\mathbf{kk} - \mathbf{ii} - \mathbf{jj}), \dfrac{1}{\sqrt{2}}(\mathbf{ii} - \mathbf{jj}), \dfrac{1}{\sqrt{2}}(\mathbf{jk} + \mathbf{kj}), \dfrac{1}{\sqrt{2}}(\mathbf{ki} + \mathbf{ik}), \dfrac{1}{\sqrt{2}}(\mathbf{ij} + \mathbf{ji})\right]$	
...	...	...		
Γ_j	$2j + 1$	... $\dfrac{\sin(j + \frac{1}{2})\phi}{\sin\frac{1}{2}\phi}$	...	
...	...	...		

A.3
Characters for the Primitive Symmetry Species Introduced into Rotation Groups by Extending the Period of Rotation from 2π to 4π.

A system with half-integral spin about a symmetry axis is not transformed into itself by a rotation of 2π about this axis. Instead, a rotation by 4π is required.

To treat the effects of this condition, a person may introduce the rotation by 2π as R with the property

$$R \neq I,$$

$$R^2 = I.$$

The resulting group contains twice as many elements as the corresponding simple geometric group. Consequently, it is often called a *double group*. Properties of the symmetry species added by thus extending the most common rotation groups are given in the following tables.

C_1^*	I	R	
$B_{1/2}$	1	-1	$e^{i\phi/2}, e^{-i\phi/2}$

C_2^*	I	C_2	R	C_2R	
$C_{1/2}$	1	i	-1	$-i$	$e^{-i\phi/2}$
$D_{1/2}$	1	$-i$	-1	i	$e^{i\phi/2}$

C_3^*	I	C_3	C_3^2	R	C_3R	C_3^2R	
$C_{1/2}$	1	$-\omega^2$	ω	-1	ω^2	$-\omega$	$e^{-i\phi/2}$
$D_{1/2}$	1	$-\omega$	ω^2	-1	ω	$-\omega^2$	$e^{i\phi/2}$
$D_{3/2}$	1	-1	1	-1	1	-1	$e^{3i\phi/2}, e^{-3i\phi/2}$

$\omega = \exp(2\pi i/3)$

$\mathbf{D_2^*}$	I	$2C_2(z)$	R	$2C_2(x)$	$2C_2(y)$
$E_{1/2}$	2	0	−2	0	0

$\mathbf{D_3^*}$	I	$2C_3$	$2C_3^2$	R	$3C_2'$	$3C_2'R$
$E_{1/2}$	2	1	−1	−2	0	0
$C_{3/2}$	1	−1	1	−1	i	$-i$
$D_{3/2}$	1	−1	1	−1	$-i$	i

$\mathbf{D_4^*}$	I	$2C_4$	$2C_2$	$2C_4^3$	R	$4C_2'$	$4C_2''$
$E_{1/2}$	2	$\sqrt{2}$	0	$-\sqrt{2}$	−2	0	0
$E_{3/2}$	2	$-\sqrt{2}$	0	$\sqrt{2}$	−2	0	0

$\mathbf{D_6^*}$	I	$2C_6$	$2C_3$	$2C_2$	$2C_3^2$	$2C_6^5$	R	$6C_2'$	$6C_2''$
$E_{1/2}$	2	$\sqrt{3}$	1	0	−1	$-\sqrt{3}$	−2	0	0
$E_{3/2}$	2	$-\sqrt{3}$	1	0	−1	$\sqrt{3}$	−2	0	0
$E_{5/2}$	2	0	−2	0	2	0	−2	0	0

$\mathbf{T_d^*}$	I	$6S_4$	$6C_2$	$6S_4^3$	R	$8C_3$	$8C_3^2$	$12\sigma_d$
$\mathbf{O^*}$	I	$6C_4$	$6C_2$	$6C_4^3$	R	$8C_3$	$8C_3^2$	$12C_2'$
$E_{1/2}$	2	$\sqrt{2}$	0	$-\sqrt{2}$	−2	1	−1	0
$E_{5/2}$	2	$-\sqrt{2}$	0	$\sqrt{2}$	−2	1	−1	0
$G_{3/2}$	4	0	0	0	−4	−1	1	0

A.4
Characters and Spin Bases for the Primitive Symmetry Species of Full Permutation Groups

Many physical systems contain sets of identical entities. The permutations within each of these sets form a group. The group for a set of n objects is the symmetric group $\mathscr{S}_n$.

Characters for the simplest such groups are described in the following tables. The classes are identified by their cycle structures; the symmetry species by their Young diagrams. Conjugate species are joined by tie lines.

$\mathscr{S}_1$	(1)		
{1}	1		Doublet

$\mathscr{S}_2$	(1^2)	(2)	
{2}	1	1	Triplet
{1^2}	1	−1	Singlet

$\mathscr{S}_3$	(1^3)	3(12)	2(3)	
{3}	1	1	1	Quartet
{21}	2	0	−1	Doublet
{1^3}	1	−1	1	

$\mathscr{S}_4$	(1^4)	$6(1^2 2)$	$3(2^2)$	8(13)	6(4)	
{4}	1	1	1	1	1	Quintet
{31}	3	1	−1	0	−1	Triplet
{2^2}	2	0	2	−1	0	Singlet
{21^2}	3	−1	−1	0	1	
{1^4}	1	−1	1	1	−1	

$\mathscr{S}_5$	(1^5)	$10(1^3 2)$	$15(1^2 2)$	$20(1^2 3)$	20(23)	30(14)	24(5)	
{5}	1	1	1	1	1	1	1	Hextet
{41}	4	2	0	1	−1	0	−1	Quartet
{32}	5	1	1	−1	1	−1	0	Doublet
{31^2}	6	0	−2	0	0	0	1	
{$2^2 1$}	5	−1	1	−1	−1	1	0	
{21^3}	4	−2	0	1	1	0	−1	
{1^5}	1	−1	1	1	−1	−1	1	

B.1
Answers to Problems

Chapter 1

1.1 $\begin{pmatrix} \frac{1}{2} & -\frac{\sqrt{3}}{2} & 0 \\ \frac{\sqrt{3}}{2} & \frac{1}{2} & 0 \\ 0 & 0 & 1 \end{pmatrix}.$

1.2 $\begin{pmatrix} \cos\phi & 0 & \sin\phi \\ 0 & 1 & 0 \\ -\sin\phi & 0 & \cos\phi \end{pmatrix}.$

1.4 $\frac{2\pi}{\sqrt{3}d}(\sqrt{3}\mathbf{i} + \mathbf{j}),\ \frac{2\pi}{\sqrt{3}d}(-\sqrt{3}\mathbf{i} + \mathbf{j}),\ \frac{2\pi}{e}\mathbf{k}.$

1.6 (a) $\mathbf{T}_d$, (b) $\mathbf{D}_{\infty h}$, (c) $\mathbf{D}_{2d}$, (d) $\mathbf{D}_{3h}$.

1.8 $\mathbf{C}_{2h}$, $\mathbf{D}_2$.

1.9 $A^4 = I.$

1.10 $\begin{pmatrix} -\frac{1}{2} & \frac{\sqrt{3}}{2} & 0 \\ -\frac{\sqrt{3}}{2} & \frac{1}{2} & 0 \\ 0 & 0 & -1 \end{pmatrix}$

356 Answers to Problems

1.11 $C_4^2 = -\sigma_h$.

1.13 $\dfrac{2\pi}{d}(1 - \dfrac{e}{f}\mathbf{j})$, $\dfrac{2\pi}{f}\mathbf{j}$, $\dfrac{2\pi}{g}\mathbf{k}$.

1.14 C_5^{-1}, 5.

1.17 (a) $\mathbf{C}_{\infty v}$, (b) $\mathbf{D}_{\infty h}$, (c) $\mathbf{D}_{2h}$, (d) $\mathbf{D}_{2d}$.

1.18 3.

Chapter 2

2.2 $c_{1jk} = \begin{pmatrix} 1 & 0 & 0 \\ 0 & 0 & 1 \\ 0 & 1 & 0 \end{pmatrix}$, $c_{2jk} = \begin{pmatrix} 0 & 0 & 1 \\ 0 & 1 & 0 \\ 1 & 0 & 0 \end{pmatrix}$, $c_{3jk} = \begin{pmatrix} 0 & 1 & 0 \\ 1 & 0 & 0 \\ 0 & 0 & 1 \end{pmatrix}$.

2.3 $1, 1, 1$; $1, \omega, \omega^2$; $1, \omega, \omega^2$ with $\omega = \exp(2\pi i/3)$.

2.6 Table for $\mathbf{C}_4$.

2.9 $\begin{pmatrix} 1 & 0 & 0 & 0 \\ 0 & 1 & 0 & 0 \\ 0 & 0 & 1 & 0 \\ 0 & 0 & 0 & 1 \end{pmatrix} \begin{pmatrix} 0 & 1 & 0 & 0 \\ 2 & 0 & 1 & 0 \\ 0 & 1 & 1 & 0 \\ 0 & 0 & 0 & 2 \end{pmatrix} \begin{pmatrix} 0 & 0 & 1 & 0 \\ 0 & 1 & 1 & 0 \\ 2 & 1 & 0 & 0 \\ 0 & 0 & 0 & 2 \end{pmatrix} \begin{pmatrix} 0 & 0 & 0 & 1 \\ 0 & 0 & 0 & 2 \\ 0 & 0 & 0 & 2 \\ 5 & 5 & 5 & 0 \end{pmatrix}$.

2.10
$1, 1, 1, 1$; $2, 2, \tfrac{1}{2}(-1 \pm \sqrt{5})$; $2, 2, \tfrac{1}{2}(-1 \pm \sqrt{5})$; $0, 0, 5, -5$.

2.13 Table for $\mathbf{T}$.

Chapter 3

3.1 $D_{\infty h}$ and subgroups thereof.

3.3 $A: u_1 - u_7;$ $B_1: u_1 - 2u_4 + u_7;$
       $B_2: u_2 - 2u_5 + u_8;$ $B_3: u_3 - 2u_6 + u_9.$

3.5 $A: \omega^2 = k_1/m_0.$

$$B_1: \omega^2 = k_1\left(\frac{1}{m_O} + \frac{2}{m_C}\right); \quad B_2 \text{ and } B_3: \omega^2 = k_2\left(\frac{1}{m_O} + \frac{2}{m_C}\right).$$

3.7 $\dfrac{2m_O}{m_C}S.$

3.8 $A_1: r_1 + r_2 + r_3 + r_4;$ $B_2: r_1 - r_2 + r_3 - r_4;$
       $E: r_1 - r_3, r_2 - r_4;$ $B_1: \alpha_1 - \alpha_2 + \alpha_3 - \alpha_4.$

3.9 $g = \alpha_5 + \alpha_6;$
       $\alpha_5 =$ upper angle subtended by 1st and 3rd corners at center of mass,
       $\alpha_6 =$ lower angle subtended by 2nd and 4th corners at center of mass.

3.10 $A_1: u_5 + u_6 + u_7 + u_8;$ $A_2: u_1 + u_2 + u_3 + u_4;$
        $B_1: u_5 - u_6 + u_7 - u_8;$ $B_2: u_1 - u_2 + u_3 - u_4;$
        $E: u_1 - u_3, u_2 - u_4, u_5 - u_7, u_8 - u_6.$

3.11   Rotation: $A_2;$
        Translation: E in $u_1 - u_6 - u_3 + u_8, u_2 - u_7 - u_4 + u_5;$
        Vibration: E in $u_1 + u_6 - u_3 - u_8, u_2 + u_7 - u_4 - u_5;$
        also $A_1, B_1, B_2.$

358 Answers to Problems

3.12 Employ C_{3v}. Radial and tangential vectors.

3.13

A_1: $\mathbf{r}_1 + \mathbf{r}_2 + \mathbf{r}_3$;

E: $2\mathbf{r}_1 - \mathbf{r}_2 - \sqrt{3}\mathbf{t}_2 - \mathbf{r}_3 + \sqrt{3}\mathbf{t}_3,\ 2\mathbf{t}_1 - \mathbf{t}_2 + \sqrt{3}\mathbf{r}_3 - \mathbf{t}_3 - \sqrt{3}\mathbf{r}_3$.

3.15 A_1: $\omega^2 = 3k$; A_2: $\omega^2 = 0$; E: $\omega^2 = \frac{3}{2}k,\ \omega^2 = 0$.

Chapter 4

4.5 C: $\sigma_{23} + i\sigma_{31}$; D: $\sigma_{23} - i\sigma_{31}$.

4.6 $\frac{1}{\sqrt{2}}\Big[-2\sin\phi\cos\phi\,\mathbf{ii} + 2\sin\phi\cos\phi\,\mathbf{jj} + (\cos^2\phi - \sin^2\phi)\mathbf{ij} + (\cos^2\phi - \sin^2\phi)\mathbf{ji}\Big]$.

4.7 $\frac{\sqrt{6}}{4}\sigma_{11} - \frac{\sqrt{6}}{4}\sigma_{22} - \frac{1}{2}\sigma_{12};\ -\frac{\sqrt{6}}{4}\sigma_{11} + \frac{\sqrt{6}}{4}\sigma_{22} - \frac{1}{2}\sigma_{12}$.

4.9 $\mp\frac{\sqrt{3}}{2}\sigma_{31} - \frac{1}{2}\sigma_{23};\ -\frac{1}{2}\sigma_{31} \pm \frac{\sqrt{3}}{2}\sigma_{23}$.

4.12 A: $\sigma_{11} + \sigma_{22},\ \sigma_{33}$; $C + D$: $(\sigma_{11} - \sigma_{22},\ \sigma_{12}),\ (\sigma_{23},\ \sigma_{31})$.

Chapter 5

5.4 $\frac{1}{\sqrt{2}}(\mathbf{e}_1 \pm i\mathbf{e}_{22})$.

5.5 B.

5.8 $A_{1g} + B_{1g} + E_u, A_{2u} + B_{2u} + E_g, A_{2g} + B_{2g} + E_u$.

5.10 $\tfrac{1}{2}(\mathbf{r}_1 + \mathbf{r}_2 + \mathbf{r}_3 + \mathbf{r}_4), \tfrac{1}{2}(\mathbf{r}_1 - \mathbf{r}_2 + \mathbf{r}_3 - \mathbf{r}_4),$
$\dfrac{1}{\sqrt{2}}(\mathbf{r}_1 - \mathbf{r}_3), \dfrac{1}{\sqrt{2}}(\mathbf{r}_2 - \mathbf{r}_4)$.

5.11

Vibrational modes: $\tfrac{1}{2}(\mathbf{r}_1 - \mathbf{t}_2 - \mathbf{r}_3 + \mathbf{t}_4), \tfrac{1}{2}(\mathbf{r}_2 + \mathbf{t}_1 - \mathbf{r}_4 - \mathbf{t}_3)$.

5.13 n even.

5.14 $1, \omega^5, \omega^4, \omega^3, \omega^2, \omega$.

5.18 $A + C + D + 3F$.

5.19

$F: \tfrac{1}{2}(\mathbf{u}_1 - \mathbf{u}_4 - \mathbf{u}_7 + \mathbf{u}_{10})$, translation;

$\dfrac{1}{\sqrt{8}}(\mathbf{u}_1 - \mathbf{u}_2 + \mathbf{u}_4 - \mathbf{u}_5 - \mathbf{u}_7 + \mathbf{u}_8 - \mathbf{u}_{10} + \mathbf{u}_{11})$, rotation;

$\dfrac{1}{\sqrt{8}}(\mathbf{u}_1 + \mathbf{u}_2 + \mathbf{u}_4 + \mathbf{u}_5 - \mathbf{u}_7 - \mathbf{u}_8 - \mathbf{u}_{10} - \mathbf{u}_{11})$, vibration.

5.20 $\dfrac{1}{\sqrt{3}}(\mathbf{e}_1 + \mathbf{e}_2 + \mathbf{e}_3), \dfrac{1}{\sqrt{3}}(\mathbf{e}_1 + \omega \mathbf{e}_2 + \omega^2 \mathbf{e}_3),$
$\dfrac{1}{\sqrt{3}}(\mathbf{e}_1 + \omega^2 \mathbf{e}_2 + \omega \mathbf{e}_3)$.

Chapter 6

6.1 $\frac{1}{\sqrt{12}}(2|1\rangle + |2\rangle - |3\rangle - 2|4\rangle - |5\rangle + |6\rangle);$

$\frac{1}{2}(|2\rangle + |3\rangle - |5\rangle - |6\rangle).$

6.2 $\frac{h_{11} + h_{12} - h_{13} - h_{14}}{1 + s_{12} - s_{13} - s_{14}}.$

6.3 No.

6.4 A: p_z, $d_{3z^2-r^2}$;
B: $d_{x^2-y^2}$, d_{xy};
C: $p_x - ip_y$, $d_{zx} - id_{zy}$;
D: $p_x + ip_y$, $d_{zx} + id_{zy}$.

6.5 A: $|1\rangle + |2\rangle + |3\rangle + |4\rangle;$
B: $|1\rangle - |2\rangle + |3\rangle - |4\rangle;$
C: $|1\rangle - i|2\rangle - |3\rangle + i|4\rangle;$
D: $|1\rangle + i|2\rangle - |3\rangle - i|4\rangle.$

6.6

A: $\frac{h_{11} + 2h_{12} + h_{13}}{1 + 2s_{12} + s_{13}}$; C and D: $\frac{h_{11} - h_{13}}{1 - s_{13}}$; B: $\frac{h_{11} - 2h_{12} + h_{13}}{1 - 2s_{12} + s_{13}}$.

6.8 $A + B + C + D.$

6.9 A: $\frac{h_{11} + 2h_{12}}{1 + 2s_{12}}$; C and D: $\frac{h_{11} - h_{12}}{1 - s_{12}}$.

6.11 $\dfrac{h_{11} + \frac{1}{3}h_{23} - \frac{4}{3}h_{12}}{1 + \frac{1}{3}s_{23} - \frac{4}{3}s_{12}}$; $\dfrac{h_{11} - h_{23}}{1 - s_{23}}$.

6.12 A_1: Y_0^0, Y_2^0; A_2: Y_1^0; E: (Y_1^1, Y_1^{-1}), (Y_2^1, Y_2^{-1}), (Y_2^2, Y_2^{-2})

6.13 A: Y_0^0, Y_1^0, Y_2^0;

C: Y_1^{-1}, Y_2^{-1}, Y_2^2;

D; Y_1^1, Y_2^1, Y_2^{-2}.

6.14 A: $|1\rangle + |2\rangle + |3\rangle + |4\rangle + |5\rangle$;

C_1: $|1\rangle + \omega^4|2\rangle + \omega^3|3\rangle + \omega^2|4\rangle + \omega|5\rangle$;

C_2: $|1\rangle + \omega^3|2 + \omega|3\rangle + \omega^4|4\rangle + \omega^2|5\rangle$.

6.15 A: $\dfrac{h_{11} + 2h_{12} + 2h_{13}}{1 + 2s_{12} + 2s_{13}}$;

C_1 and D_1: $\dfrac{h_{11} + 0.6180\,h_{12} - 1.6180\,h_{13}}{1 + 0.6180\,s_{12} - 1.6180\,s_{13}}$;

C_2 and D_2: $\dfrac{h_{11} - 1.6180\,h_{12} + 0.6180\,h_{13}}{1 - 1.6180\,s_{12} + 0.6180\,s_{13}}$.

6.17 $A + C_1 + D_1 + C_2 + D_2$.

6.18 A: $\dfrac{1}{\sqrt{3}}(|1\rangle + |2\rangle + |3\rangle)$, $|4\rangle$;

C: $\dfrac{1}{\sqrt{3}}(|1\rangle + \omega^2|2\rangle + \omega|3\rangle)$;

D: $\dfrac{1}{\sqrt{3}}(|1\rangle + \omega|2\rangle + \omega^2|3\rangle)$.

6.19 $h_{11} \pm \sqrt{3}\, h_{14}$.

Chapter 7

7.1 2

7.2 $C \times C = D$, $D \times D = C$.

7.3 C_2.

7.5 $c_1 = 1/\sqrt{2}$, $c_2 = -1/\sqrt{2}$.

7.6 $(1/\sqrt{3})|0, 0\rangle + (1/\sqrt{2})|1, 0\rangle + (1/\sqrt{6})|2, 0\rangle$.

7.7 $c_1 = -\dfrac{1}{\sqrt{2j+1}}$, $c_2 = \dfrac{\sqrt{2j}}{\sqrt{2j+1}}$.

7.8 $-\sqrt{j}$.

7.9 $\begin{pmatrix} 1 & 0 \\ 0 & -1 \end{pmatrix} \times \begin{pmatrix} -\frac{1}{2} & -\frac{\sqrt{3}}{2} \\ \frac{\sqrt{3}}{2} & -\frac{1}{2} \end{pmatrix}$.

7.10 B_3, B_1, B_2.

7.11 B_2, B_1, B_3.

7.13 $c_1 = 1/\sqrt{3}, c_2 = \sqrt{2/3}$.

7.14 $c_1 = \sqrt{3/10}, c_2 = -2/\sqrt{10}, c_3 = \sqrt{3/10}$.

7.15 $-\dfrac{\sqrt{2j-1}}{\sqrt{j(j+1)}}$.

7.16 $\dfrac{j-1}{\sqrt{j(j+1)}}$.

Chapter 8

8.1 $\mathbf{C}_{3v}, \mathbf{D}_3$.

8.2 (135).

8.3 $(1/\sqrt{24})(2|1>_1|2>_2|3>_3|4>_4 + 2|1>_2|2>_1|3>_4|4>_3$
$+ \ldots -|1>_1|2>_3|3>_4|4>_2 - \ldots)$.

8.4 $\dfrac{1}{120}\{[\chi(R)]^5 - 10[\chi(R)]^3\chi(R^2) + 15\chi(R)[\chi(R^2)]^2$
$+ 20[\chi(R)]^2\chi(R^3) - 20\chi(R^2)\chi(R^3) - 30\chi(R)\chi(R^4)$
$+ 24\chi(R^3)\}$.

8.5 $^5A_g, {}^3F_{1g}, {}^3F_{2g}, {}^3G_g, {}^3H_g, {}^1A_g, {}^1G_g, 3{}^1H_g$.

Answers to Problems

8.6 $\mathbf{T}_d$ and $\mathbf{O}$.

8.7 $(2^1 3^1 4^1)$.

8.8 $\{1^4\} + \{2^2\} + \{4\}$.

8.9 $\frac{1}{30}\{[\chi(R)]^5 - 5[\chi(R)]^3\chi(R^2) + 5[\chi(R)]^2\chi(R^3)$
           $+ 5\chi(R^2)\chi(R^3) - 6\chi(R^5)\}$.

8.10 $^3\Sigma_g^-, {}^1\Sigma_g^+, {}^1\Delta_g$.

Chapter 9

9.1 e^{ika}.

9.2 iM.

9.3 $x^2 + y^2 = a^2$, $\phi = \tau + b$, $z = \lambda\tau + c$.

9.4 $u' = u + a$, $v' = v$.

9.5 $\begin{pmatrix} 0 & -1 & 0 \\ 1 & 0 & 0 \\ 0 & 0 & 0 \end{pmatrix}$.

9.6 $\frac{1}{2}n(n+1)$.

9.9 iM.

9.10 ik.

9.11 $y = a\tau + b$.

9.12 $\begin{pmatrix} i & 0 & 0 \\ 0 & -i & 0 \\ 0 & 0 & 0 \end{pmatrix}.$

Chapter 10

10.3 $-\sin\phi \dfrac{\partial}{\partial\theta} - \cot\theta \cos\phi \dfrac{\partial}{\partial\theta}.$

10.8 $\cos\phi \dfrac{\partial}{\partial\theta} - \cot\theta \sin\phi \dfrac{\partial}{\partial\phi}.$

Index

Abelian group, 45
Active transformation, 3, 246
Addition,
  of group elements, 51–52, 241
Amorphous materials,
  point groups for, 107
  stress-strain relations for, 122–123
Angular momenta,
  in quantum mechanics, 198–200
  and **SO**(3) group, 276–283
Annihilation operator a_j^-, 295
Anticommutation relations,
  for fermions, 296
Anti-Hermitian operator, 248
Antiparticle, 319–320
Antiquarks, 311
Antisymmetrizing operator, 220
Array,
  of displacements, 76
    symmetry-adapted, 77–78
  of mass elements, 74
  of stresses, 108–110
    symmetry-adapted, 111–112
Associative law, 18
Axis,
  coordinate, 3, 325
  rotational, 4, 6

Baryon number B, 299, 302
Baryons,
  properties of, 312
  structure of, 319–320
Bases,
  for symmetry operations, 2–4
Basis,
  standard expressions as, 75, 128–129, 325
  correspondence between, 326
  multiplication of, 191–192
  reduction of, 142
Benzene molecule,
  pi kets for, 165–166
  symmetry-adapted
    combinations of, 167–168
  energy levels of, 170–172
  symmetry group for, 164–165
Bilinear products,
  as Lie algebra elements, 297–298
Binary combination,
  of elements, 17–18
Boson states,
  relations connecting, 295
Boundary,
  of quark-composite lattice, 303–305, 315

Canonical coordinates, 249–251
Canonical parameter, 244, 251–252
Casimir operator, 278, 290
Causality, 1
Cayley diagram, 21–22
Character table,
  construction of, 57–58
  for C_{3v} group, 60–63
  for translational groups, 37–38
  for 2-D rotational groups, 31–32, 274–276
  notation in, 63
  orthogonality in, 64–65
Character vector $\chi^{(r)}$, 58
  composition of, 142–143
  multiplication of,
    in orthogonality condition, 65
    for product representation, 192, 196
Characteristic differential equation, 247
Class, 45–48
  character components for, 66–68, 142–143
  exclusively of, 48
  matrix traces for, 139–142
  in permutation group, 215
Class sum $\mathscr{C}_j$, 52
  character component $y_j^{(r)}$ for, 58–59

Class sum $\mathscr{C}_i$ (continued)
  eigenoperator $\mathscr{E}$ of, 56–57
  eigenvalue $\lambda_j^{(r)}$ for, 56–57
  elements in, h_j, 53
  splitting or coalescing of, 97
Class sum product $\mathscr{C}_j\mathscr{C}_k$,
  commutativity of, 52
  composition of, 53–54
  in C_{3v} group, 54–55
Clebsch-Gordon coefficient,
  200–205
Closure, 17
Coefficient c_{jkl}, 53
Color field,
  representation of, 318–321
Common eigenarray,
  of displacements, 77–78
  of stresses, 110
Commutation relations,
  for bosons, 295, 297–298
  for fermions, 298
Commutativity,
  of class sums, 52
  of group elements, 45
Commutator,
  of infinitesimal operators,
  258–260
Commutators,
  for SU(2) group,
    for isospin operators, 300
  for SU(3) group,
    for unitary spin operators, 313
Compact group, 252–253
Completeness of basis set,
  of displacements, 75–76
  of kets, 191
  of vectors, 129
Component,
  of jkth class sum product, c_{jkl},
    53–54
  of rth character vector, $y_j^{(r)}$, 58
Conjugate,
  of symmetry species, 227
Connectivity, 253
Constitutive relations,
  between stress and rate of strain,
    in liquids, 122–123
  between stress and strain,
    in crystals, 116–120
    in glasses, 122

Continuous group, 240–241
Continuum,
  for group elements, 240
Continuum approximation,
  for condensed phases, 106
Conversion operation, 3–4
Covariance,
  under symmetry operations,
    76–77
Covering group,
  universal, 253
  for SO(3, r) group, 287
Creation operator a_j^+, 295
Crystalline materials,
  point groups for, 106–107
  stress-strain relations for,
    115–120
Cubic material,
  point groups for, 107
  stress-strain relations for, 119
Cycle,
  of group elements, 22–30
  of permutations, 13–14, 214
    from transpositions, 217–218
Cyclic group C_n, S_n, 22, 26, 30–31
Cyclic structure,
  of permutation group, 215
Cyclopropenyl radical,
  C_{2v} structure for, 182–183
  D_{3h} structure for, 177, 182
  pi kets for, 176
    basis combinations of, 178–181

Decuplet,
  of quark triplets, 310–312
Defining relationship, 21
Degeneracy, 1–2
  and character component $y_1^{(r)}$,
    63, 78
  molecular, 163, 165–166
    reduction of, 182–183
  nuclear, 298–299
  particle, 301–302
    in proton octet, 317–318
  rotational, 283
  stress-strain relational, 111–113
  vibrational, 78–79, 94
Degree,
  of cycle, 214

Derivative,
  of continuous-group element, 242–243, 252
Dihedral group $\mathbf{D}_n$, $\mathbf{D}_{nd}$, $\mathbf{D}_{nh}$, 23–25
Dimensionality,
  of continuum, 1
  of manifold, 241, 251–253
  of representation, 130
  of symmetry species, 163
Direct product,
  of groups, 195–198
  of matrices, 192–194
  of representations, 191–192
Distortion,
  interactions introduced by, 97
  stabilization resulting from, 183–184
Double groups, 352
  character tables for, 352–353
Dyads,
  as basis, 325–326
  correlation with stress arrays of, 120
  transformation of, 120–121

Eigenarray,
  of displacement vectors, 77–78
  of stress elements, ϕ_{lm}, 110
Eigenfunction,
  for rotation, Ψ, 273–274
Eigenoperator,
  for class sum, $\mathscr{E}^{(r)}$, 56–59
Eigenvalue,
  for class sum, $\lambda_j^{(r)}$, 56–59, 141–142
  for $\mathbf{C}_{3v}$ group, 59–60
Eigenvalue equation,
  for class sums, 56–57, 141
  for group elements, 33–34, 37–38
  for infinitesimal operators,
    of $\mathbf{SO}(3, r)$ group, 283
Electronic states,
  for single shell, 233–234
  for multiple shells, 235
Elements of group,
  distinguishment of, 19–24, 45
Equilibrium positions,
  in condensed phase, 106
  in molecule, 74–75

Equivalent particles,
  states of, 229–231
Equivalent system, 2, 38
Exponential form,
  for group element, 30–32, 244–246, 252
Exponentiation, 246

Factor group, 253
Fermion states,
  relations connecting, 296
Finite group, 21
Fluid media,
  point groups for, 107
  stress-strain relations for, 122–123
Foldness (multiplicity),
  of rotation, 6
Form-preserving transformation, 246–247
Full rotation group, 18
  subgroups of, 19–20

Generalized coordinates,
  elements of strain as, 113
Generation of representations,
  by kets, 175–176
  by vectors, 129–130
Generator,
  finite, 19
    table of, 24
  infinitesimal, 243–245, 252
Geometric group, 18–19, 51
Glasses,
  point groups for, 107
  stress-strain relations for, 122
Group, 18, 240
  choice of,
    for molecular orbitals, 164–166
    for vibrational modes, 78–80
  identification of,
    through generators, 24–25
    by key bases, 19–20
  structure of, 21–22

Hermitian operator, 248
Hexagonal material,
  point groups for, 107
  stress-strain relations for, 119

Hilbert space, 161
  for particle physics, 294
Hypercharge Y, 303

Icosahedral group $\mathbf{I}$, 30
Identicalness,
  consequences of, 218-220
  for three particles, 221-224
Identity operation I, 3-5
Index order,
  in direct product, 192
Infinitesimal generators,
    243-245, 252
  commutator of, 258-259
  linear combinations of,
    252, 260-261
  product of, 261
Infinitesimal operator,
  characteristic equations for, 247
    finite transformation from,
      247-249
Interaction-line model,
  of color field, 318-321
Invariant function, 250
Invariant subgroup, 253
Inverse,
  of element, A^{-1}, 17, 46
  of combination, $(AB)^{-1}$, 48
Inverse class sum $\mathscr{C}_j$, 52
Inversion, i, 4-5
Irreducible representations,
  contributions of, 142-143
  matrix,
    general, 130
    second-order, 33-36
    numerical, 26-32
    properties of, 149-153
Isospin (isotopic spin) T, I,
    301, 303-304
  operators for, 299-300, 302

Jahn-Teller distortion, 183-184
Ket, 161-162
  symmetry-adapted, 162-163
    for benzene molecule, 166-168
    for cyclopropenyl radical,
      182-184

Lagrange equation, 76
  covariance of, 76-77

Lattice,
  for condensed phases,
    in $\mathbf{k}$-space, 9-10
    in $\mathbf{r}$-space, 8-9, 107
  for quark composites, 301-303
    occupation of, 304-305
    symmetry operations for,
      303-304
Lie algebra, 260-261
  from creation-annihilation
    operators, 297-298
Lie groups, 242
  classification of, 253-258
Linearity condition,
  in elasticity, 115
  for quantized states, 161
  for vibration, 76
Linear group,
  general $\mathbf{GL}(n, ..)$, 254
  special $\mathbf{SL}(n, ..)$, 254
Lorentz group, 257

Manifold,
  for group parameters,
    241, 252-253
  of $\mathbf{SO}(3, r)$, 278-280
  of $\mathbf{SU}(2, c)$, 287
Matrix element,
  for scalar operator, 172-173
    vanishing of, 173-174
  for tensorial operator, 205-206
Matrix representation, 129-130
  from basis kets, 175-176
  matrix-diagonal form for,
    138-139
Mesons,
  properties of, 312
  structure of, 320-321
Modes of motion,
  equations for, 76-77
  interactions among, 94-95
  separation of, 88-94
Monoclinic material,
  point groups for, 107
  stress-strain relations for,
    116-117
Multiparameter group, 251-253
Multiplet,
  with given angular momentum,
    283

with given Casimir operators, 290
  for proton-neutron systems,
    299-301
  for quark systems, 301-305
Multiplication,
  binary combining operation as,
    18, 51
  commutator form as, 261
Multiplication table,
  atomic bra with atomic ket,
    170-171
Multiplicity,
  for Young diagram, 228

Near-symmetry, 2
Normal subgroup, 253
Nuclear forces, 298-299

Observer,
  possibility of, 1
Octahedral group O, O_h, 28-29
Octet,
  of quark triplets, 309, 316-317
  degeneracy in, 317-318
Operator,
  geometric, 3-6, 13-14
  projection, 79
  quantum mechanical, 161
  scalar, 172-173
  tensorial, 206
Order,
  in cycle, 13-14
  of generator, 21
  of group, 18
Orthogonal group,
  general $O(n,..)$, 254-255
  special $SO(n,..)$, 255
Orthogonality,
  of character vectors, 65
  of class sums, 67-68
  of eigenoperators, 64
Orthorhombic material,
  point groups for, 107
  stress-strain relations for, 117
Oscillator,
  harmonic,
    differential equation for, 92

Parameter,
  for group elements,
    241-244, 251-253
Partitioning,
  in symmetric group, 218
Passive transformation, 3, 246
Patterns,
  in nature, 1
Pauli exclusion principle, 227
Pauli spin matrices, 264
Permutation matrix, 13
Permutation operation, 213
  symbol for, 214-215
Permutations,
  number of, 214
  with given cyclic structure, 216
Pi bonding,
  in benzene molecule, 166-172
  in cyclopropenyl radical, 177-183
Poincare group, 257
Point groups, 18-19, 326-327
  character tables for, 328-351
Polymers,
  point groups for, 107
  stress-strain relations for, 122
Polynomial ket, 224-225
  reorientation of, 225-226
Primitive cell, 10
Primitive symmetry species, 75
  coalescing of, 97
  splitting of, 97
Product,
  of group elements, 17-18, 51
Product group, 195-196
Product representation, 191-192
Projection operator, 79
Proper transformation, 255
Proton-neutron systems,
  relations among, 299-301
Proton octet, 309, 316-317
  degeneracy in, 317-318

Quantum angular momenta,
  198-200
Quantum eigenstate,
  symmetry composition of,
    162-163
Quarks, 301, 311
  charge on, 307-309
Quasirotation groups, 289-290

Reciprocal lattice, 10

Reciprocal vector, 9
Reciprocity,
  between stress and strain, 114
Reducibility,
  of representation, 33-36, 142, 149
Reducing symmetry,
  effects of, 97, 181-184
Reflection,
  by plane, σ, 4-5
  through point, i, 4-5
Reorientation matrices, 3-7
Representation,
  of group elements, 128
    by complex numbers, 26, 30-32
    by matrices, 33-36
      composition of, 147
      freedom in, 148-149
      generation of,
        129-130, 175-176
    of group structure, 21-22
Rotation C_n, 6
Rotoreflection S_n, 6
Rotation group, 271
  general, 289-290
  2-dimensional, 272-276
  3-dimensional, 276-280
Row,
  of primitive symmetry species, 163

Schur's lemma, 149-153
Secular equation, 57
Separation of symmetry species, 77-79, 147
  modes of motion, 88-94
  molecular orbitals, 162-163
  stress and strain arrays, 111-112
Similarity transformation, 48
  of class sum, 52
  in permutation group, 215
$SO(2, r)$ group,
  infinitesimal operators for, 272
$SO(3, r)$ group,
  infinitesimal operator for, 276-277
  manifold for, 278-280
Space, physical, 1
Spanning,
  of group-element space, 148-149

Spin factor,
  characterization of, 227-228
Spinorial transformation, 285-287
Splitting of energy level,
  by reducing symmetry, 165, 182-184
Square assembly, 80-82
  frequencies for, 92-94
  normal modes for, 84-87
Step operator,
  commutation relation for, 287-289
Strain,
  element ε_{jk} of, as generalized coordinate, 113
  symmetry-adapted arrays of, 111-112
Strain energy,
  density W of, 113
  as analytic function, 113-114
Strangeness S, 301-303
Stress,
  component σ_{jk} of, 108-110
  dependence on strain of, 114-115
  symmetry-adapted arrays of, 111-112
Strong interaction, 298-299
Structure constant c_{jk}^l, 259-261
Structure of group,
  Cayley diagram for, 21-22
$SU(2, c)$ group, 264
  finite operators for, 285-287
  nuclear manifestation of, 301
$SU(3, c)$ group,
  quark manifestation of, 313
  Young diagram for, 314-316
Subtraction,
  of group elements, 241
Symmetric groups, 214, 353
  character tables for, 354
Symmetrizing operator, 220
Symmetry operation, 2-3
  kinds of, 4
Symmetry species,
  for geometric groups, 45-46, 75
  generation of, 77-79
  labeling of, 63, 219
  making up representation, 147
Symplectic group,
  general $Sp(n, ..)$, 256-257

special $\mathbf{SSp}(n,..)$, 257
System, 2

Tensor operator,
  like rotational eigenket, 284
Tetragonal material,
  point groups for, 107
  stress-strain relations for, 117–118
Tetrahedral group $\mathbf{T}$, $\mathbf{T}_h$, $\mathbf{T}_d$, 26–28
Time, 1
Trace of matrix, 140
  for class sum, 140–142
  for geometric operation, 142
  as character component, 143
  for reorientation,
    of Young-diagram ket, 226
Transformation,
  of coordinates, 3
  in translation, 8
  of dyads, 120–121
  of kets, 175–176
  of operators, 46–48
  of stress arrays, 121–122
  of vector arrays, 129–130
Translation,
  by displacement $\mathbf{a}$, $\mathbf{b}$, 8
  representation of, 37–38
Transposition, 217
Triangular molecule,
  base vectors for, 130–131
    linear combinations of, 131–136
  pi kets for, 176
    linear combinations of, 177–183

Triclinic material,
  point groups for, 107
  stress-strain relations for, 116
Trigonal material,
  point groups for, 107
  stress-strain relations for, 118
Twin cyclic group $\mathbf{C}_{nh}$, $\mathbf{C}_{nv}$, 22–24

Unit cells, 8–9
$\mathbf{U}(1, c)$ group, 262
$\mathbf{U}(2, c)$ group, 263–265
Unitary group,
  general $\mathbf{U}(n,..)$, 255–256
  special $\mathbf{SU}(n,..)$, 256
Unitary spin,
  operators for, 302–303

Vacuum state, 294
Vibrating system,
  linearization of, 87–88
Viscosities,
  in fluid, 122–123

Wavevector $\mathbf{k}$, 9, 38
  lattice for, 10
Wigner-Eckart theorem, 206

Young diagrams, 219–220
  for quark composites, 314–316
Young tables, 220
  for orbital motions,
    of three particles, 222–224
  for proton octet, 316–317
  for spin motions, 227–228
    of five particles, 228–229